Large-Scale Adsorption and Chromatography

Volume I

Author

Phillip C. Wankat, Ph.D.
Professor
Department of Chemical Engineering
Purdue University
West Lafayette, Indiana

CRC Press, Inc.
Boca Raton, Florida

Library of Congress Cataloging-in-Publication Data

Wankat, Phillip C., 1944-
Large-scale adsorption and chromatography.

Includes bibliographies and indexes.
1. Chromatographic analysis. 2. Adsorption.
I. Title.
QD79.C4W36 1986 543'.089 86-13668
ISBN 0-8493-5597-4 (v. 1)
ISBN 0-8493-5598-2 (v. 2)

This book represents information obtained from authentic and highly regarded sources. Reprinted material is quoted with permission, and sources are indicated. A wide variety of references are listed. Every reasonable effort has been made to give reliable data and information, but the author and the publisher cannot assume responsibility for the validity of all materials or for the consequences of their use.

Direct all inquiries to CRC Press, Inc., 2000 Corporate Blvd., N.W., Boca Raton, Florida, 33431.

International Standard Book Number 0-8493-5597-4 (Volume I)
International Standard Book Number 0-8493-5598-2 (Volume II)

Library of Congress Card Number 86-13668
Printed in the United States

PREFACE

My major goal in writing this book has been to present a unified, up-to-date development of operating methods used for large-scale adsorption and chromatography. I have attempted to gather together the operating methods which have been used or studied for large-scale applications. These methods have been classified and compared. The main unifying principle has been to use the same theory, the solute movement or local equilibrium theory, to present all of the methods. Mass transfer and dispersion effects are included with the nonlinear mass transfer zone (MTZ) and the linear chromatographic models. More complex theories are referenced, but are not discussed in detail since they often serve to obscure the reasons for a separation instead of enlightening. Liberal use has been made of published experimental results to explain the operating methods.

Most of the theory has been placed in Chapter 2. I recommend that the reader study Sections II and IV.A and IV.B carefully since the other chapters rely very heavily on these sections. The rest of Chapter 2 can be read when you feel motivated. The remaining chapters are all essentially independent of each other, and the reader can skip to any section of interest. Considerable cross-referencing of sections is used to guide the reader to other sections of interest.

I have attempted to present a complete review of the open literature, but have not attempted a thorough review of the patent literature. Many commercial methods have been published in unconventional sources such as company brochures. Since these may be the only or at least the most thorough source, I have referenced many such reports. Company addresses are presented so that interested readers may follow up on these references. Naturally, company brochures are often not completely unbiased. The incorporation of new references ceased in mid-May 1985. I apologize for any important references which may have been inadvertently left out.

Several places throughout the text I have collected ideas and made suggestions for ways to reduce capital and/or operating expenses for different separation problems. Since each separation problem is unique, these suggestions cannot be universally valid; however, I believe they will be useful in the majority of cases. I have also looked into my cloudy crystal ball and tried to predict future trends; 5 years from now some of these predictions should be good for a laugh.

Much of this book was written while I was on sabbatical. I wish to thank Purdue University for the opportunity to take this sabbatical, and Laboratoire des Sciences du Genie Chimique, Ecole Nationale Superieure des Industries Chimiques (LSGC-ENSIC) for their hospitality. The support of NSF and CNRS through the U.S./France Scientific Exchange Program is gratefully acknowledged. Dr. Daniel Tondeur, Dr. Georges Grevillot, and Dr. John Dodds at LSGC-ENSIC were extremely helpful in the development of this book. My graduate level class on separation processes at Purdue University served as guinea pigs and went through the first completed draft of the book. They were extremely helpful in polishing the book and in finding additional references. The members of this class were Lisa Brannon, Judy Chung, Wayne Curtis, Gene Durrence, Vance Flosenzier, Rod Geldart, Ron Harland, Wei-Yih Huang, Al Hummel, Jay Lee, Waihung Lo, Bob Neuman, Scott Rudge, Shirish Sanke, Jeff Straight, Sung-Sup Suh, Narasimhan Sundaram, Bart Waters, Hyung Suk Woo, and Qiming Yu. Many other researchers have been helpful with various aspects of this book, often in ways they are totally unaware of. A partial listing includes Dr. Philip Barker, Dr. Brian Bidlingmeyer, Dr. Donald Broughton, Dr. Armand deRosset, Dr. George Keller, Dr. C. Judson King, Dr. Douglas Levan, Dr. Buck Rogers, Dr. William Schowalter, and Dr. Norman Sweed. The typing and help with figures of Connie Marsh and Carolyn Blue were invaluable and is deeply appreciated. Finally, I would like to thank my parents and particularly my wife, Dot, for their support when my energy and enthusiasm plummeted.

THE AUTHOR

Phillip C. Wankat is a Professor of Chemical Engineering aat Purdue University in West Lafayette, Ind. Dr. Wankat received his B.S.Ch.E. from Purdue University in 1966 and his Ph.D. degree in Chemical Engineering from Princeton University in 1970. He became an Assistant Professor at Purdue University in 1970, an Associate Professor in 1974, and a Professor in 1978. Prof. Wankat spent sabbatical years at the University of California-Berkeley and at LSGC, ENSIC, Nancy, France.

His research interests have been in the area of separation processes with an emphasis on operating methods for adsorption and large-scale chromatography. He has published over 70 technical articles, and has presented numerous seminars and papers at meetings. He was Chairman of the Gordon Research Conference on Separation and Purification in 1983. He is on the editorial board of Separation Science. He is active in the American Institute of Chemical Engineers, the American Chemical Society, and the American Society for Engineering Education. He has consulted with several companies on various separation problems.

Prof. Wankat is very interested in good teaching and counseling. He earned an M.S.Ed. in Counseling from Purdue University in 1982. He has won several teaching and counseling awards, including the American Society for Engineering Education George Westinghouse Award in 1984.

TABLE OF CONTENTS

Volume I

Volume II

Chapter 1

INTRODUCTION

The purpose of this book is to provide a unified picture of the large number of adsorption and chromatographic operating methods used for separation. The macroscopic aspects of the processes differ, but on a microscopic scale all of these separation methods are based on different velocities of movement of solutes. The solute velocities in turn depend upon the phenomena of flow through a porous media, sorption equilibria, diffusion, mass transfer, and sorption/desorption kinetics.

Since I do not read books serially from cover to cover, but instead skip to those sections I am most interested in, this book has been written for this type of selective reading. Except for Chapter 2, the chapters are essentially independent so that the reader can start anywhere. All of the chapters do rely heavily on the local equilibrium or solute movement theory. Thus, a review of Chapter 2 (Sections III.A and B, plus possibly Section IV) would be helpful before reading other parts of the book. The remainder of Chapter 2 can be picked up as needed.

We will first look (in Chapter 2) at a physical picture of solute movement in a packed column. For most systems the separation can be predicted by combining the average rate of solute movement and zone spreading effects. The average rate of solute movement will be derived for both linear and nonlinear isotherms. This average solute wave velocity depends upon the bed porosity, solvent velocity, and equilibrium conditions, and is essentially the fraction of time the solute is in the mobile phase times the fluid velocity. The solute velocity is easily calculated and easy to use to explain the macroscopic aspects of different operating methods. Nonlinear adsorption, thermal waves, changing gas velocities, and coupled systems will all be studied. The spreading of the solute zones depends on diffusion, mass transfer rates, and sorption/desorption kinetics. The amount of zone spreading is easily determined from theories for systems with linear isotherms. From these theories one obtains the familiar rule that zone spreading is proportional to the square root of the distance traveled. For nonlinear systems which form constant patterns, the mass transfer zone (MTZ) approach will be developed.

The pictures of solute movement and of zone spreading will be combined to explain the operating methods in Chapters 3 to 8. Where necessary, the results from more detailed theories will be used to explain experimental results. Chapters 3 to 8 describe different operating methods and use the theories from Chapter 2 to explain these methods. The division of different separation methods into chapters is somewhat arbitrary. Essentially, Chapters 3 to 5 cover fixed-bed systems while Chapters 6 to 8 cover moving or simulated moving beds. These six chapters are all independent and can be read in any order, although they are cross-referenced. The development of mathematical theories is mainly restricted to Chapter 2 and, to a lesser extent, Chapter 6.

The adsorption of a single solute with simple cycles is discussed in Chapter 3. The basic type of operating cycle used is shown in Figure 1-1. The adsorption of solute occurs for some period and then the solute is desorbed either with a hot fluid or a desorbent. This is a batch process with a large number of possible variations. The method has been applied for cleaning up gas streams using a hot gas for desorption, for solvent recovery from a gas stream using activated carbon and steam desorption, for liquid cleanup using either a hot liquid or a desorbent for the desorption step, and for wastewater treatment systems. General considerations are covered in Section II of Chapter 3 and specific separations are covered in the rest of the chapter. Section II.D in Chapter 3, on the effect of particle size, will probably be of interest to all readers. Many of the common commercial adsorption processes are briefly reviewed in this chapter.

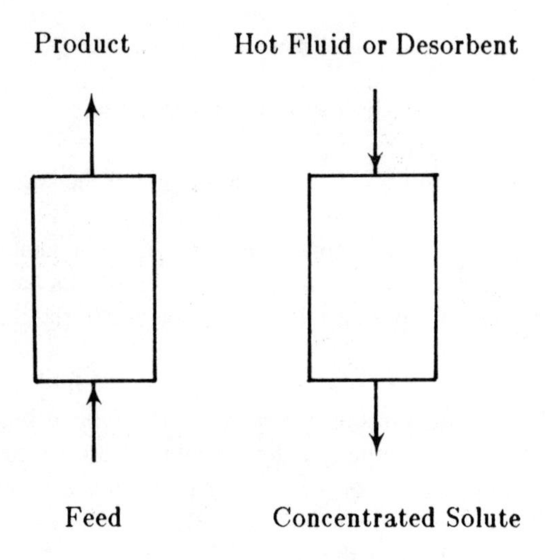

FIGURE 1-1. Basic cycle for adsorption of a single solute. (A) Adsorption step. (B) Desorption.

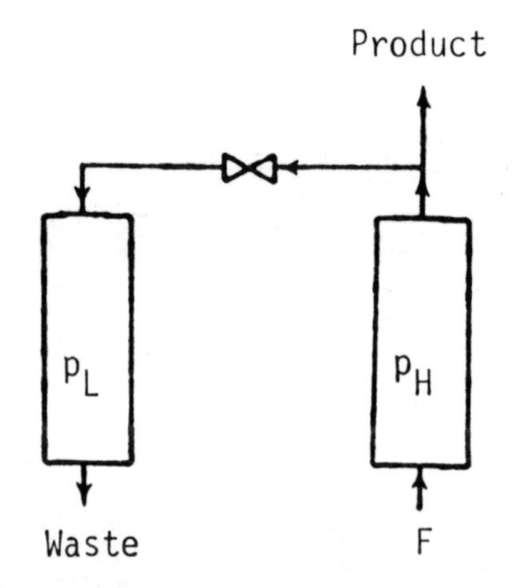

FIGURE 1-2. Basic pressure swing adsorption apparatus.

Chapter 4 covers cyclic operations which are somewhat more complex than those shown in Figure 1-1. Pressure swing adsorption (PSA) first adsorbs solute from a gas stream at elevated pressure and then desorbs the solute using a purge at much lower pressure. A very simple system is shown schematically in Figure 1-2. Since the volume of gas expands when depressurized, a larger volume but fewer moles of gas can be used for the purge step. Every few minutes the columns change functions. For liquid systems, parametric pumping and cycling zone adsorption are based on the shift in the equilibrium isotherm when a thermodynamic variable such as temperature is changed. Although this change in concentration is often small, a large separation can be built up by utilizing many shifts. A variety of cycles will be explored for both gas and liquid systems.

The separation or fractionation of more than one solute by large-scale chromatography is the subject of Chapter 5. The basic method and typical results are illustrated in Figure 1-3. Solvent or carrier gas is continuously fed into a packed column and a pulse of feed is injected intermittently. Since different solutes travel at different velocities, they exit the column at

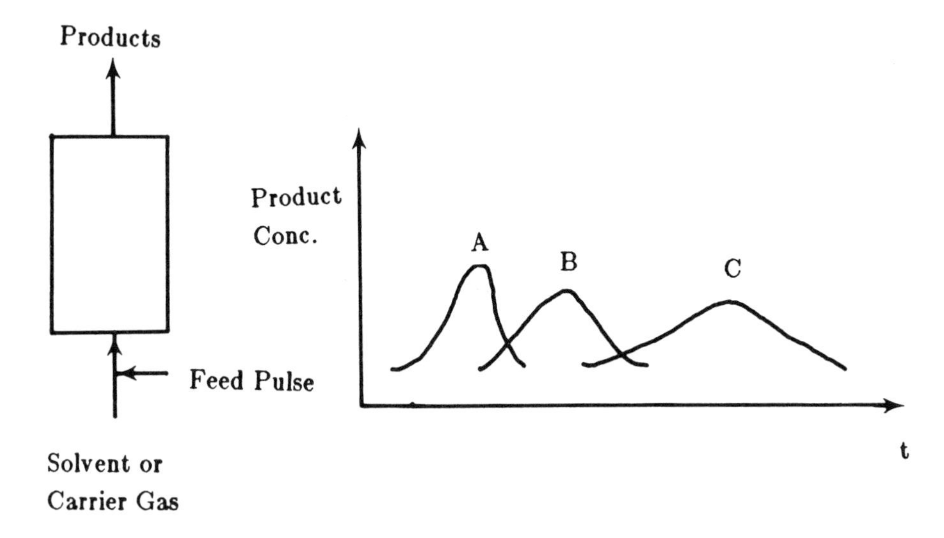

FIGURE 1-3. Apparatus and results for chromatographic separation.

different times. Large-scale liquid chromatography, size exclusion chromatography (SEC), gas-liquid chromatography (GLC), biospecific affinity chromatography, and ion-exchange chromatography will be explored.

Countercurrent moving bed systems and simulated countercurrent systems are the subject of Chapter 6. For single solute removal the basic apparatus is shown in Figure 1-4. The function of the sorption and desorption chambers is the same as in Figure 1-1, but the countercurrent apparatus operates at steady state. To fractionate two solutes the moving bed arrangement shown in Figure 1-5 could be used. This is a steady-state apparatus for binary separation. Since it is difficult to move solids in a uniform plug flow, simulated moving bed (SMB) systems have been developed commercially. In an SMB the solid does not move. Instead, the location of each product and feed port is switched in the direction of fluid flow every few minutes. When a port location moves, an observer at that port sees the solid move in the opposite direction. Thus, countercurrent motion is simulated.

Chromatographic and simulated countercurrent processes both have advantages. These two processes are combined in hybrid chromatographic processes, which are discussed in Chapter 7. With column-switching procedures the products are removed at different locations in the column. In a moving-feed chromatograph the input location of the feed pulse moves up the column to follow the movement of solute. Then chromatographic development is used to completely separate the solutes. Moving port chromatography combines these two methods.

Two-dimensional and rotating methods are discussed in Chapter 8. The prototype two-dimensional system is the rotating annulus apparatus shown in Figure 1-6. The annulus is packed with sorbent while carrier gas or solvent flows continuously upward. Feed is added continuously at one point. The result is a steady-state separation similar to the results shown in Figure 1-3, but with the angular coordinate, θ, replacing time. Many other two-dimensional arrangements have been developed. Centrifugal chromatography, the chromatofuge, is also discussed.

Most real separation problems can be solved using any of several different operating methods. Hopefully, this compendium of operating methods will provide the designer with ideas for creating new schemes. The "best" scheme will vary depending on all the facets of the problem being solved.

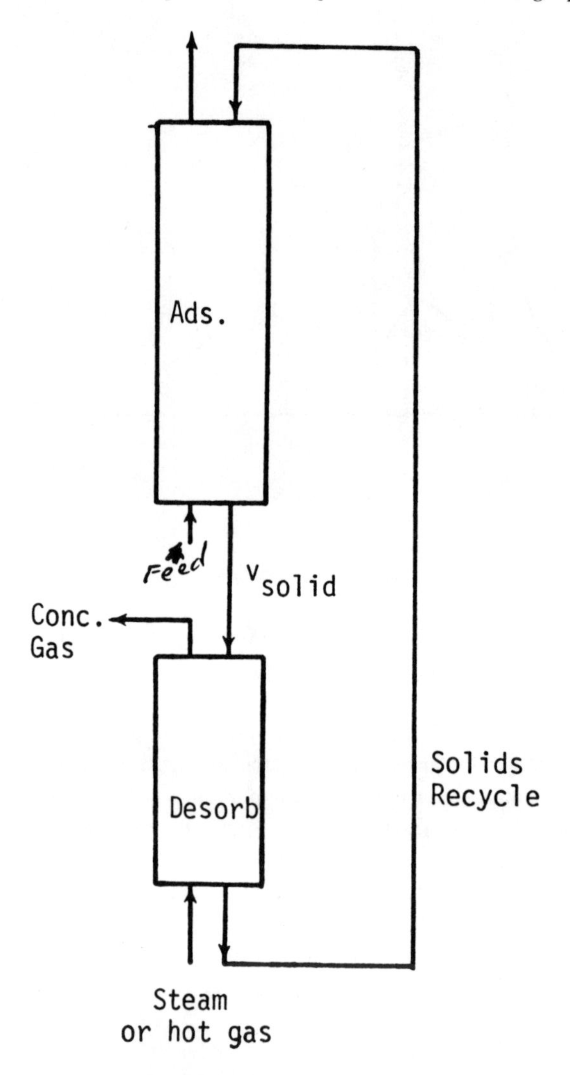

FIGURE 1-4. Countercurrent system for single solute.

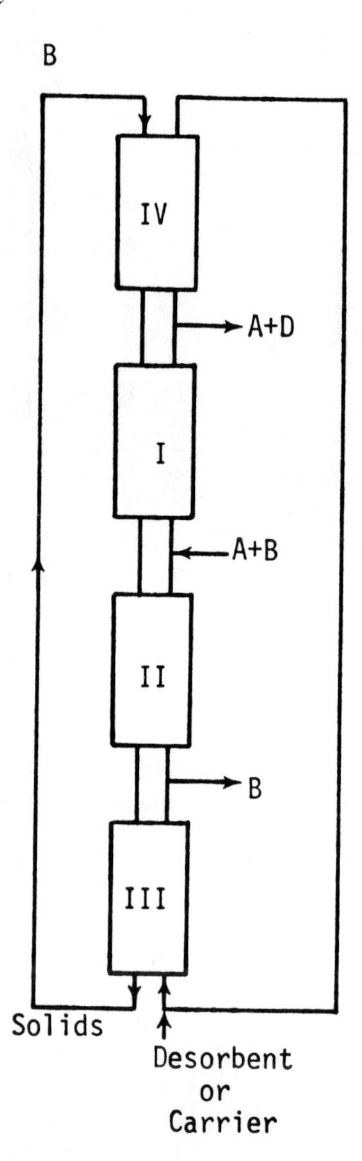

FIGURE 1-5. Countercurrent system for fractionation of two solutes.

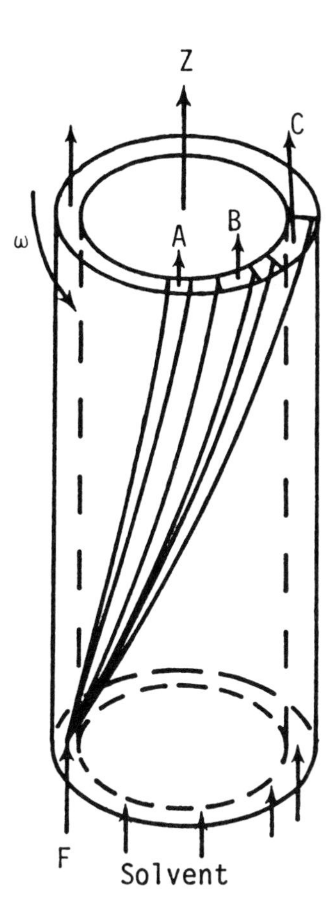

FIGURE 1-6. Apparatus and results for two-dimensional system.

Chapter 2

PHYSICAL PICTURE AND SIMPLE THEORIES FOR ADSORPTION AND CHROMATOGRAPHY

I. INTRODUCTION

What phenomena are involved in chromatographic and adsorptive separations? How do these phenomena combine to produce the desired separation? How can we simply predict the separation which will occur? In this chapter we will try to answer these and other questions. First, a physical picture of sorption will be presented. Then, equilibrium isotherms will be discussed. A physical argument will be used to explain movement of solute and energy in an adsorption column followed by a rigorous mathematical development of the equations. Zone spreading by mass transfer and diffusion will be explained mathematically for linear systems. The mass transfer zone (MTZ) or length of unused bed (LUB) approach will be introduced for nonlinear systems. The purpose of this chapter is to provide a physical understanding and relatively simple mathematical theories which will be useful in later chapters. This chapter is not a design manual; other sources should be consulted for design equations.[165,865,901,1015-1017]

II. PHYSICAL PICTURE

Commercially significant sorption operations (including adsorption, chromatography, ion exchange, ion exclusion, etc.) use sorbents which are highly porous and have large surface areas per gram of sorbent. The sorbent particles are commonly packed in a column as illustrated in Figure 2-1. In general, the particles will be of different sizes and shapes. They will pack in the column and have an average interparticle (between different particles) porosity of α. In a poorly packed bed α may vary considerably in different parts of the column. This can lead to poor flow distribution or channeling and will decrease the separation. Since the particles are porous, each particle has an intraparticle (within the particle) porosity, ϵ, which is the fraction of the particle which is void space. If the packing is manufactured uniformly, ϵ will be the same for all particles. Approximately 2% of the surface area is on the outer surface of the packing; thus, most of the capacity is inside the particles. An alternate model using a single porosity is also commonly used and is discussed in Section V.B.

The pores are not of uniform size. Large molecules such as proteins or synthetic polymers may be sterically excluded from some of the pores. The fraction of volume of pores which a molecule can penetrate is called K_d. Very small molecules can penetrate all the pores and $K_d = 1.0$ while very large molecules can penetrate none of the pores and $K_d = 0$. For a nonsorbed species, K_d can be determined from a simple pulse experiment. Very small nonsorbed species will have available both the external void volume V_o and the internal void volume V_i. Thus, small molecules will exit at an elution volume, V_e, of

$$V_e = V_o + V_i = \alpha V_{col} + (1 - \alpha)\epsilon V_{col} \tag{2-1}$$

where V_{col} is the volume of the packed column. Since large molecules have available only the external void volume V_o, their elution volume is

$$V_e = V_o = \alpha V_{col} \tag{2-2}$$

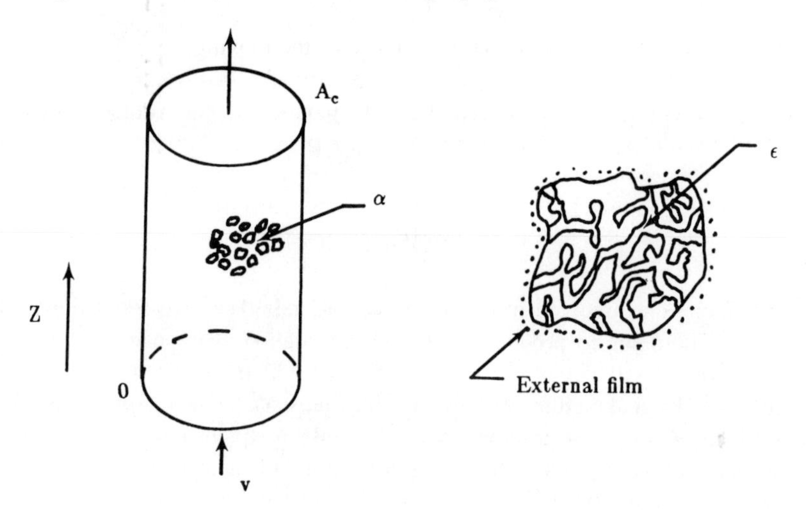

FIGURE 2-1. Particles packed in a bed.

Equations 2-1 and 2-2 allow determination of V_o,α, V_i, and ϵ from one experiment with large molecules and one experiment with small molecules. Molecules of intermediate size can penetrate some of the pores. For these nonsorbed molecules, K_d can be determined from:

$$K_d = \frac{V_e - V_o}{V_i} \tag{2-3}$$

Size exclusion (gel permeation or gel filtration) separations are based entirely on differences in K_d.

This picture is somewhat too simple for some sorbents. Some activated carbons[792,1073] and ion exchange resins[18] have two types of pores: macropores and micropores. These particles have two internal porosities: one for macropores and one for micropores. The two internal and one external porosity can be measured by three experiments. Very small molecules will permeate all pores; large molecules will permeate only the macropores; very large molecules will stay in the external void volume. The solute movement theory presented later could be adjusted for these more complex systems, but K_d can be used to approximately include the two types of internal pores.

Molecular sieve zeolites differ from both these pictures.[105,168,645,865] The zeolite crystals form a porous three-dimensional array and have a highly interconnected, regular network of channels and cavities of very specific sizes. Thus the crystal geometry is well defined. Commercial zeolite adsorbents are pelleted agglomerates of zeolite crystals and binder. The binders have large pores and relatively little sorption capacity compared to the zeolite crystals. Typical values for void fractions are[645] interpellet = 32%, intercrystal = 23%, and intracrystal = 19% based on the fraction of the entire bed.

For the system shown in Figure 2-1, the processes which occur during a separation are as follows: Fluid containing solute flows in the void volume outside the particles. The solute diffuses through an external film to the particle. Here the solute may sorb on the external surface or (more likely) diffuse into the stagnant fluid in the pores. If the pores are tight for the solute this diffusion will be hindered. The solute finds a vacant site and then sorbs by physical or electrical forces or by a chemical reaction. While sorbed the solute may diffuse along the surface. The solute desorbs and diffuses through the pores, back across the external film, and into the moving fluid. A given molecule may sorb and desorb many times during its stay inside a single particle. Once in the moving fluid the solute is carried along at the fluid velocity until the solute diffuses into another particle and the whole process is repeated. As far as migration down the column is concerned, the particle is either moving at the interstitial velocity, v, of the fluid or it has a velocity of zero when it is inside a particle.

A large number of adsorbents, ion exchange resins, partition chromatography supports, and size exclusion packings have been developed. The properties of these are available in the following sources: ion exchangers,[18,202,549,1016,1017,1076] size exclusion media,[549,827,1076,1109] activated carbon,[616,792,1016,1017] silica gel,[112,616,1016,1017] activated alumina,[112,438,616,1016,1017] chromatographic packings,[827,930,1076,1126] and molecular sieves.[105,112,168,616,645,865,1017] An extremely complete annotated bibliography of adsorption up to 1953 was compiled by Dietz.[321,322]

III. EQUILIBRIUM ISOTHERMS

A wide variety of equilibrium isotherms have been published.[18,105,168,645,792,825,865,1016,1017] A few of these will be reviewed for gas and liquid systems.

A. Gas Systems

For gas systems the adsorbed phase has essentially the same density as a liquid. This makes pure component equilibrium data very easy to obtain. The weight of the sorbent or the pressure can easily be measured. Many different adsorption isotherms have been developed to fit the results obtained. One of the simplest is the Langmuir isotherm.[589,629,692,809,865,1114] This isotherm is also appealing because either a simple physical picture[629,865] or a statistical mechanical argument[809,865] can be used to develop the isotherm. Langmuir assumed that at most a monolayer of adsorbate could cover the solid surface. If q is the adsorbent loading and $\bar{p}$ is the partial pressure, the Langmuir isotherm is

$$q = \frac{q_{max} K_A \bar{p}}{1 + K_A \bar{p}} \tag{2-4}$$

where K_A is the sorption equilibrium constant and q_{max} is capacity at monolayer coverage. The shape of Equation 2-4 is shown in Figure 2-2A. This is known as a "favorable" isotherm, because this shape leads to sharp breakthrough curves. Since K_A is a reaction equilibrium constant, it should follow the Arrhenius relationship:

$$K_A = K_o \exp[-\Delta H/RT] \tag{2-5}$$

The Langmuir isotherm is applicable for dilute, single-component adsorption with an inert gas present or for pure gases at low pressures.

At very low partial pressure the Langmuir equations takes the linear form:

$$q = K_A q_{max} \bar{p} \tag{2-6}$$

The linear isotherm is extremely important not only because it is a limiting form of more complex isotherms when the solute is dilute, but also because the mathematical theories become much more tractable when the isotherm is linear.

If several solutes can adsorb the Langmuir isotherm can be extended to:[589,692,809,1114]

$$\frac{q_i}{q_{imax}} = \frac{K_i \bar{p}_i}{1 + \sum_{j=1}^{n} (K_j \bar{p}_j)} \tag{2-7}$$

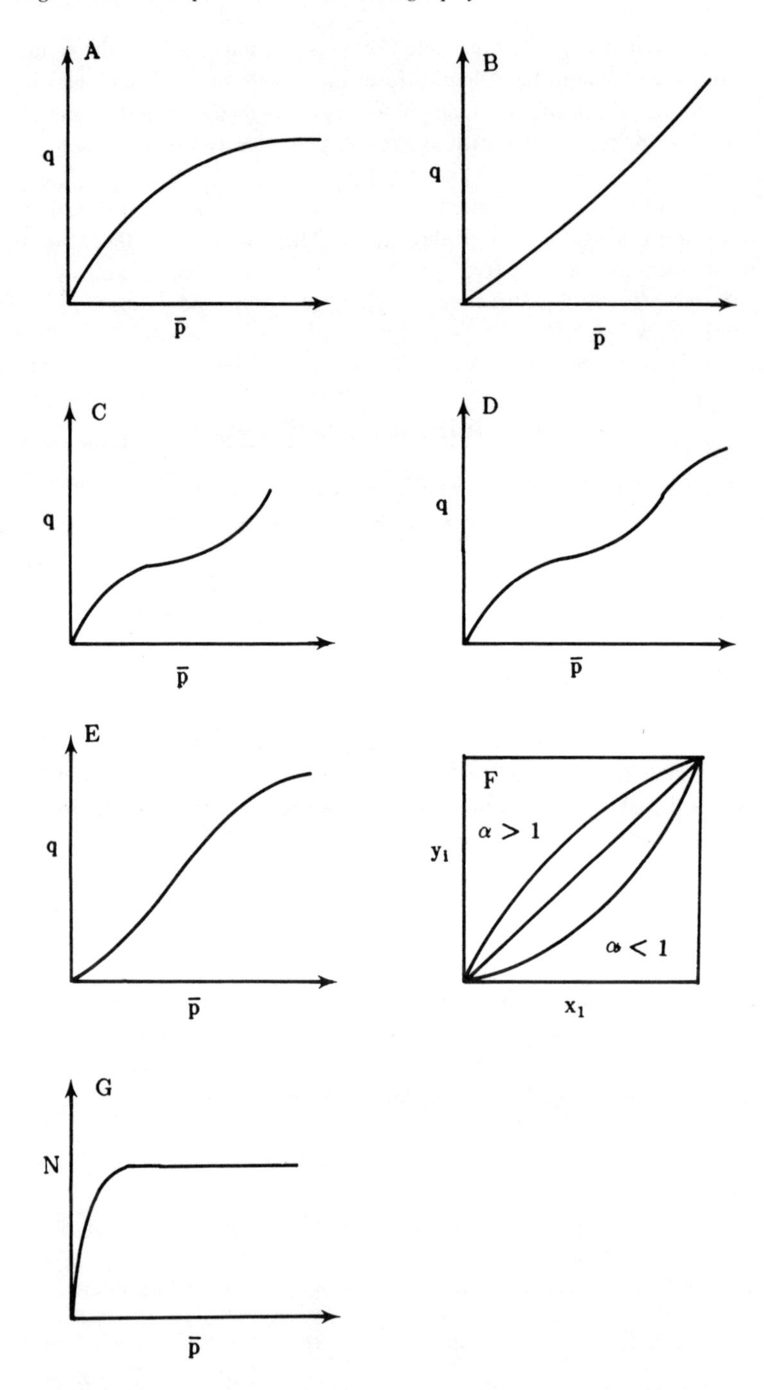

FIGURE 2-2. Equilibrium isotherms. (A) Langmuir; (B) BET type 2 behavior (K > 1); (C) BET type 3 behavior (K < 1); (D) BET type 4 behavior; (E) BET type 5 behavior; (F) Lewis correlation, constant α; (G) molecular sieve isotherm.

In addition to the previous assumptions, it is necessary to assume that the only interaction of solutes is competition for sites on the adsorbent. In order for Equation 2-7 to be thermodynamically consistent, the monolayer coverages, q_{imax}, of all components must be equal.[652] If the monolayer coverages differ, extra terms are required for thermodynamic consistency.[652] Equation 2-7 predicts that less solute is adsorbed when other solutes are present. A few systems will show cooperative adsorption where the presence of other solutes aids adsorption.[1100] Competitive adsorption is much more common.

Langmuir's isotherm is based on a specific physical picture of adsorption. The Langmuir isotherm agrees with data for some systems but not for others. (This is generally true of all isotherms. There are none that fit all systems.) When different physical pictures are used, different isotherms result. For example, by assuming that the forces involved in adsorption and in condensation are the same, and allowing for more than one molecular layer, Brunauer et al.[188,189] derived the BET isotherm. The simplest form of the BET equation is

$$\frac{q}{q_{mono}} = \frac{K\,\bar{p}}{[p^o + (K - 1)\,\bar{p}][1 - \bar{p}/p^o]} \tag{2-8}$$

where q_{mono} is the adsorbate concentration for monolayer coverage and p^o is the vapor pressure of pure solute at the adsorption temperature. The shape of Equation 2-8 depends on the values of the constants. If $K < 1$ the isotherm is unfavorable throughout and is shown in Figure 2-2B, while if $K > 1$ the isotherm is favorable at low concentrations and unfavorable at higher concentrations (Figure 2-2C). Other possible shapes which require a more complex equation[188,809,1114] are shown in Figures 2-2D and E. If $K \gg 1$ and $\bar{p} \ll p^o$ Equation 2-8 reduces to the Langmuir form. In the limit of very dilute systems Equation 2-8 becomes linear. The adsorption isotherms for many gas systems can be fit by one of the BET forms. This is true for both adsorption of a pure gas and adsorption of a gas when an inert carrier is present. Of course, just because the isotherms fit a BET form is not proof that the mechanism postulated in the derivation is correct.

As layers build up, eventually capillary condensation will occur. This will cause an inflection point in the isotherm as in Figures 2-2C and D. In the capillary condensation regime, results for adsorption and desorption will differ. Several plausible physical pictures for this hystersis have been suggested.[865]

For binary systems where both gases adsorb, Lewis et al.[589,651] developed a correlation for adsorption from the mixture compared to pure component adsorption. This equation is

$$\frac{q_1}{q_1^o} + \frac{q_2}{q_2^o} = 1 \tag{2-9}$$

where q_i^o is the amount adsorbed from a pure gas and q_i is the amount adsorbed from the mixture. With the additional assumption that the ratio

$$\alpha_{12} = \frac{y_1/x_1}{y_2/x_2} = \frac{y_1/q_1}{y_2/q_2} = \frac{\bar{p}_1/q_1}{\bar{p}_2/q_2} \tag{2-10}$$

is a constant, Equations 2-9 and 2-10 can be used for predictions; y_i and x_i are the mole fractions in the gas and solid phase, respectively; α_{12} is a separation factor which is essentially the same as the relative volatility of a liquid vapor mixture. The shape of these isotherms is shown in Figure 2-2F.

Isotherms can also be derived from a thermodynamic argument.[809] For example, for the two-dimensional surface the pressure is replaced by the spending pressure and the volume by the area. For an ideal adsorbed solution Myers and Prausnitz[752] showed that mixture equilibrium isotherms could be predicted from pure component isotherms. The procedure has been extended to multilayer adsorption.[865,914] The ideal adsorbed solution theory has also been applied to heterogeneous surfaces.[749,1125] Unfortunately, these theories do not give a single equation which can be substituted into theories of adsorption.

Molecular sieve zeolite adsorbents behave differently than other adsorbents. Instead of forming a surface layer the entire pore fills with adsorbed material. In addition, some solutes may be totally excluded since they are too large to fit in the pores. The loading ratio correlation (LRC)[639,645,1112] extends the Langmuir expression but replaces the monolayer capacity by the maximum attainable loading. For a pure component the loading ratio is

$$\frac{N}{N_o} = \frac{(k\bar{p})^{1/m}}{1 + (k\bar{p})^{1/m}} \tag{2-11}$$

where N is the loading on the adsorbent and N_o is the maximum attainable loading. Temperature effects can be included by using an Arrehenius relation for k and rearranging Equation 2-11:

$$\ln \bar{p} = A_1 + A_2/T + n \ln N/(N_o - N) \tag{2-12}$$

where n may also be temperature dependent. The shape of a typical molecular sieve isotherm is shown in Figure 2-2G. Equation 2-12 can be extended to multicomponent systems in the same way the Langmuir equation was extended. More detailed theories applicable to molecular sieve zeolites are discussed elsewhere.[105,168,188,589,809,865]

In gas-liquid chromatography (GLC) the phenomena is essentially absorption of the solute into the stationary liquid phase, not adsorption at a surface. At low concentrations the isotherms follow a Henry's law or linear relationship. At higher concentrations the isotherm shape is similar to Figure 2-2B. Absorption is favored as more solute dissolves in the stationary phase.

B. Liquid Systems

Adsorption from liquid systems is more complex than from gas systems. Extensive reviews are available.[589,671,750,865] Both solute and solvent compete for adsorbent surface. For dilute solutions of solid dissolved in a liquid the solvent effects can often be ignored and the Langmuir isotherm can be derived by the same procedure used for gases.[589]

$$q = \frac{q_{max} K_A c}{1 + K_A c} \tag{2-13}$$

For liquids it is common to generalize this to

$$q = \frac{a c}{1 + b c} \tag{2-14}$$

where a and b do not have to be $q_{max} K_A$ and K_A, respectively. The constants a and b can be fit to experimental data by plotting c/q vs. c. Extension to multisolute systems is analogous to Equation 2-7.

A better fit to experimental data at low concentrations can often be obtained with the Freundlich equation:[387,589]

$$q = k c^{1/n} \qquad n > 1 \tag{2-15}$$

Unfortunately, the Freundlich equation does not approach a linear isotherm for very dilute solutions, and it does not approach a limiting asymptotic value observed for many real systems.[589] Use of the Freundlich equation is discouraged.

Neither the Langmuir nor the Freundlich isotherms fit data for adsorption of organic mixtures from aqueous solution on activated carbon. An empirical equation for multicomponent systems which fits this data is[392,865]

$$q_i = \frac{a_{io}\, c_i^{b_{io}}}{\beta_i + \sum_{i=1}^{n} a_{ij}\, c_i^{b_{ij}}} \qquad (2\text{-}16)$$

of course, with five constants for one solute it is easier to fit any data. The Langmuir and Freundlich isotherms are special cases of Equation 2-16. Adsorption of organics on activated carbon has generated a ridiculously huge literature. The reader who feels compelled to attack this area can start with several reviews.[119,251,526,714,750]

For completely miscible binary liquid mixtures adsorption is complicated.[364,365,589] Over the entire range of mole fractions the component which is "solvent" must change. Except for molecular sieves, both components will compete for the adsorption surface. Experimental data is also much more difficult to obtain and interpret. The adsorbed layer has essentially the same density as the bulk liquid, and it is difficult to separate adsorbed material from liquid in the pores. A simple method[589] is to measure the decrease in mole fraction in component 1 (more strongly adsorbed) in the liquid. From this a composite isotherm can be calculated

$$\frac{n_o\, \Delta x_1}{m} = n_1^s\, x_2 - n_2^s\, x_1 \qquad (2\text{-}17)$$

where n_o is the original number of total moles of 1 plus 2, m is the mass of adsorbent, Δx_1 is the decrease in mole fraction of 1 in the liquid, n_1^s and n_2^s are the moles of 1 and 2 adsorbed per mass of adsorbent, and x_1 and x_2 are the mole fractions of 1 and 2 in the liquid. Three different classes of composite isotherms are shown in Figure 2-3.[589] In case A, component 1 is always most strongly adsorbed while in case B the preference switches. A negative Δx_1 on Figure 2-3B means that there is an increase in the mole fraction of component 1 in the liquid. For molecular sieves where component 2 is completely excluded from the pores $n_2^s = 0$ and Equation 2-17 becomes a straight line. This is illustrated in Figure 2-3C. In practice, the small amount of adsorption on the external surface will cause a slight curvature. The effect of temperature on the composite isotherm is shown in Figure 2-3D.

The composite isotherm does not give the individual isotherms directly except in the case of molecular sieves where component 2 is excluded. If it is assumed that the surface is completely covered by an adsorbed layer and that this layer is one molecule thick, then

$$\frac{n_1^s}{(n_1^s)_m} + \frac{n_2^s}{(n_2^s)_m} = 1 \qquad (2\text{-}18)$$

where $(n_1^s)_m$ and $(n_2^s)_m$ are the number of moles of the pure components required to cover the surface of a unit weight of solid. This equation is essentially the same as Equation 2-9. From Equations 2-17 and 2-18, n_1^s and n_2^s can be calculated from the experimental data. Individual isotherms are shown in Figure 2-4A and B, while plots in terms of mole fractions are shown in Figure 2-4C and D. The use of Equation 2-18 does involve the assumption of a monolayer coverage. Unfortunately, for nonideal mixtures the monolayer coverage assumption is not thermodynamically consistent.[364] In these cases one must use a multilayer theory where there is a gradual change from surface to bulk properties, or a thermodynamic approach which does not postulate a physical model.[364,365]

The equations for adsorption of a solute from a very dilute solution are consistent with

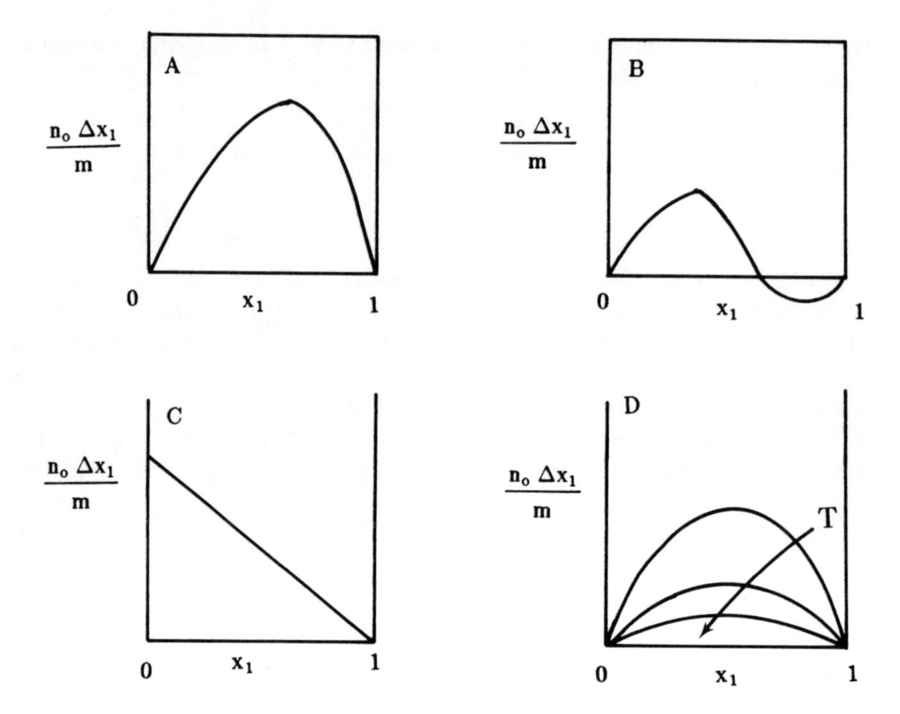

FIGURE 2-3. Composite isotherms for miscible liquids. (A) U-shaped, component 1 always preferred; (B) S-shaped, preferred component changes; (C) linear, for molecular sieves, component 2 excluded; (D) temperature effects for U-shaped composite isotherms.

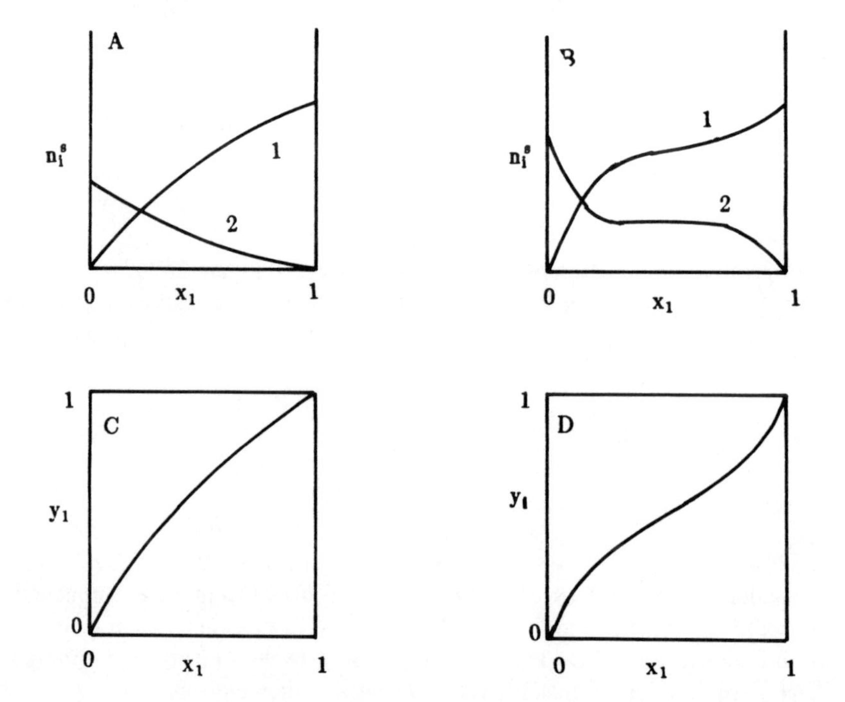

FIGURE 2-4. Individual isotherms for miscible liquids. (A,C) Component 1 always preferred; (B,D) preferred component changes.

the more general development given in Equations 2-17 and 2-18. For very dilute solutions $x_1 \simeq 0$, $x_2 \simeq 1.0$, and Equation 2-17 becomes:

$$\frac{n_o \, \Delta x_1}{m} = n_1^s \tag{2-19}$$

Thus the amount adsorbed can be determined directly from the change in concentration of the solution.

For ion exchange systems[18,484,1016] the situation is again different since ion exchange follows a stoichiometric material balance. For monovalent ion exchange

$$A^+ + R^-B^+ + X^- \rightleftarrows R^-A^+ + B^+ + X^-$$

If the activity coefficients are all 1.0, the law of mass action gives an equilibrium expression

$$c_{RA} = \frac{c_{Rtotal} \, K_{AB} \, c_A}{c + (K_{AB} - 1) \, c_A} \tag{2-20}$$

where c_{Rtotal} is the total resin capacity in equivalents per bulk volume and c is the total concentration of ions in the liquid in equivalents per liter. It is convenient to work in terms of fractions:

$$y_i = \frac{c_{Ri}}{c_{Rtotal}} \quad , \quad x_i = \frac{c_i}{c} \tag{2-21}$$

Thus, for monovalent ions Equation 2-20 becomes:

$$y_A = \frac{K_{AB} \, x_A}{1 + (K_{AB} - 1)x_A} \tag{2-22}$$

For monovalent ions Equation 2-22 does not depend upon either the total ionic strength of the solution or the resin capacity (except for secondary effects due to the activity coefficients). Note that Equation 2-22 is a Langmuir-type form which will be favorable for A if $K_{AB} > 1$. The isotherms will have the same shape as those in Figure 2-2F if K_{AB} is constant. Selectivity values for a number of systems are available.[18,202,203,484,1016] Because the equilibrium forms are the same, solutions for Langmuir adsorption are also valid for binary ion exchange with constant selectivities.

For removal of a divalent ion the situation is different. Now the reaction is

$$D^{++} + 2R^-B^+ + 2X^- \rightleftarrows D^{++}R_2^- + 2B^+ + 2X^-$$

and the equilibrium expression is

$$\frac{y_B^2}{1 - y_B} = \left[\frac{1}{K_{DB} \dfrac{c_{Rtotal}}{c}}\right] \frac{x_B^2}{1 - x_B} \tag{2-23}$$

The general Langmuir form has been lost but the isotherm will be favorable for D if $(K_{DB}c_{Rtotal}/c) > 1$. Note that the isotherm now depends on both the total resin capacity and the liquid concentration. This is important in the chemical regeneration of these systems. In real systems K_{DB} may not be constant.

IV. MOVEMENT OF SOLUTE AND ENERGY WAVES IN THE COLUMN

Separation occurs because different solutes move at different velocities. This solute movement is mainly controlled by the equilibrium, and solute movement can be predicted with a simple physical picture and some algebra. In Section IV.A a physical argument will be presented for the movement of uncoupled solutes. In Section IV.B the movement of pure energy waves will be explored and their effect on solute concentrations will be developed. In Section IV.C the "sorption effect" in gas systems will be explained. In Section IV.D, coupling effects between two solutes and between temperature and a solute are considered. The formal mathematical development will be presented in Section V.

A. Movement of Solute Waves in the Column

Solute diffuses between the moving mobile fluid and the stagnant fluid in the pores. While in the pores, solute may also adsorb on the solid (or dissolve into the stationary fluid coating the solid in GLC). Solute in the mobile phase moves at the interstitial velocity, v, of the mobile phase. Solute in the stagnant fluid or sorbed onto the solid has a zero velocity. Each solute molecule spends some time in the mobile fluid and then diffuses back into the stagnant fluid and so forth. Thus the movement of a given molecule is a series of random steps. The average velocity of all solute molecules of a given species is easily determined, but the randomness causes zone spreading.

If we consider a large number of solute molecules, this average velocity can be determined from the fraction of time they are in the mobile phase. Thus,*

$$u_{solute} = (v)\,(\text{fraction solute in mobile phase}) \tag{2-24}$$

If we consider an incremental change in solute concentration, Δc, which causes an incremental change in the amount sorbed, Δq, the fraction of this Δc in the mobile phase is

$$\text{Fraction incremental } \Delta c \text{ in mobile phase} = \frac{\text{Amount in mobile phase}}{\text{Amount in (mobile + stagnant fluid + solid)}} \tag{2-25}$$

For the system shown in Figure 2-1, each of the terms in Equation 2-25 is easily calculated. The incremental amount of solute in the mobile phase is

$$\text{Amount mobile} = (\Delta z\ A_c)\alpha\ \Delta c \tag{2-26a}$$

In Equation 2-26a ($\Delta z\ A_c$) is the volume of the column segment and α is the fraction of that volume which is mobile phase. The incremental amount of solute in the stagnant fluid is

$$\text{Amount in stagnant fluid} = (\Delta z\ A_c)\,(1 - \alpha)\epsilon\ K_d\ \Delta c \tag{2-26b}$$

Here $(1 - \alpha)$ is the fraction of the fluid volume, $\Delta z\ A_c$, which is not mobile phase and ϵ is the fraction of this which is stagnant fluid. K_d tells what fraction of these pores are available to the solute. For the solid we have

$$\text{Amount on solid} = (\Delta z\ A_c)\,(1 - \alpha)\,(1 - \epsilon)\rho_s\ \Delta q \tag{2-26c}$$

* This development is similar to but more complete than that in References 1051 to 1053, and 1056.

The first three terms give the volume which is solid. Since Δq is measured in kilogram moles per kilogram solid, the solid density ρ_s is required to convert from volume to weight. The solid density ρ_s is the structural density of the solid, i.e., of crushed solid without pores. Putting Equation 2-26 into Equation 2-25, we have

$$\text{Fraction} = \frac{(\Delta z\, A_c)\alpha\, \Delta c}{\Delta z\, A_c[\alpha\, \Delta c + (1 - \alpha)\epsilon\, K_d\, \Delta c + (1 - \alpha)(1 - \epsilon)\rho_s\, \Delta q]} \tag{2-27}$$

Substituting Equation 2-27 into Equation 2-24 and rearranging we obtain:

$$u_s = \frac{v}{1 + \left(\frac{1 - \alpha}{\alpha}\right)\epsilon\, K_d + \left(\frac{1 - \alpha}{\alpha}\right)(1 - \epsilon)\rho_s \frac{\Delta q}{\Delta c}} \tag{2-28}$$

The solute wave velocity, u_s, is the average velocity of the incremental amount of solute.

To use Equation 2-28 we must relate Δq to Δc. If we assume that solid and fluid are in equilibrium, then any of the equilibrium expressions developed in Section III can be used. Although a number of assumptions are now inherent in the physical derivation of u_s (see Section V), the result agrees quite well with experiment.

1. Solute Movement with Linear Isotherms

The simplest isotherm to use is the linear isotherm:

$$q_i = k_i(T)c_i \tag{2-29}$$

Then $\Delta q/\Delta c = k_i(T)$, and Equation 2-28 becomes:

$$u_{s_i} = \frac{v}{1 + \left(\frac{1 - \alpha}{\alpha}\right)\epsilon\, K_{d_i} + \left(\frac{1 - \alpha}{\alpha}\right)(1 - \epsilon)\rho_s\, k_i} \tag{2-30}$$

Equation 2-30 says that in the low concentration limit where the linear isotherm is valid there is a limiting solute velocity which does not depend on concentration. This solute velocity does depend upon temperature and other thermodynamic variables which change k_i. If we plot axial distance in the column z vs. time t, each solute will have a slope equal to its solute velocity. For an isothermal system this is illustrated in Figure 2-5A for a pulse of feed. The solute waves are drawn at the beginning and end of the feed pulse. Each solute moves at a constant velocity given by Equation 2-30, and the results for different solutes can be superimposed since we have assumed independent isotherms. The predicted outlet concentrations are the square waves shown in Figure 2-5B. Essentially the same results were first obtained by DeVault[320] using the equations first developed by Wilson.[1094] Obviously, the model is too simple since experimental results show zone spreading. This zone spreading will be discussed in Section VI. What is important here is that the simple theory accurately predicts where the peak maximum exit, and thus predicts whether or not a separation will occur (but not how good the separation is). The solute movement theory is simple enough to use with very complex operating methods, and it can easily be extended to nonlinear isotherms.

2. Solute Movement with Nonlinear Isotherms

In most large-scale adsorption and chromatography applications the concentrations are high enough that isotherms are nonlinear. For systems where there is only one solute $\Delta q/$

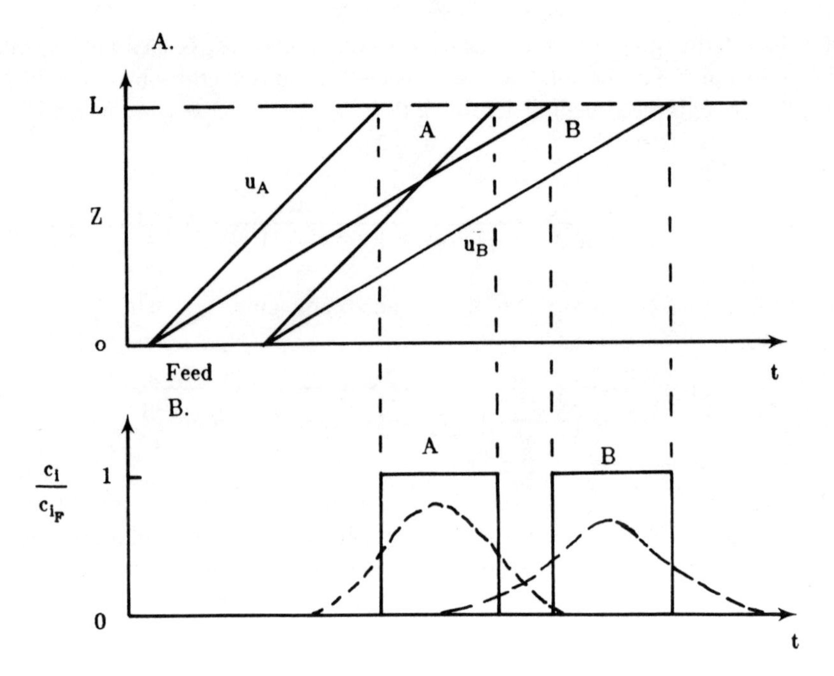

FIGURE 2-5. Solute movement theory for pulse input in an isothermal system with linear isotherms. (A) Solute movement in column; (B) product concentrations. — Predicted; ---- experimentally observed.

Δc will be a function of the solute concentration. Hence, the solute wave velocity given by Equation 2-28 will also depend upon concentration. The specific effects actually depend upon the isotherm form used. For dilute systems the Langmuir isotherm (Equations 2-4, 2-13, or 2-14) is often a good approximation of the equilibrium data. Suppose that a column is first saturated with a concentrated solution, c_h and this is then displaced with a dilute solution, c_l (see Figure 2-6A). The value of $\Delta q/\Delta c$ now increases monotonically as concentration decreases from c_h to c_l. Thus $\Delta q/\Delta c$ can be approximated by the derivative, where Equation 2-14 has been used for the isotherm form. Equation 2-31 can be substituted into Equation 2-28:

$$\frac{\Delta q}{\Delta c} \simeq \left(\frac{\partial q}{\partial c}\right)_T = \frac{a}{(1 + bc)^2} \tag{2-31}$$

$$u_s = \frac{v}{1 + \frac{(1 - \alpha)}{\alpha} \epsilon K_d + \frac{(1 - \alpha)}{\alpha} (1 - \epsilon)\rho_s \frac{a}{(1 + bc)^2}} \tag{2-32}$$

This equation shows that u_s is now a function of concentration and that the solute velocity decreases monotonically as concentration decreases.

This result is easily shown on a distance vs. time diagram. Figure 2-6A shows the feed to the column when a high concentration fluid is displaced by a fluid of zero concentration. Figure 2-6B shows the z vs. t diagram. Until time t_o, all solute moves at a solute velocity u_s (c_h) which can be calculated from Equation 2-32. At t_o a "diffuse wave" or "fan" is generated. This occurs because the concentration at this point ($z = 0$, $t = t_o$) varies from c_h to c_l. Each concentration generates a solute wave with a slope u_s (c) as shown in Figure 2-6B. By choosing arbitrary concentrations between c_h and c_l, a number of solute waves in

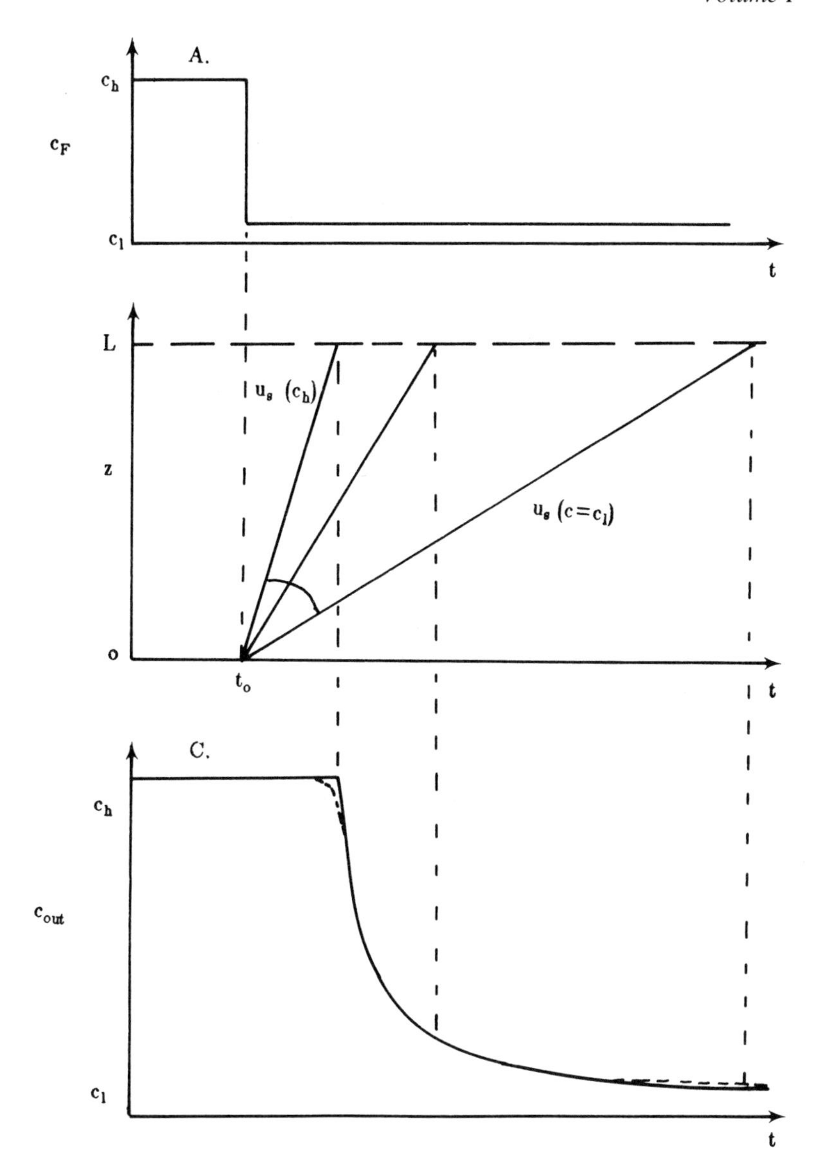

FIGURE 2-6. Solute movement theory for displacement of a concentrated solution by a dilute solution for Langmuir isotherm. (A) Feed composition; (B) solute movement in column; (C) product concentration. — Predicted; ---- experimentally observed. Note that scales in parts A and C differ.

the diffuse wave can be drawn. In Figure 2-6C these are projected into the product concentration vs. time diagram to predict the outlet concentration profile. If the column is made longer, the spread between the fastest and slowest waves increases and the diffuse wave grows in direct proportion to the column length. This is called "proportional pattern behavior". Similar results were obtained by DeVault,[320] Walter,[1031] and Weiss[1081] using different developments, and are reviewed in several sources.[165,865,901,1015-1017]

Comparison of experimental data and predictions for the effluent concentration show that there is often excellent agreement for both adsorption[901] and for chromatography.[385] This agreement is illustrated schematically in Figure 2-6C. The reason for the agreement between theory and experiments for diffuse waves is the isotherm effect controls. Additional zone spreading caused by mass transfer resistances and axial dispersion are almost negligible.

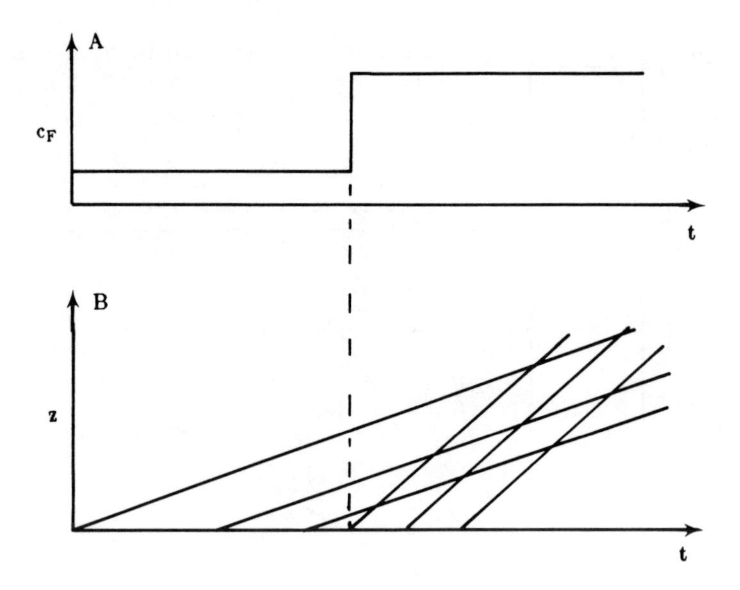

FIGURE 2-7. Concentrated material displacing dilute material for Langmuir isotherm. (A) Feed; (B) solute movement in column, Equation 2-32, impossible case.

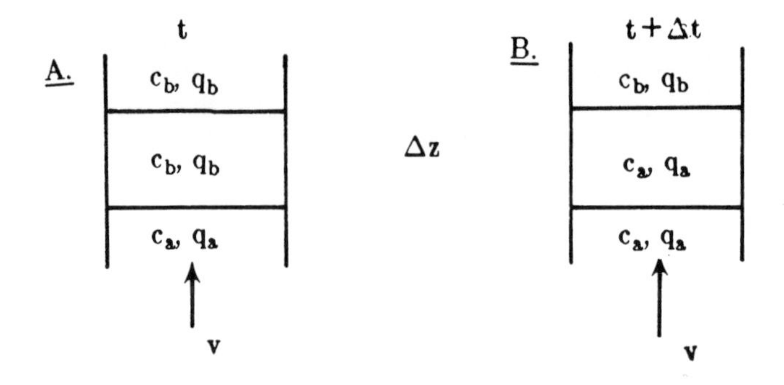

FIGURE 2-8. Control volume for mass balance for shock wave. (A) Shock wave enters control volume; (B) shock wave leaves control volume.

Any equilibrium data which is "favorable" and has the same general shape as the Langmuir isotherm will have similar diffuse wave behavior. The equation for other isotherms is easily derived by determining $\partial q/\partial c$ and substituting this for $\Delta q/\Delta c$ in Equation 2-28. If an isotherm equation is not available, the solute wave velocity at any concentration can be predicted by determining the isotherm slope either graphically or numerically and using this slope as the local value of $\partial q/\partial c$.

For favorable isotherms, if a dilute solution is displaced by a concentrated solution the situation is very different. Now Equation 2-32 for a Langmuir isotherm predicts that the more concentrated, faster-moving waves overtake the less concentrated, slower-moving waves. As shown in Figure 2-7B, this equation predicts a series of intersections of two waves of different concentrations. This implies that there is a region of the column where two different concentrations occur simultaneously. This is obviously physically impossible. The reason why this physically impossible prediction occurs is we assumed in Equation 2-31 that c and q were continuous functions. This is apparently not the case, and Equations 2-31 and 2-32 are not valid for the feed shown in Figure 2-7A.

Since concentration is not continuous inside the column, we must have a "shock" or discontinuous wave. To analyze this a macroscopic mass balance can be used. Figure 2-8 shows a macroscopic section of the column of height Δz. We wish to do a mass balance for the time period, Δt, during which the shock wave passes through this section of the column. If the shock wave velocity is u_{sh}, then the time required for the shock wave to move a distance Δz is

$$\Delta t = \frac{\Delta z}{u_{sh}} \tag{2-33}$$

In Figure 2-8, c_a and q_a are the fluid concentration and amount adsorbed *after* the shock wave has passed, where c_b and q_b refer to *before* the shock wave. We assume solid and fluid are in equilibrium at each point. For the feed shown in Figure 2-7A, $c_a = c_{low}$ and $c_b = c_{high}$. The mass balance for period Δt over segment Δz is

$$\begin{aligned} &\alpha v c_a \Delta t - \alpha v c_b \Delta t - [\alpha + K_d \epsilon (1 - \alpha)](c_a - c_b)\Delta z \\ &\quad - [(1 - \alpha)(1 - \epsilon)]\, \rho_s (q_a - q_b)\Delta z = 0 \end{aligned} \tag{2-34}$$

The first term in Equation 2-34 is input, the second term is minus the output, and the third and fourth terms are minus the accumulation in segment Δz. Substituting in Equation 2-33 and solving for the shock wave velocity we obtain:

$$u_{sh} = \frac{v}{1 + \dfrac{1 - \alpha}{\alpha}\epsilon K_d + \dfrac{(1 - \alpha)}{\alpha}(1 - \epsilon)\rho_s\left(\dfrac{q_a - q_b}{c_a - c_b}\right)} \tag{2-35}$$

Comparing Equations 2-28 and 2-35, we see that they are essentially the same. In Equation 2-35, $\Delta q/\Delta c$ has been defined as the difference across the shock wave. The shock wave velocity depends on $c_a = c_{high}$ and $c_b = c_{low}$ plus the isotherm. The term $(q_a - q_b)/(c_a - c_b)$ is the slope of the chord of the isotherm from the initial condition (c_b, q_b) to the final condition (c_a, q_a). This term can be calculated from the isotherm equation or from the equilibrium data. If the Langmuir isotherm is used, Equation 2-14 is substituted in for q_a and for q_b in Equation 2-35. For favorable isotherms:

$$u_s(c_{high}) > u_{sh} > u_s(c_{low}) \tag{2-36}$$

The shock wave calculation is shown in Figure 2-9. Note in Figure 2-9B that Equation 2-36 is satisfied. The product concentration shown in Figure 2-9C is a sharp jump which will be the same regardless of the column length. The shock wave shown in Figure 2-9C is different from the linear isotherm result shown in Figure 2-5B. The linear isotherm predicts no change in shape of the added pulse. If the input wave is not sharp it will stay this way with a linear isotherm. With a shock wave the diffuse input will be sharpened into a shock. This difference is illustrated in Figure 2-10. With a nonlinear isotherm the feed in Figure 2-10A will generate a series of shocks which intersect each other and eventually form a single shock wave (if the column is long enough) which gives the product concentration shown in Figure 2-10C. The linear isotherm transmits the input without change in shape (Figure 2-10B). Other developments for the shock wave are given in several reviews.[165,901,1016]

In actual practice, finite mass transfer rates, diffusion, and dispersion will spread out the shock wave as shown by the dotted line in Figure 2-9C. The sharpening effect of the shock wave illustrated in Figure 2-10 and the dispersive effects reach a balance so that a "constant pattern" is eventually obtained. Then, no matter how long the column, the shape shown by the dotted line in Figure 2-9C does not change. This S-shaped region of changing concen-

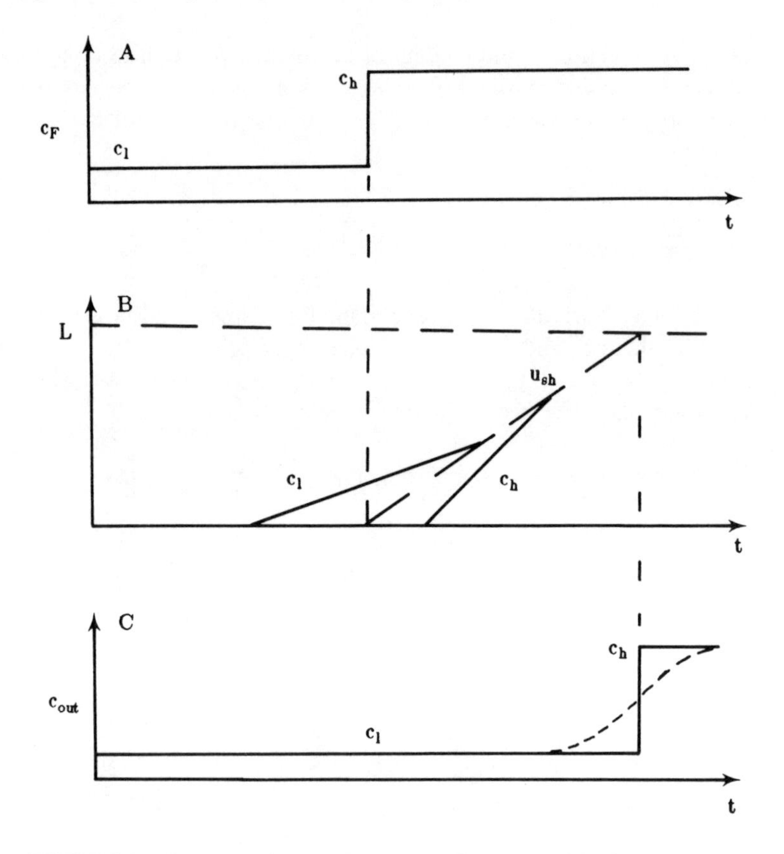

FIGURE 2-9. Concentrated material displacing dilute material for favorable isotherm. (A) Feed; (B) shock wave solution for solute movement in column; (C) predicted product concentration.

tration is called the "mass transfer zone". This constant pattern greatly simplifies the difficulty of including mass transfer in the theory and is discussed in Section VII. Once the constant pattern has been formed the movement of the mass transfer zone (MTZ) can be shown on a solute movement diagram.

This solution method for nonlinear isotherms can also be applied to monovalent, binary ion exchange[1015-1017] and to GLC.[1006] With a constant equilibrium constant the ion exchange equilibrium is given by Equation 2-22, which has the same form as the Langmuir expression. Thus, Figures 2-6, 2-9, and 2-10C could also be for binary ion exchange. Concentrations are now fractions, x_A (see Equation 2-22), of ion A. Fraction of ion B can be found as $1 - x_A$.

For GLC, the isotherms have the opposite curvature of a Langmuir isotherm and are shown schematically in Figure 2-2B. Diffuse and shock waves still occur, but in the reverse situations as for Langmuir isotherms. Thus in GLC a shock wave occurs when a dilute gas displaces a concentrated gas. The methods of constructing the diagrams are the same, but the diagrams look "backwards" compared to adsorption. With more complex isotherms such as BET type 3, 4, and 5 isotherms (Figures 2-2C to E), interacting shock and diffuse waves can occur.

B. Movement of Pure Energy Waves

If the heat of adsorption and the heat of mixing are negligible and the column is adiabatic, then we have a pure energy wave. That is, the adsorption and desorption of solutes will not affect the column temperature although temperature does affect the isotherms. These as-

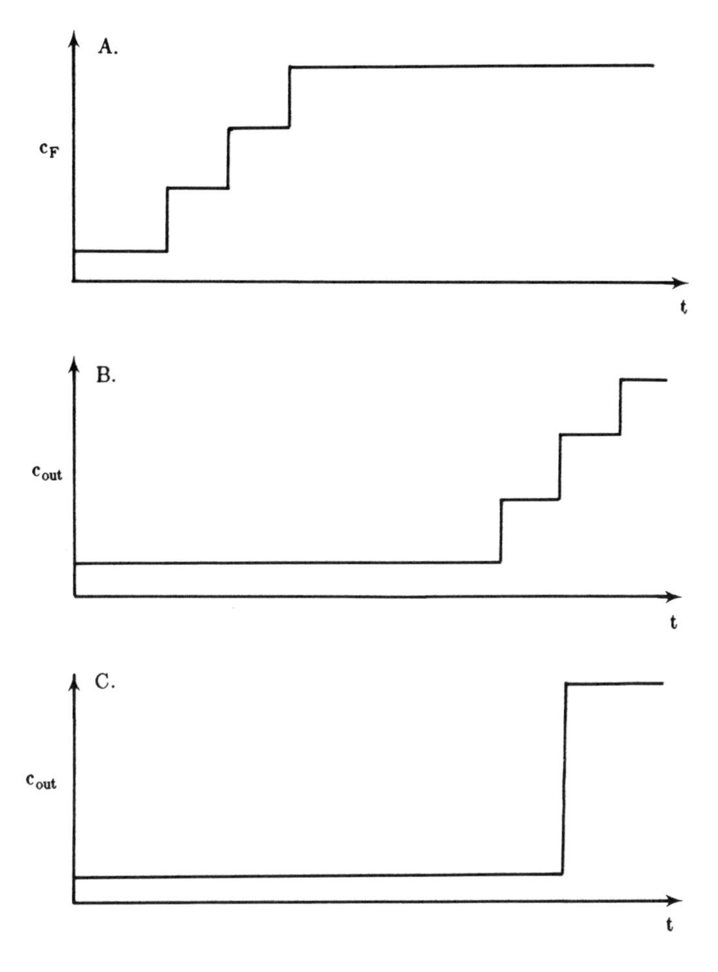

FIGURE 2-10. Effect of changing input wave. (A) Feed; (B) predicted outlet concentration for linear isotherm; (C) predicted outlet concentration for favorable isotherm.

sumptions are reasonable in many liquid systems and in very dilute gas systems. The movement of a pure energy wave can be determined from an analysis similar to that used previously for solute movement.

1. Thermal Wave Velocity

The velocity of the energy wave, u_{th}, can be found from the fraction of energy in the mobile phase times the interstitial fluid velocity.

$$u_{th} = \left(\frac{\text{Energy in mobile phase}}{\text{Total energy in segment}}\right) v \tag{2-37}$$

The fraction of energy stored in the mobile phase is

$$\frac{\text{Energy in mobile phase}}{\text{Total energy in segment}} = \frac{\text{Energy in mobile phase}}{\text{Energy in: (mobile + stagnant fluid + solid + wall)}} \tag{2-38}$$

or

$$\frac{\text{Energy in mobile phase}}{\text{Total energy in segment}} =$$

$$\frac{(\Delta z A_c)\alpha\rho_f C_{P_f}(T_f - T_{ref})}{\{\Delta z A_c[(\alpha + (1-\alpha)\epsilon)\rho_f C_{P_f}(T_f - T_{ref}) + (1-\alpha)(1-\epsilon)\rho_s C_{P_s}(T_s - T_{ref})] + (\Delta z W)C_{P_w}(T_w - T_{ref})\}} \tag{2-39}$$

The C_P values are the heat capacities while W is the weight of column wall per length and T_w is the wall temperature. The energy storage in the wall is important in laboratory-scale columns, but not in large commercial-scale columns. If we have local thermal equilibrium

$$T = T_f = T_s = T_w \tag{2-40}$$

and the term $(T - T_{ref})$ divides out. Then combining Equations 2-37 and 2-39, we obtain the thermal wave velocity:

$$u_{th} = \frac{v\, \rho_f\, C_{P_f}}{\left\{\left[1 + \left(\frac{1-\alpha}{\alpha}\right)\epsilon\right]\rho_f C_{P_f} + \frac{(1-\alpha)(1-\epsilon)}{\alpha} C_{P_s}\rho_s + \frac{W}{\alpha\, A_c} C_{P_w}\right\}} \tag{2-41}$$

Because of the simplifying assumptions made here the thermal wave velocity is independent of temperature and concentration. Comparison of Equations 2-30 and 2-41 is instructive. The solute and thermal wave velocities have similar forms. The thermal wave velocity has an additional term for the wall effect. For energy the equilibrium expression given in Equation 2-40 is linear with an equilibrium constant of 1.0. Also, since all pores are accessible to energy, K_d is implicitly equal to 1.0 in Equation 2-41. Equation 2-41 represents the average rate of movement of the thermal wave. An alternate approach including the heat of adsorption is discussed in Section IV.D.2. More exact analyses will include the effects of dispersion and the rate of heat transfer.

Temperature changes in the column can be analyzed on a axial distance z vs. time t diagram in the same way solute movement was analyzed. This is illustrated in Figure 2-11. The pure thermal wave moves through the column at a velocity u_{th}. The predicted shape of the temperature pulse is totally unchanged as it passes through the column. Of course, experimental results show dispersion and heat transfer effects as illustrated in Figure 2-11C. Comparison of Figures 2-11 and 2-5 shows that the thermal wave moves through the column in the same way as a solute with a linear isotherm.

2. *Effect of Thermal Wave on Solute*

Next, let us look at how the temperature changes affect the solute concentration and movement. Suppose we put a pulse of solute into a long column and then follow this with a step increase in temperature as shown in Figure 2-12A. If the solute is fairly strongly adsorbed the thermal wave will move faster than the solute wave, $u_{th} > u_s(T_c)$, and eventually overtake it. At the higher temperature the equilibrium constant k_i decreases and thus the solute velocity increases. This situation is illustrated in Figure 2-12B for the case where $u_{th} > u_s(T_h)$. Note that after the temperature increase the solute waves are closer together. The predicted outlet concentrations are shown in Figure 2-12C. Since the concentration wave takes less time to exit than to enter, it must have a higher concentration.

The effect of the thermal wave on the solute concentration can be determined by a mass balance. This mass balance calculation is similar to the balance done for the shock wave. A column segment of length Δz is selected. The thermal wave will pass through this segment in a time:

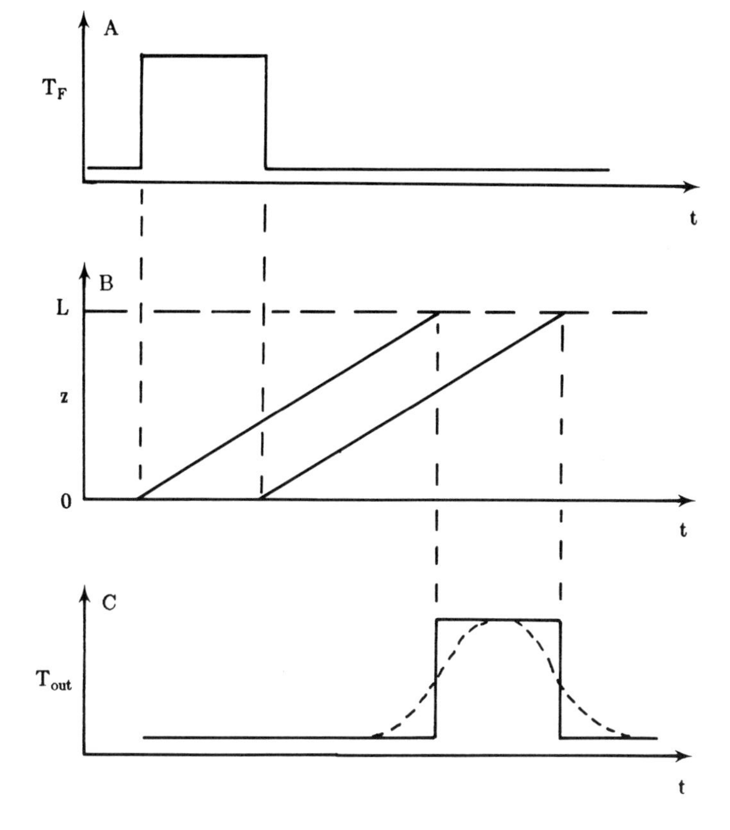

FIGURE 2-11. Movement of pure energy waves in a column. (A) Feed; (B) energy waves in column; (C) outlet temperatures. — Predicted; ---- experimental.

$$\Delta t = \frac{\Delta z}{u_{th}} \tag{2-42}$$

The mass balance over length Δz and time Δt is

$$\alpha v(c_a - c_b)\Delta t - [\alpha + K_d\,\epsilon(1 - \alpha)]\,(c_a - c_b)\Delta z$$
$$-[(1 - \alpha)(1 - \epsilon)]\rho_s(q_a - q_b)\Delta z = 0 \tag{2-43}$$

Substituting in Equation 2-42 and rearranging we have

$$\left[\alpha + \epsilon(1 - \alpha)K_d - \frac{\alpha\, v}{u_{th}}\right](c_a - c_b) + (1 - \alpha)(1 - \epsilon)\rho_s(q_a - q_b) = 0 \tag{2-44}$$

Now u_{th} is known from Equation 2-41 and the conditions before the intersection with thermal wave, c_b and q_b, are also known. We wish to determine the values c_a and q_a after the intersection. To do this we assume local equilibrium and solve Equation 2-44 simultaneously with the isotherm equation.

For linear isotherms (Equation 2-29) this simultaneous solution is simple. After some rearrangement the result is

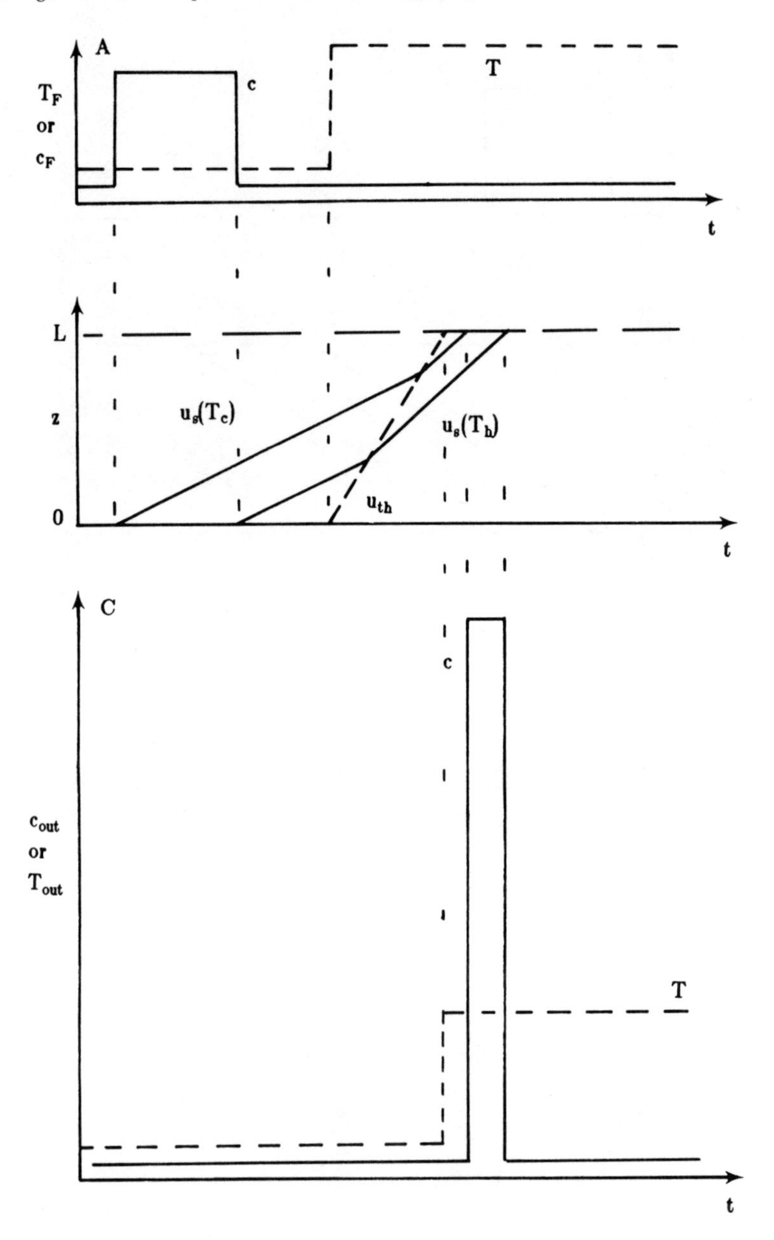

FIGURE 2-12. Effect of thermal wave on solute with a linear isotherm. (A) Feed; (B) waves in column; (C) predicted outlet concentration and temperature.

$$\frac{c(T_h)}{c(T_c)} = \frac{\dfrac{1}{u_s(T_c)} - \dfrac{1}{u_{th}}}{\dfrac{1}{u_s(T_h)} - \dfrac{1}{u_{th}}} \tag{2-45}$$

Equation 2-45 was first developed by Baker and Pigford.[74] In a typical dilute liquid system $u_{th} > u_s(T_h) > u_s(T_c)$ and Equation 2-45 predicts $c(T_h) > c(T_c)$. This is shown in Figure 2-12C.

For gas systems we may have $u_s(T_h) > u_s(T_c)$ and $c(T_c) > c(T_h)$. This is opposite to our intuition but does agree with some experimental results. If $u_s(T_h) > u_{th} > u_s(T_c)$, Equation 2-45 predicts a negative concentration which is physically impossible. In this case the nonlinear mass balance Equation 2-44 must be solved.

An alternate way to heat or cool the column is to use a jacket or heating coils. This is called the direct mode. Now the entire length of the column is heated or cooled simultaneously. Thus, u_{th} is essentially infinite, and Equation 2-45 becomes:

$$\frac{c(T_h)}{c(T_c)} = \frac{u_s(T_h)}{u_s(T_c)} \tag{2-46}$$

If the equilibrium constant $k_i(T)$ decreases as temperature increases, Equation 2-46 predicts that $c(T_h) > c(T_c)$. That is, as solute desorbs it diffuses into the fluid and increases the fluid concentration. Equation 2-46 is valid for both gases and liquids.

For nonlinear equilibrium a simple closed form result is usually not attainable. The mass balance (Equation 2-44), and the isotherm equation often have to be solved simultaneously.

This analysis is also applicable if a thermodynamic variable other than temperature (e.g., pH) is changed. Equations 2-44 to 2-46 are valid but with u_{th} replaced by the velocity of this thermodynamic variable.

C. Effect of Velocity Changes

In the preceding two sections we assumed that the interstitial fluid velocity v was a constant. For liquids the density of adsorbed material is approximately the same as the density of the liquid. If the liquid density does not change drastically with composition, then the overall density of the liquid is roughly constant. This implies that v will be constant if the porosities are constant. A significant density change will cause v to vary; however, for liquids this is the exception, not the rule. For gases the density of the adsorbed phase is approximately the same as the liquid density. Thus when gas adsorbs there must be a decrease in volume which will cause a decrease in the velocity. When the solute desorbs, the opposite happens and v increases. This effect produces a shock wave when gas adsorbs since the faster-moving material is displacing the slower-moving material. A diffuse wave results during desorption. These effects will occur for any type of isotherm and are called "the sorption effect".

Quantitatively we can study the sorption effect by doing a mass balance for a control volume. We will assume the pores are readily available to all species. The solute mass balance over time Δt and length Δz is

$$\begin{aligned} &\alpha v_a c_a \Delta t - \alpha v_b c_b \Delta t - [\alpha + \epsilon(1 - \alpha)](c_a - c_b)\Delta z \\ &\quad - (1 - \alpha)(1 - \epsilon)\rho_s(q_a - q_b)\Delta z = 0 \end{aligned} \tag{2-47}$$

which differs from Equation 2-34 since v will have different velocities before and after the sorption wave. Equation 2-47 can be written for each solute. We can also write an overall balance

$$\begin{aligned} &\alpha v_a \bar{\rho}_{f_a} \Delta t - \alpha v_b \bar{\rho}_{f_b} \Delta t - [\alpha + \epsilon(1 - \alpha)](\bar{\rho}_{f_a} - \bar{\rho}_{f_b})\Delta z \\ &\quad - (1 - \alpha)(1 - \epsilon)\rho_s(\bar{q}_a - \bar{q}_b)(\Delta z) = 0 \end{aligned} \tag{2-48}$$

where $\bar{\rho}_{f_a}$ and $\bar{\rho}_{f_b}$ are molar densities and $\bar{q}_i$ are total amounts adsorbed. To keep the situation as simple as possible we will assume an ideal gas at constant temperature and pressure. Then the molar density is constant and $\bar{\rho}_{f_a} = \bar{\rho}_{f_b} = \bar{\rho}_f$. We will also assume that the isotherm is linear, Equation 2-29, and that there is only one species which will adsorb plus a non-adsorbed carrier.

The analysis is different for diffuse and for shock waves. For diffuse waves we can solve both Equation 2-47 and 2-48 for $\Delta z/\Delta t$ and take the limit as $\Delta t \to 0$. This limit does exist for diffuse waves, and $\Delta c \to 0$ as $\Delta t \to 0$. The results for linear systems are

$$\lim_{\Delta t \to 0} \frac{\Delta z}{\Delta t} = \frac{\alpha}{\alpha + \epsilon(1-\alpha) + (1-\alpha)(1-\epsilon)\rho_s k} \frac{d(vc)}{dc} \tag{2-49}$$

and

$$\lim_{\Delta t \to 0} \frac{\Delta z}{\Delta t} = \frac{\bar{\rho}_f \alpha}{(1-\alpha)(1-\epsilon)\rho_s} \frac{1}{k} \frac{dv}{dc} \tag{2-50}$$

Setting these equations equal and rearranging we eventually obtain

$$\frac{dv}{dy} = \frac{[(1-\alpha)(1-\epsilon)\rho_s k]v}{\alpha + \epsilon(1-\alpha) + [(1-\alpha)(1-\epsilon)\rho_s k](1-y)} \tag{2-51}$$

where y is the mole fraction of adsorbate

$$y = \frac{c}{\bar{\rho}_f} \tag{2-52}$$

Equation 2-51 can be rearranged and integrated.

$$\int_{v_{in}}^{v} \frac{dv}{v} = \int_{y_{in}}^{y} \frac{[(1-\alpha)(1-\epsilon)\rho_s k]dy}{\alpha + \epsilon(1-\alpha) + [(1-\alpha)(1-\epsilon)\rho_s k](1-y)} \tag{2-53}$$

The result of the integration is

$$v = \frac{v_{in}[\alpha + \epsilon(1-\alpha) + (1-\alpha)(1-\epsilon)\rho_s k(1-y_{in})]}{\alpha + \epsilon(1-\alpha) + (1-\alpha)(1-\epsilon)\rho_s k(1-y)} \tag{2-54}$$

For a diffuse wave where $y_{out} > y_{in}$, Equation 2-54 predicts that $v_{out} \geqslant v_{in}$ since adsorbate is desorbing.

The wave velocity can be determined by substituting Equation 2-54 into either Equation 2-49 or 2-50 and realizing that[865,1051]

$$u_{wave} = \lim_{\Delta t \to 0} \frac{\Delta z}{\Delta t} \tag{2-55}$$

The result is

$$u_{wave} = \frac{\alpha[\alpha + \epsilon(1-\alpha) + (1-\alpha)(1-\epsilon)\rho_s k]v_{in}}{[\alpha + \epsilon(1-\alpha) + (1-\alpha)(1-\epsilon)\rho_s k(1-y)]^2} \tag{2-56}$$

This predicts that the wave velocity increases as y increases. Equations 2-55 and 2-56 agree with results derived by a different method.[865]

The sorption effect for a diffuse wave is shown in Figure 2-13. Note that the resulting

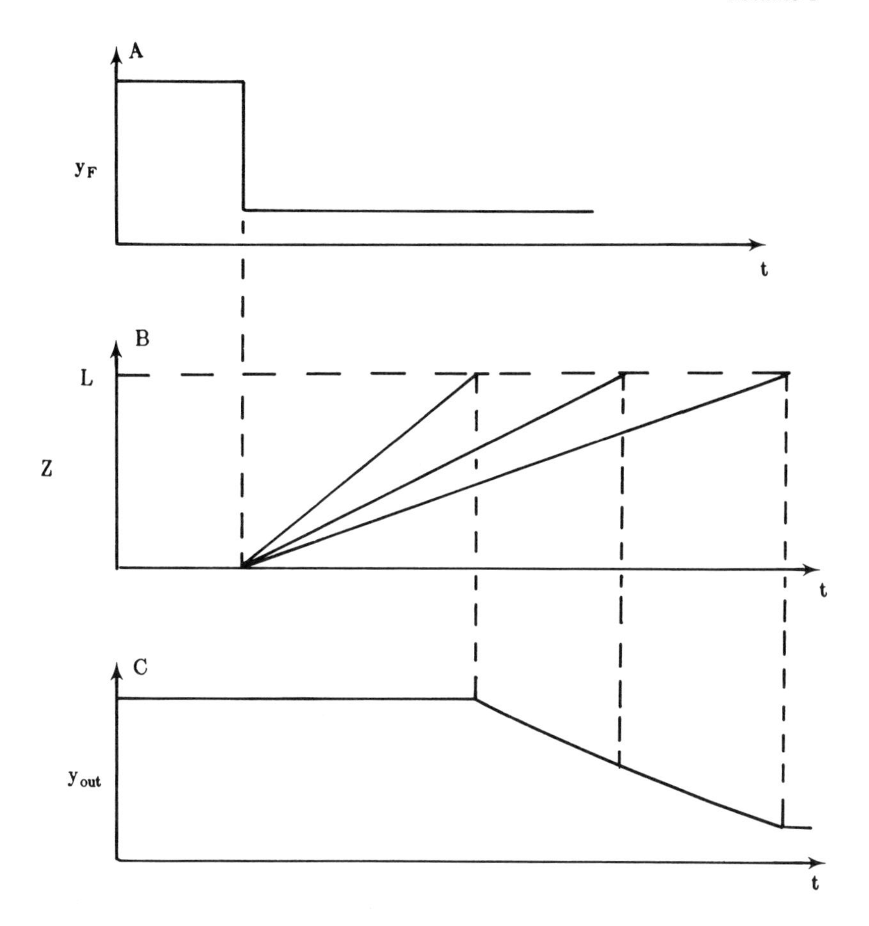

FIGURE 2-13. Sorption effect for dilute wave displacing a concentrated wave for linear isotherms. (A) Feed; (B) solution using Equation 2-56; (C) predicted product concentration.

outlet concentration is very similar to that predicted for a diffuse wave caused by a nonlinear isotherm.

When a concentrated solution displaces a dilute solution, Equation 2-56 predicts that solute waves cross. This will be similar to Figure 2-7B and is physically impossible. In this case the limit in Equations 2-49 and 2-50 does not exist. Now the balances used in Equations 2-47 and 2-48 must be over finite Δz and Δt. The time increment Δt must be related to Δz:

$$\Delta t = \frac{\Delta z}{u_{wave}} \tag{2-57}$$

Now we can solve Equations 2-47, 2-48, and 2-57 for the fluid velocity and the wave velocity. The results are

$$v_b = v_a \left[\frac{\alpha + \epsilon(1 - \alpha) + (1 - \alpha)(1 - \epsilon)\rho_s k(1 - y_a)}{\alpha + \epsilon(1 - \alpha) + (1 - \alpha)(1 - \epsilon)\rho_s k(1 - y_b)}\right] \tag{2-58}$$

and

$$u_{wave} = \frac{\alpha v_a}{\alpha + (1 - \alpha)\epsilon + (1 - \alpha)(1 - \epsilon)\rho_s k(1 - y_b)} \tag{2-59}$$

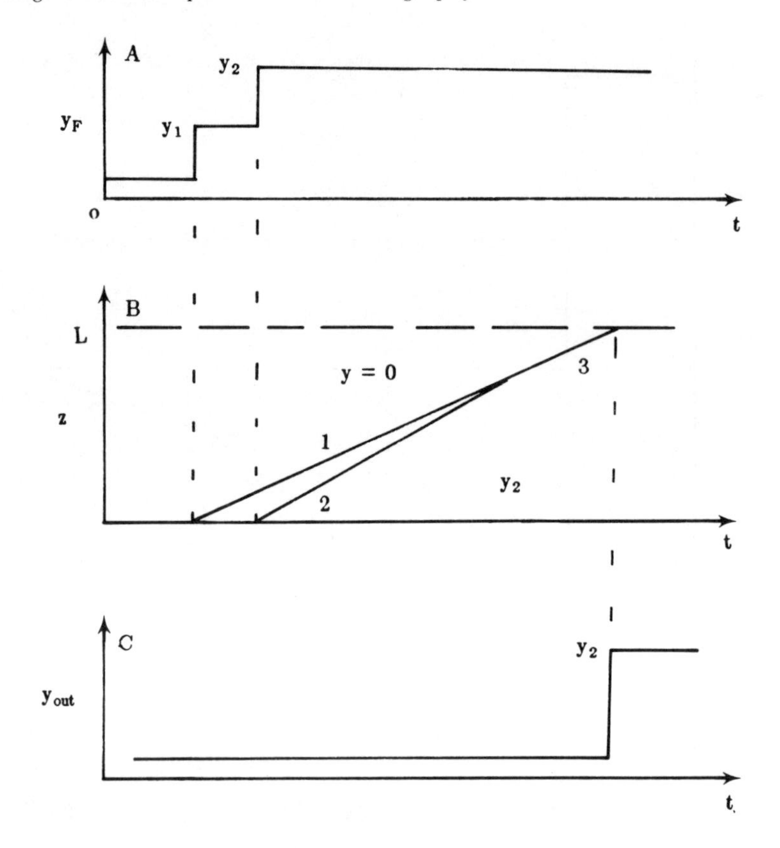

FIGURE 2-14. Sorption effect for concentrated wave displacing a dilute wave for linear isotherms for series of steps. (A) Feed; (B) solution using Equation 2-59; (C) predicted product concentration.

where $y_a \geqslant y_b$. These results can also be obtained from a balance around the entire column. Note that the wave velocity depends on the conditions before the shock while the downstream velocity depends on conditions on both sides of the shock wave.

The sorption effect for a shock wave is shown in Figure 2-14. A series of steps are shown so that the shock wave effect will be obvious. Wave 1 sees $y_b = 0$ and has the wave velocity which would normally be expected for a linear isotherm. However, the downstream velocity $v_b < v_a$ since $y_a = y_1$. Wave 2 sees $y_b = y_1$ and thus moves faster than wave 1. The two waves eventually intersect to form wave 3. This wave sees $y_b = 0$ and has the same velocity as wave 1. The downstream fluid velocity will be different. The outlet mole fraction shown in Figure 2-14C appears as a shock wave.

For nonlinear isotherms the sorption and isotherm effects may reinforce or oppose each other. If the isotherm is favorable (Langmuir shape), both the isotherm and the sorption effects predict a shock wave when a concentrated solution displaces a dilute solution. When a dilute solution displaces a concentrated solution, both effects predict a diffuse wave. Thus the two effects reinforce each other for favorable isotherms. For unfavorable isotherms (e.g., GLC) the isotherm and sorption effects are opposite and oppose each other. Because of this opposition there is an optimum operating temperature for preparative GLC which will give quite sharp peaks.

Temperature and pressure changes will also cause velocity changes for gases. Temperature increases will increase the gas velocity. When this effect is important the molar density is not constant. The molar density can be calculated from the appropriate PVT relationship such as the ideal gas law. The effect is easily visualized for a very simple case. Assume the ideal gas law holds, the system is isobaric, and the change in amount adsorbed is negligible compared to the total gas flow rate. Thus ρv = constant. Applying the ideal gas law, this leads to the conclusion that

$$v = T(\text{constant}) \quad (2\text{-}60)$$

Therefore, a temperature increase increases the velocity. This means that hot waves are compressive since the upstream portion of the wave moves faster than the downstream portion. For a temperature increase from 20 to 80°C the velocity increase is about 20%. Cooling waves will be diffuse and thus increase zone spreading. For dilute isothermal systems which follow the ideal gas law:

$$v = \frac{\text{Constant}}{p} \quad (2\text{-}61)$$

Thus a pressure drop increases the fluid velocity downstream which is dispersive. Pressure waves will also change the fluid velocity.

Changes in porosity will also change the velocity. Generally speaking, if α decreases the velocity increases. With soft gels such as Sephadex® (a polydextran) and Biogel P® (a polyacrylamide) used for size exclusion chromatography (SEC), temperature changes change the porosity if the bed is constrained. The exact effect this has on solute velocites depends on the gel used.[612] Ion exchange resins commonly swell or shrink when exchange occurs.[18,202,484,693,1016] For instance, with a carboxylic acid cation exchange resin going from the H^+ to Na^+ form the resin swells. If the bed is constrained so that expansion is not allowed the porosities decrease. The high Na^+ wave will move faster than the low Na^+ wave. This is then compressive. The reverse situation occurs and a diffuse wave results if the resin goes from the H^+ to the Na^+ form. The compressive or diffusive wave effects will be added onto the effect caused by isotherm curvature.

D. Coupled Systems

1. Two or More Solutes

When there are two or more solutes which adsorb, they will usually compete with each other for sites on the adsorbent. This was illustrated by the multicomponent Langmuir isotherm, Equation 2-7, which we will write as:

$$q_i = \frac{a_i c_i}{1 + \sum_{j=1}^{n} (b_j c_j)} \quad (2\text{-}62)$$

Since the equilibrium for a solute depends upon the concentration of all other solutes, the movement of each solute is coupled to the movement of all other solutes which are present. For dilute solutions with constant velocity the mass balance for each solute and the solute movement equations are unchanged by the addition of additional solutes. Thus Equations 2-28 and 2-35 are the same, except $\Delta q/\Delta c$ becomes $\Delta q_i/\Delta c_i$. However, since Δq_i depends on the concentrations of all solutes present, the solute wave velocity u_{si} will depend on all c_i. The mathematical solution requires uncoupling the equations. This uncoupling can be done several different ways and is discussed in detail elsewhere.[59,385,426,485,486,536,596,657,832,836,865,982] In this section we will look at the problem qualitatively, and solve the one situation where the solutes uncouple themselves.

In Section IV.A we saw that a shock wave is produced if a concentrated solution displaces

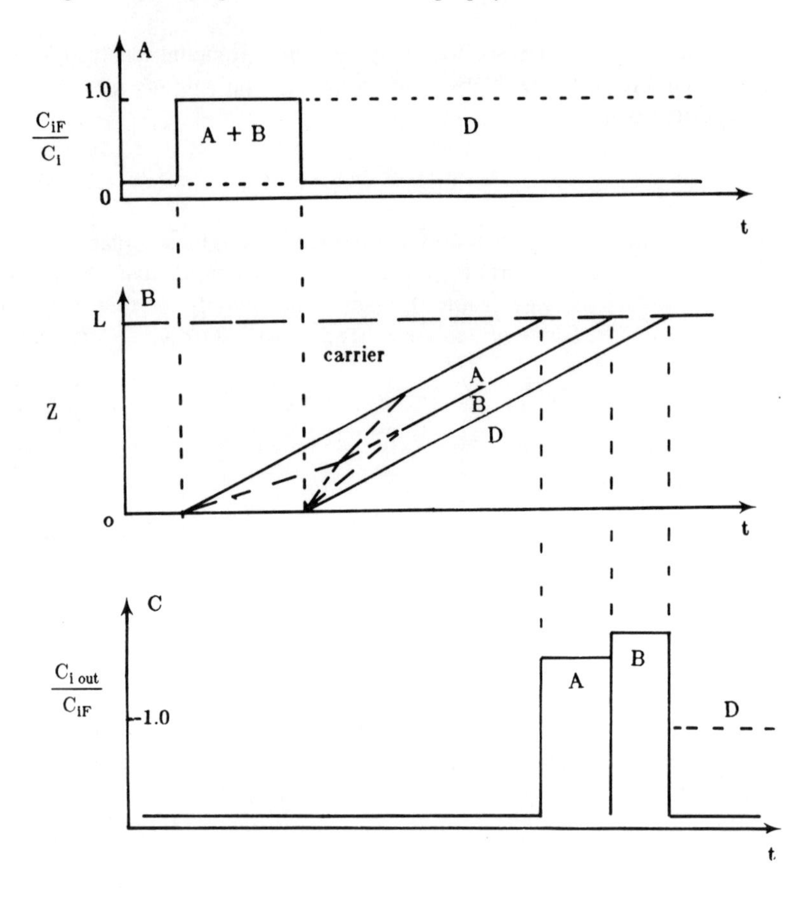

FIGURE 2-15. Displacement development with coupled solutes. (A) Feed; (B) local equilibrium theory. Solid lines can be predicted from Equations 2-35, 2-63, and 2-64. Dotted lines require more detailed theories; (C) predicted outlet concentrations.

a dilute solution when the solute has a Langmuir isotherm. In that case there really were two components present: the solute and a carrier. The solute was much more strongly adsorbed. Extending this situation to several solutes whose isotherms follow Equation 2-62 we would expect to observe a shock wave (constant pattern behavior) when a weakly adsorbed solute is displaced by a more strongly adsorbed solute. This is observed experimentally and is predicted by several theories.[59,426,486,832,865,982] If we reverse the situation and have a strongly adsorbed solute displaced by a weakly adsorbed solute a diffuse wave results.

In the case of shock waves the solutes will eventually decouple themselves. Suppose a pulse of two solutes, A and B, are displaced by a third solute, D, which is strongly adsorbed. (An example would be methane, A, and ethane, B, in hydrogen carrier being displaced by butane, D, on activated carbon adsorbent). Since the displacer, D, is pushing out less strongly adsorbed A and B, a shock wave will result. The pulse of A and B is pushing out a less strongly adsorbed material, the solvent or carrier gas, and a shock will form here also. Finally, since solute B is more strongly adsorbed than solute A, the B will displace the A. The result will eventually be a band of pure B displacing a band of pure A. Thus the final result which is observed experimentally[427,518] is four pure components separated by three shock waves. Since mass balances on A and B must be satisfied, the band widths for these solutes must become constant. This means that the shock waves must all be parallel. This condition is observed experimentally and is predicted by the theories. The solute movement diagram is shown in Figure 2-15. Prediction of the dotted lines in Figure 2-15B requires the more complex theories[59,385,486,832,865,982] while the solid lines can be predicted based on the argument presented here.[262,320,425,518,835]

The shock wave velocities must all be equal once the pure bands are obtained. The concentration of D upstream of shock 3 is the inlet concentration of D and the concentration of D downstream of shock 3 is zero. For this particular case where the shock separates pure component D from a solution containing no D, the equations are uncoupled and u_{sh3} (Equation 2-35) does not depend on c_A or c_B. Now we know that

$$u_{sh_1} = u_{sh_2} = u_{sh_3} \tag{2-63}$$

Shock wave 1 depends only upon the concentration of A in the A band and u_{sh2} only on the concentration of B in the B band. Equations 2-63 and 2-35 can now be used to find the concentrations of A and B in their respective bands. Figure 2-15C shows that the solutes are concentrated in displacement development. Finally, the band widths can be found from a mass balance over an entire cycle.

$$(c_{iF})(t_F) = (c_{i\ band})(t_{band}) \qquad i = A,B \tag{2-64}$$

In displacement development the final products can be predicted without knowing the details of the interactions inside the column. However, calculation of the length of column necessary to get pure bands does require these details and hence the more detailed theories.

Displacement behavior only occurs if the displacing agent is strongly enough adsorbed and has a high enough feed concentration. If the weakest adsorbed solute, A, moves faster than u_{sh}, a dilute band or a separate peak of A will be formed instead of a concentrated band. This can be checked. If

$$\left.\frac{\Delta q_A}{\Delta c_A}\right|_{c_A<c_{A_{band}}} < \left.\frac{\Delta q_D}{\Delta c_D}\right|_{c_{D_F}}$$

then $u_{sh}(c_A) > u_{sh3}$ and the A band will be diluted. If

$$\left.\frac{\Delta q_A}{\Delta c_A}\right|_{c_{A_F}} < \left.\frac{\Delta q_D}{\Delta c_D}\right|_{c_{D_F}}$$

$u_{sh_A}(c_{AF}) > u_{sh3}$ and a separate A peak will form. If

$$\left.\frac{\Delta q_A}{\Delta c_A}\right|_{c_A=0} < \left.\frac{\Delta q_D}{\Delta c_D}\right|_{c_{D_F}}$$

then $u_s(c_A \rightarrow 0) > u_{sh3}$ and the A peak will have the diffuse tail normally expected and will be unaffected by the displacer.

Analysis of diffuse wave patterns requires the detailed theories to uncouple the equations. However, the characteristic diagrams can be understood without these theories. The characteristic diagram for three ions, which must be coupled because electroneutrality must hold, is shown in Figure 2-16A for a pulse input. The eluant, H^+, is the least strongly sorbed ion. The less-sorbed ion in the feed pulse, A^+, is pushed ahead and concentrated to form a region of pure A^+. This is shown in Figure 2-16A. This region is bounded by shocks. Then comes a region at the feed concentration. This is separated from a plateau region by a fast diffuse wave. In the fast diffuse wave the A^+ concentration drops to zero. The plateau

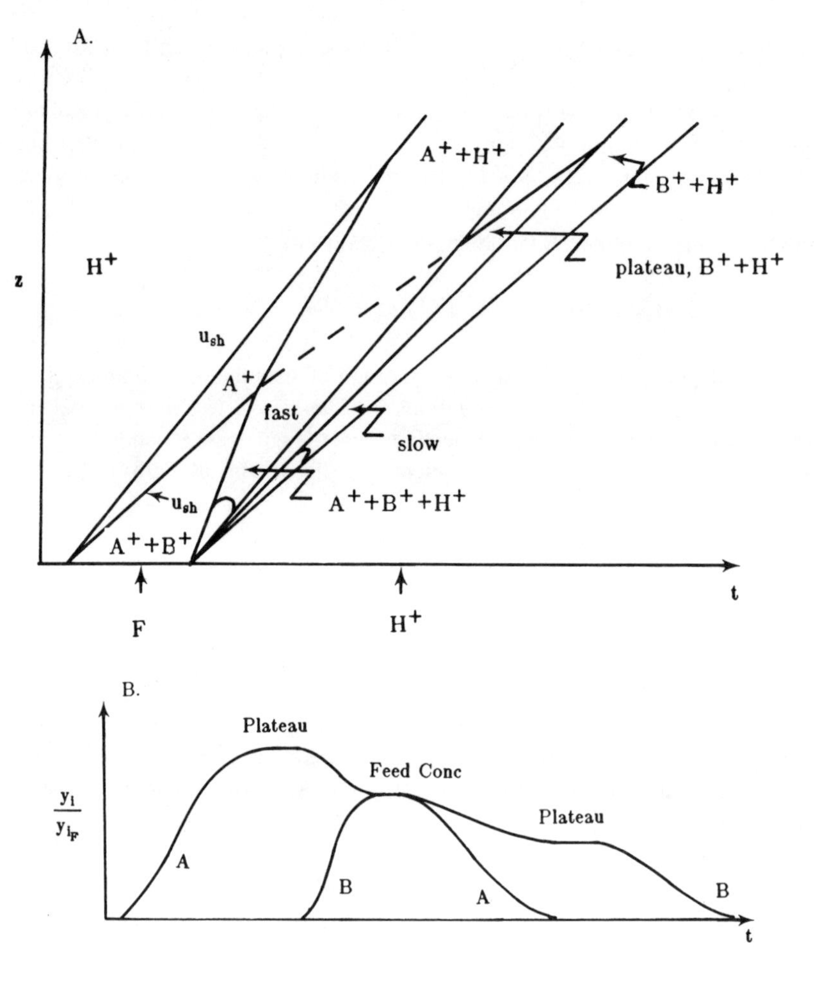

FIGURE 2-16. Coupled solutes with shock and diffuse waves. (A) Characteristic diagram for ternary ion exchange; (B) outlet concentration profiles for coupled adsorption for input of large pulse.

region contains no A^+. This plateau is separated from pure eluant by a slow diffuse wave across which the B^+ concentration drops to zero. This ordering of solutes according to their sorption characteristics demonstrates "coherence".[485,486] If given a long enough period, coupled species will always order themselves according to affinity.

Binary systems with two adsorbed species in a carrier will show a similar coupling. The observed concentration profiles for input of a long pulse are shown in Figure 2-16B for an isothermal system. Usually,[114] the fronts for both compounds form constant patterns. The concentration of component A is increased to above its feed concentration because B serves as a desorbent. After breakthrough of B, both compounds exit at their feed concentrations. When the column is eluted with a nonadsorbed species, both solutes form disperse waves. Solute B does this in two steps with a plateau in between. The results shown in Figure 2-16B can be approximately predicted using the effective equilibrium pathway method.[114] For adiabatic systems with two adsorbates the behavior is much more complex and numerical integration is required.[472]

2. Coupling with Heat of Adsorption

A second type of coupling occurs when heat of adsorption effects are important. The amount adsorbed affects the heat released, which changes the fluid tempeature. This in turn changes the equilibrium and hence the rate of solute movement. For very dilute mixtures the amount of heat released is small and the mass and energy balances can be treated as if they were uncoupled. In exchange adsorption the net heat of adsorption is usually close to zero and the balances are uncoupled. In liquid systems the volumetric heat capacity is large and the energy is usually carried down the column past the mass transfer zone (MTZ). This effectively uncouples the balances. In gas systems which are not very dilute the coupling between the amount adsorbed and the heat of adsorption is usually important.

The mathematical techniques used to decouple solutes for equilibrium analysis will also decouple the mass and energy balances.[109,112,513a,538,865,958] To rigorously solve the problem, including mass transfer and axial dispersion, numerical solutions are required.[281,472,865,912] The equilibrium theories predict a longer breakthrough time than actually occurs and thus caution is required in their use. The approximate behavior of gas systems can be predicted by comparing the velocity of a pure thermal wave (Equation 2-41) to the solute wave velocity (Equation 2-28). For gas systems the first term in the denominator of Equation 2-41 will often be small compared to the second. In large-diameter columns the wall effect will be negligible. If adsorption is strong the first two terms in the denominator of Equation 2-28 will be small compared to the last term. Then, taking the ratio of u_{th} to u_s we have

$$\frac{u_{th}}{u_s} \simeq \frac{\Delta q/\Delta c}{(C_{P_s}/C_{P_f})\,\dfrac{1}{\rho_f}} = R_T \tag{2-65}$$

R_T is the cross-over ratio.[403] If $R_T \ll 1$ the solute velocity is significantly greater then the thermal wave velocity, and the thermal wave will be well behind the solute wave. If $R_T \gg 1$ the thermal wave runs ahead of the solute wave. In these two cases the temperature in the region of the solute wave can be calculated. When R_T is near 1.0, the thermal and solute waves travel together and a simple solution is not possible. Operation with R_T near one may be operationally desirable since much of the thermal energy of adsorption is stored in the column where it is available for desorption (see Chapter 3 [Section III.B]).

However, the calculation will be more difficult. Basmadjian[109] uses a different method to develop similar but more exact criteria for when the thermal wave will run ahead of the concentration wave. When shock waves are predicted, constant pattern solution methods can be used (see Section VII.A).

V. FORMAL MATHEMATICAL DEVELOPMENT OF SOLUTE MOVEMENT THEORY

In this section the simple solute movement theory will be formally developed from the complete mass and energy balances. This development is useful since it clearly spells out the assumptions inherent in the more physical development presented in Section IV. Since the basics have been presented, this section will be brief.

A. Two Porosity Model

We will first assume that the packing is homogeneous, that radial gradients in temperature, velocity, and concentration are negligible, that thermal diffusion and pressure diffusion can be neglected, that there are no phase changes other than adsorption, and that chemical reactions do not occur (reversible chemisorption is allowed). Then the solute balance on fluid and solid is[1051]

$$\alpha \frac{\partial c_i}{\partial t} + K_{d_i}(1 - \alpha)\epsilon \frac{\partial c_i^*}{\partial t} + \rho_s(1 - \alpha)(1 - \epsilon)\frac{\partial q_i}{\partial t}$$

$$+ \alpha \frac{\partial(vc_i)}{\partial z} - \alpha(E_D + D_M)\frac{\partial^2 c_i}{\partial z^2} = 0 \qquad (2\text{-}66)$$

where c_i^* is the concentration of solute i in the stagnant fluid contained inside the pores and E_D and D_M are the eddy and molecular diffusivities. Assuming that a lumped parameter mass transfer expression is adequate, we obtain the following solute balance on the solid phase:

$$\rho_s(1 - \epsilon)(1 - \alpha)\frac{\partial q_i}{\partial t} + K_{d_i}\epsilon(1 - \alpha)\frac{\partial c_i^*}{\partial t} = -k_T a_p(c_i^* - c_i) \qquad (2\text{-}67)$$

If the lumped parameter expression is not adequate, intraparticle diffusion terms replace the right-hand side in Equation 2-67.[865,901] For the local equilibrium assumptions the form of Equation 2-67 will be immaterial.

For the energy balance we make the additional assumptions that no electrical or magnetic fields are present, radiant heat transfer is negligible, viscous heating can be neglected, kinetic and potential energy changes are small, and either density is independent of temperature or pressure is constant. Then the energy balance for both phases is

$$\rho_f C_{P_f}\alpha\frac{\partial T}{\partial t} + \rho_f C_{P_f}\epsilon(1 - \alpha)\frac{\partial T^*}{\partial t} + \rho_s C_{P_s}(1 - \epsilon)(1 - \alpha)\frac{\partial T_s}{\partial t} + \rho_f C_{P_f}\alpha\frac{\partial(vT)}{\partial z}$$

$$-(E_{DT} + D_T)\rho_f C_{P_f}\alpha\frac{\partial^2 T}{\partial z^2} = h_w A_w(T_{amb} - T_w) - C_{P_w}\frac{W}{A_c}\frac{\partial T_w}{\partial t} \qquad (2\text{-}68)$$

where T^* is the temperature of the stagnant fluid, T_{amb} is the ambient temperature, and h_w is the heat transfer coefficient for the column walls. If we assume that heat transfer into the solid can be represented by an overall linear driving force, the energy balance for the solid is

$$\rho_s C_{P_s}(1 - \epsilon)(1 - \alpha)\frac{\partial T_s}{\partial t} + \rho_f C_{P_f}\epsilon(1 - \alpha)\frac{\partial T^*}{\partial t} = -h_p a_p(T^* - T)$$

$$+ (1 - \epsilon)(1 - \alpha)\rho_s \Delta H_{ads}\frac{\partial q}{\partial t} \qquad (2\text{-}69)$$

where h_p is the heat transfer coefficient and a_p is the surface area per unit volume.

Although already somewhat simplified, Equations 2-66 to 2-69 or the equivalent single porosity forms (see Equations 2-73, 2-74, 2-77, and 2-78) along with the isotherm expression are the usual starting points for modeling adsorption and chromatography. Now we are ready to make the additional assumptions necessary to obtain simple solutions. The first additional assumption is that stagnant fluid and solid are in equilibrium. Thus, $T_s = T^*$ and q_i and c_i^* are related by the isotherm equation. Next we assume that all parameters (e.g., α,ϵ, D_M, C_{Pf}) except density are constant. To obtain simple solutions, density is also assumed to be constant, although this assumption can be relaxed. This set of additional assumptions are often included even for numerical solutions.

To obtain analytical solutions additional assumptions are required. Usually the fluid velocity is assumed to be constant. For nonisothermal systems we usually assume that axial dispersion and diffusion terms can be neglected. The rates of heat and mass transfer are assumed to be very high. Thus $c = c^*$ and $T = T^*$, and liquid and solid are locally in equilibrium. This removes Equations 2-67 and 2-69. The column is usually assumed to be

Table 1
MODEL ASSUMPTIONS

1.	Homogeneous packing
2.	Thermal and pressure diffusion are negligible
3.	No chemical reactions other than adsorption
4.	No phase changes other than adsorption
5.	No electrical or magnetic fields
6.	Radiant heat transfer is negligible
7.	Viscous heating is negligible
8.	Kinetic and potential energy changes negligible
9.	Constant α, ϵ, ρ_s, E_D, D_M, D_T, k_T, a_P, C_{P_f}, K_d, C_{P_s}, C_{Pw}, h_w, h_p, ΔH_{ads}
10.	Radial gradients in q, c, T, and v are negligible
11.	Lumped parameter expressions for heat and mass transfer are adequate
12.	Fluid density is constant
13.	Fluid velocity is constant
14a.	Column is adiabatic or
14b.	Column is isothermal
15.	Axial thermal dispersion can be neglected
16.	Axial mass dispersion can be neglected
17.	Heat and mass transfer rates are very high and local equilibrium occurs
18.	The heat of adsorption is negligible
19.	Solutes are independent
20.	The isotherm is linear

adiabatic or if the column is heated and cooled by a jacket (the direct mode) the heat transfer rates are assumed to be very high and radial gradients do not occur. For simple solutions the solutes are assumed to be independent and the heat of adsorption is neglected. With these assumptions, Equation 2-66 becomes:

$$[\alpha + K_d(1 - \alpha)\epsilon]\frac{\partial c}{\partial t} + \rho_s(1 - \alpha)(1 - \epsilon)\frac{\partial q}{\partial t} + \alpha v\frac{\partial c}{\partial z} = 0 \qquad (2\text{-}70)$$

The amount adsorbed, q, is related to c and T by the isotherm. By applying the chain rule and rearranging, Equation 2-70 simplifies to:

$$\frac{\partial c}{\partial t} + u_s\frac{\partial c}{\partial z} = -\frac{u_s}{v}\left(\frac{1 - \alpha}{\alpha}\right)(1 - \epsilon)\rho_s\frac{\partial q}{\partial T}\frac{\partial T}{\partial t} \qquad (2\text{-}71)$$

For isothermal systems, or for systems with instantaneous temperature changes or for systems with square wave changes in the feed temperature $\partial T/\partial t = 0$. Then Equation 2-71 is easily solved by the method of characteristics.[865,1051] The results are the same as shown in Figures 2-5 (for linear isotherm assumption), 2-6, and 2-9.

The energy balance Equation 2-68 simplifies to:

$$\frac{\partial T}{\partial t} + u_{th}\frac{\partial T}{\partial z} = 0 \qquad (2\text{-}72)$$

When solved by the method of characteristics, the solution to this equation is the same as Figure 2-11.[865,1051]

For easy reference the assumptions are listed in Table 2-1. This is a rather staggering list of assumptions. Fortunately, many of them are valid in most situations. As we have already seen, some of the assumptions can be relaxed. In Section VI.B we will see that by invoking assumptions 19 and 20 we can relax assumptions 16 or 17 and explore the effects of dispersion and finite mass transfer rates.

B. Single Porosity Model

Before continuing it will be useful to compare Equations 2-66 to 2-69 to the single porosity equations which are often employed. It is common to use a single porosity which we will call ϵ'. Then $(1 - \epsilon')$ includes the solid and the stagnant fluid. The mass balance equations are now

$$\epsilon' \frac{\partial c_i}{\partial t} + \rho_B(1 - \epsilon') \frac{\partial q_i'}{\partial t} + \epsilon' \frac{\partial (c_i v)}{\partial z} - \epsilon'(E_D + D_M) \frac{\partial^2 c_i}{\partial z^2} = 0 \tag{2-73}$$

and

$$\rho_B(1 - \epsilon') \frac{\partial q_i'}{\partial t} = -k_T a_p(c_i^* - c_i) \tag{2-74}$$

Where ρ_B is the bulk density of the solid including fluid in the pores. Comparison of Equations 2-73 and 2-74 with Equations 2-66 and 2-67 shows that $\epsilon' = \alpha$ and

$$\rho_B q_i' = \epsilon c_i^* + \rho_s(1 - \epsilon) q_i \tag{2-75}$$

Thus the equilibrium isotherm relating q'_i to c_i must include the fluid in the pores if the single porosity equations are used. If the two porosity equations are used the equilibrium relation between q and c must not include the fluid in the pores. Unfortunately, the literature is not always clear as to how isotherms were determined. When Equations 2-73 and 2-74 are simplified and solved in the same way as Equations 2-66 and 2-67, essentially the same result is obtained except the solute wave velocity is now

$$u_s = \frac{v}{1 + \left(\frac{1 - \epsilon'}{\epsilon'}\right) \rho_B \frac{\Delta q_i'}{\Delta c_i}} \tag{2-76}$$

With a single porosity the energy balance equations are

$$\rho_f C_{P_f} \epsilon' \frac{\partial T}{\partial t} + \rho_B C_{P_B}(1 - \epsilon') \frac{\partial T_s}{\partial t} + \rho_f C_{P_f} v \epsilon' \frac{\partial T}{\partial z} - (E_D + D_T)\rho_f C_{P_f} \epsilon' \frac{\partial^2 T}{\partial z^2} = h_w A_w (T_{amb} - T_w) - \frac{C_{pw} W}{A_c} \frac{\partial T_w}{\partial t} \tag{2-77}$$

and

$$\rho_B C_{P_B}(1 - \epsilon') \frac{\partial T_s}{\partial t} = -h_p a_p (T^* - T) + (1 - \epsilon')\rho_B \Delta H_{ads}' \frac{\partial q'}{\partial t} \tag{2-78}$$

where C_B is the bulk heat capacity of the solid particles and $\Delta H'_{ads}$ is the heat of adsorption determined as a function of q'. Comparison of Equations 2-77 and 2-78 to Equations 2-68 and 2-69 with $\epsilon' = \alpha$ shows that

$$\rho_B C_{P_B} = \rho_s C_{P_s}(1 - \epsilon) + \rho_f C_{P_f} \epsilon \tag{2-79}$$

if $T_s = T^*$. If $\partial T/\partial t = 0$, $\Delta H'_{ads}$ can be related to ΔH_{ads}.

$$\Delta H'_{ads} = \frac{(1 - \epsilon)\rho_s \Delta H_{ads}}{\epsilon \dfrac{\partial c^*}{\partial q} + \rho_s(1 - \epsilon)} \tag{2-80}$$

If Equations 2-77 and 2-78 are simplified and solved by the same method used for Equations 2-68 and 2-69, essentially the same result is obtained except the thermal wave velocity is now

$$u_{th} = \frac{v\rho_f C_{P_f}}{\rho_f C_{P_f} + \left(\dfrac{1 - \epsilon'}{\epsilon'}\right)\rho_B C_{P_B} + \dfrac{W}{\epsilon' A_c} C_{P_W}} \tag{2-81}$$

Either the single porosity or two porosity forms can be used. The two porosity forms are physically more realistic and are required if a combination of size exclusion and adsorption is occurring. The single porosity model is often mathematically more convenient. Care must be used when trying to convert data from one form to another.

VI. ZONE SPREADING EFFECTS FOR LINEAR SYSTEMS

We know that ignoring dispersion and mass transfer effects often give results which are physically unrealistic. Fortunately, when the solutes are independent and the isotherms are linear the mass transfer and dispersion effects can be included in the equations. This will allow us to study these effects for linear systems and qualitatively extrapolate the effects of dispersion and mass transfer to nonlinear systems. The linear results are exact for dilute solutions.

Several different types of theories have been used to study linear chromatography. We will first consider the historically important plate theories and then rate theories. Comparison of these theories gives the Van Deemter equation, which has been modified by stochastic theories. This equation predicts the zone spreading effects in linear chromatography. Next, the theories will be used to predict the resolution of two components in linear chromatography, and finally superposition of these results onto the solute movement theory results will be discussed.

A. Plate Theories

In their classical paper on liquid-liquid chromatography, Martin and Synge[695] developed a plate theory to explain zone spreading. Their model consisted of a series of stages with a stationary phase and a mobile phase. The mobile phase moved in discrete transfer steps followed by equilibrium of each stage. Mobile phase from different stages does not mix. Thus this theory really modeled the countercurrent distribution system which they had worked on before developing the chromatographic column. The result was a binomial expression which reduces to a Gaussian form. The number of stages or the plate height had to be determined experimentally.

Although none of the plate models are physically realistic, the continuous flow plate model shown in Figure 2-17 is more realistic than the countercurrent distribution model. In this plate model the column is divided into N equilibrium stages each of height H.[416,586,1008] Each component is assumed to be independent and the isotherms are assumed to be linear. A balance around stage j for solute i gives

$$c_{i,j-1}\, dV - c_{i,j}\, dV = V_M dc_{i,j} + M_s dq_{i,j} \tag{2-82}$$

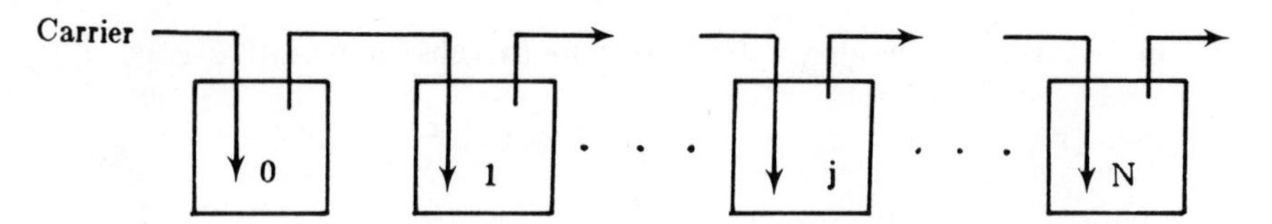

FIGURE 2-17. Continuous flow staged model for chromatography.

where V_M and M_s are the volume of mobile phase and the mass of stationary phase on each stage. V is the volume of carrier gas. Substituting in the linear isotherm (Equation 2-29) and rearranging, we have the ordinary differential equation:

$$\frac{d\,c_{i,j}}{d\,V} = \frac{c_{i,j-1} - c_{i,j}}{V_M + k_i\,M_s} \quad (2\text{-}83)$$

For a pulse of F_i moles of feed into stage 0, the solution is the Poisson distribution:

$$c_{i,j} = \frac{F_i}{V_M + k_i\,M_s}\,\frac{e^{-\nu_i}\,\nu_i^j}{j!} \quad (2\text{-}84)$$

where the dimensionless cumulative flow of mobile phase, ν_i, is

$$\nu_i = \frac{V}{V_M + M_s k_i} = \frac{u_{si}\,t}{H} \quad (2\text{-}85)$$

where Equation 2-30 and the height of a stage (Equation 2-91) have been used to relate the solute velocity to the staged terms. The outlet concentration is found by setting $j = N$. Equation 2-82 has also been solved for step inputs.

With a large number of stages, Equation 2-84 can be approximated as

$$c_{i,j} = \frac{F_i}{V_M + k_i M_s}\,\frac{1}{\sqrt{2\pi j}}\,e^{-(\nu_i - j)^2/2j} \quad (2\text{-}86)$$

which at the column outlet is

$$c_{i_{out}} = \frac{F_i}{V_M + k_i M_s}\,\frac{1}{\sqrt{2\pi N}}\,e^{-(\nu_i - N)^2/2N} \quad (2\text{-}87)$$

By differentiation it is easy to show that in dimensionless terms the peak maximum occurs at $\nu_i = N$ and thus

$$c_{imax} = = \frac{F_i}{V_M + k_i M_s}\,\frac{1}{\sqrt{2\pi N}} \quad (2\text{-}88)$$

Also, the dimensionless peak width is

$$w_i = 4\sqrt{N} \quad (2\text{-}89)$$

For a Gaussian peak the peak width equals 4σ where σ is the standard deviation. Thus in dimensionless terms

$$\sigma = \sqrt{N} \quad (2\text{-}90)$$

Equation 2-90 is easily written in terms of time by substituting in Equations 2-85, 2-88 plus a definition of plate height, H.

$$N = L/H \tag{2-91}$$

The result is

$$c_i = c_{imax}\, e^{-\left(\frac{u_{si}t}{H} - \frac{L}{H}\right)^2 / 2\left(\frac{L}{H}\right)} \tag{2-92}$$

After rearrangement this becomes

$$c_i = c_{imax}\, e^{-(t - t_{Ri})^2 / 2\left(\frac{H}{L} t_{Ri}^2\right)} \tag{2-93}$$

where the retention time for solute i is

$$t_{Ri} = L/u_{si} \tag{2-94}$$

By further algebraic manipulation the equation for solute concentration can be written in length units as

$$c_i = c_{imax}\, e^{-(z_i - L)^2/2LH} \tag{2-95}$$

where z_i is the location of the peak center.

Equations 2-87, 2-92, and 2-95 can all be written in the common form

$$c_i = c_{imax} \exp(-X^2/2\sigma^2) \tag{2-96}$$

where X is the difference from the peak maximum and σ is the standard deviation, both in appropriate units. In dimensionless units σ is given by Equation 2-90. Comparing Equations 2-93 and 2-87 in time units,

$$\sigma_t = \left(\frac{H}{L}\right)^{1/2} t_{Ri} = (LH)^{1/2}/u_{si} \tag{2-97}$$

while in length units

$$\sigma_l = \sqrt{HL} \tag{2-98}$$

This development of the Gaussian distribution and the standard deviations has been carried out at some length since different sources report different equations for the concentration profile and the standard deviation. This can be confusing, especially if the authors do not define their terms.

The staged theory correctly predicts that zone spreading is proportional to the square root of the distance traveled. This correct prediction does *not* imply that the model is "correct". Other linear theories also predict that zone spreading is proportional to $\sqrt{L}$. Since Equation 2-94 shows that the exit time for the each solute peak is proportional to L, peaks can always be separated by increasing L as long as the solute wave velocities (or distribution coefficients) differ. According to Equation 2-98, the zone spreading within the column is the same for all components except for minor differences due to different H values. Equation 2-97 shows

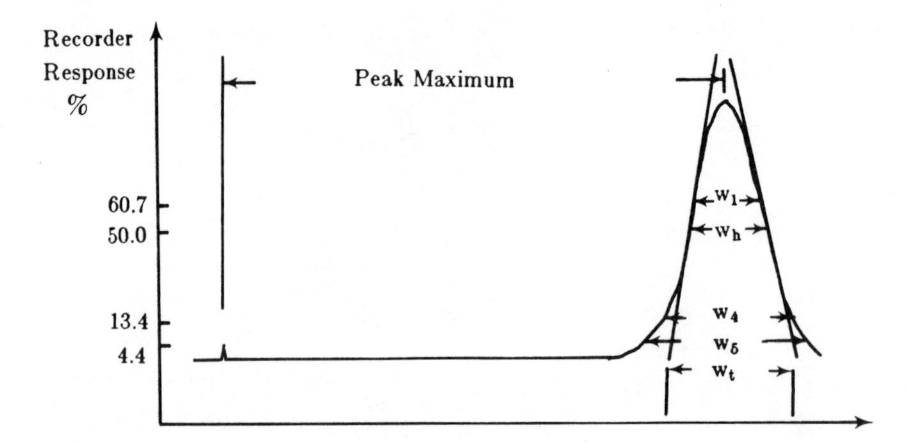

FIGURE 2-18. Single Gaussian peak. Constants for calculation of N from Equation 2-99. $w_1 = 4$, $w_h = 5.54$, $w_{4\sigma} = 16$, $w_{5\sigma} = 25$, and $w_t = 16$.

that the zone spreading with respect to time (and this is essentially what is observed on a strip chart recorder) is inversely proportional to the solute wave velocity. Slower solutes spread more. The predictions of Equations 2-97 and 2-98 are not incompatible. When the front of a solute wave just reaches the end of the column, the width of the zone inside the column is $4\sigma_l$. It takes longer to elute a slowly moving solute this distance and thus more zone spreading is observed on the strip chart.

The value of N is easily determined from an experiment. The distance (or time) that the peak maximum takes to exit and the width of the peak are measured on the strip chart as shown in Figure 2-18. Then N is

$$N = (\text{Const}) \left[\frac{\text{peak maximum}}{\text{width}} \right]^2 \tag{2-99}$$

where the constant depends on which width is used[134] and values are listed in Figure 2-18. The plate height H can then be determined from Equation 2-91. The ratio is used in Equation 2-99 so that the relationship between distance on the strip chart and time divides out. The width at half height, w_h, is most common, but the use of the 5σ width will be most accurate for asymmetric peaks.[134] All methods give the same results for Gaussian peaks.

Staged models have been used as the basis for numerical simulations for both linear and nonlinear isotherms for a variety of operating methods. Unfortunately, the staged models do not tell what affects N or equivalently H, and how the values for H and N will change as the operating conditions change. The linear theory is very useful for pulse testing of columns at low concentrations to determine how well they are packed.

B. Rate Theories

The rate theories use the solute mass balance Equation 2-66 or 2-73 and the mass transfer expression Equation 2-67 or 2-74 as the starting point and then simplify these equations with additional assumptions. Lapidus and Amundson[630] solved two cases. First they used Equation 2-73 (Equation 2-66 can also be used) and for an isothermal column made all the assumptions in Table 2-1 except that dispersion is negligible. With these assumptions Equations 2-29 and 2-73 reduce to:

$$\left[1 + \rho_B\left(\frac{1-\epsilon'}{\epsilon'}\right)k_i\right]\frac{\partial c_i}{\partial t} + v\frac{\partial c_i}{\partial z} - (E_D + D_M)\frac{\partial^2 c_i}{\partial z^2} = 0 \tag{2-100}$$

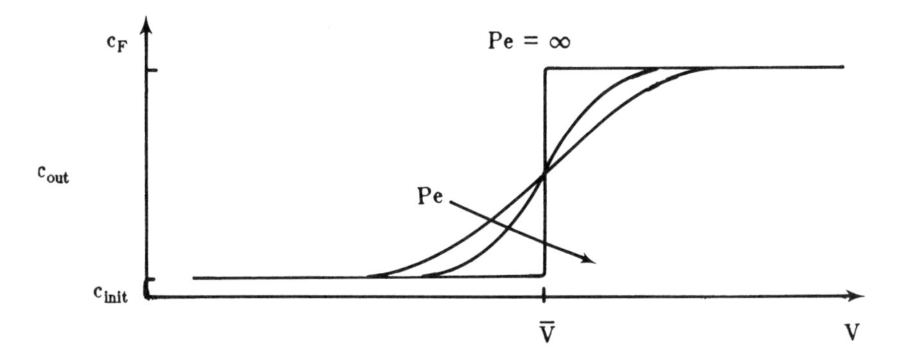

FIGURE 2-19. Response to a step input predicted by Equation 2-101.

This equation was solved for both arbitrary initial and inlet distributions and for constant initial and inlet concentrations. For constant initial and inlet concentrations a simpler, approximate solution can be obtained for columns of sufficient length.[657] For a step input this solution is

$$\left(\frac{c - c_{init}}{c_F - c_{init}}\right)_i = \frac{1}{2}\{1 + \text{erf}[(Pe_z)^{1/2} (V - \bar{V})/2(V\bar{V})^{1/2}]\} \tag{2-101}$$

where $V = \epsilon' \, v \, A_c t$ is the elution volume, $\bar{V} = A_c L[\epsilon' + (1 - \epsilon')\rho_B k_i]$ is the volume of solution required to saturate a column of length L, and the Peclet number is $Pe_z = L\, v/(D_M + E_D)$. This Peclet number is now based on an effective axial dispersion which includes mass transfer effects. The error function

$$\text{erf}(u) = -\text{erf}(-u) = \frac{2}{\pi^{1/2}} \int_0^u \exp(-\varsigma^2) d\varsigma \tag{2-102}$$

is related to the normal curve of error which is a tabulated integral. The error function is also available on many computers.

The solution given in Equation 2-101 is convenient to use. For a step input an S-shaped curve is predicted. This is shown in Figure 2-19. The outlet concentration profile becomes sharper as $(D_M + E_D)$ become smaller (larger Peclet number). In the limit of an infinite Peclet number the result reduces to the solute movement theory result for linear isotherms. The concentration profiles all intersect at $1/2\,(C_F + C_{init})$ at $V = \bar{V}$. The effective Peclet number can easily be obtained from experimental data. In linear systems the zone spreading effects of mass transfer and axial dispersion look the same. Thus an effective axial dispersion coefficient can model zone spreading which is caused by a combination of mass transfer and axial dispersion.

In addition to giving a simple solution, the use of linear equilibrium has the major advantage that superposition can be used to predict column behavior for a variety of situations.[502,657] Let X(L, V) be the breakthrough solution given in Equation 2-101 for an initially clean column.

$$X(L,V) = \left(\frac{c}{c_F}\right)_i = \text{Equation 2-101} \tag{2-103}$$

Now solutions for other cases can be determined by superposition. If a column which is initially uniformly loaded is eluted with pure solvent, we can subtract the breakthrough solution from the uniform loading,

$$\frac{c}{c_{init}} = 1 - X(L,V) = \frac{1}{2}\left\{1 - \text{erf}\left[\frac{(Pe_z)^{1/2}(V - \bar{V})}{2(V\bar{V})^{1/2}}\right]\right\} \tag{2-104}$$

If a pulse input of volume V^o is fed to an initially empty column we have first a step up and then a step down at a later time. Thus,

$$\frac{c}{c_F} = X(L,V) - X(L,V - V^0) \tag{2-105}$$

and the result is easily generated from Equation 2-101.

In linear chromatography we are interested in a differential pulse of feed. Then we want the limit of Equation 2-105 as $V_o \rightarrow 0$. After some manipulation this is

$$\left(\frac{c}{c_F}\right)_i = V_0 \frac{\partial X(L,V)}{\partial V} \tag{2-106}$$

Applying this to Equation 2-101 and noting that near the peak

$$\left(\frac{V - \bar{V}}{V}\right) \ll 1$$

then

$$\left(\frac{c}{c_F}\right)_i = \frac{1}{2}\frac{V_0}{\bar{V}}\left(\frac{Pe_z}{\pi}\right)^{1/2} \exp\left[\frac{-Pe_z(V - \bar{V})^2}{4V\bar{V}}\right] \tag{2-107}$$

The peak maximum occurs at $V = \bar{V}$ and the maximum concentration is

$$(c_i)_{max} = \frac{c_F}{2}\frac{V_0}{\bar{V}}\left(\frac{Pe_z}{\pi}\right)^{1/2} \tag{2-108}$$

Although Equations 2-108 and 2-88 are defined in different terms, they both give c_{max} for a pulse input. When these equations are set equal to each other and the terms are rationalized, we find

$$Pe_z = 2N \tag{2-109}$$

Thus the Peclet number can easily be determined from experiments from Equations 2-99 and 2-109.

In their second case, Lapidus and Amundson[630] assumed that the rate expression could be written as

$$\frac{dq'}{dt} = k_1 c - k_2 q' \tag{2-110}$$

A special case of this equation when equilibrium is linear is the lumped parameter expression for mass transfer, Equation 2-74. The solutions obtained included untabulated definite integrals and thus are not easy to use. Van Deemter et al.[1008] were able to simplify the solution for the input of a differential pulse in a system with linear equilibrium. If the column was large enough, their solution simplified to

$$c_i = \frac{F_i}{A_c\, v(\epsilon' + (1 - \epsilon')k')\sqrt{2\pi(\sigma_1^2 + \sigma_2^2)}}$$

$$\exp\left[\frac{-\left[L\left(1 + \frac{1 - \epsilon'}{\epsilon'} k_i\right)/v - t\right]^2}{2\left(1 + \frac{1 - \epsilon'}{\epsilon'} k_i\right)^2 (\sigma_1^2 + \sigma_2^2)}\right] \tag{2-111}$$

where

$$\sigma_1^2 = \frac{2\, L\, D_T}{v^3} \tag{2-112}$$

$$\sigma_2^2 = 2\left[\frac{(1 - \epsilon')k_i}{\epsilon' + (1 - \epsilon')k_i}\right]^2 \frac{L\, \epsilon'}{k_T\, a_P\, v} \tag{2-113}$$

D_T is the total effective axial diffusivity. Equation 2-111 is a Gaussian distribution whose variance is the sum of the variances for diffusion and mass transfer. This solution will be employed in the next section.

Another useful solution is Thomas' solution.[974,975] Thomas solved mass balance Equation 2-73 for a single solute under isothermal conditions with negligible dispersion. He assumed that sorption was controlled by a kinetic rate expression. The two rate expressions he used are different from the mass transfer expression Equation 2-74, but his kinetic constant can be related to the mass transfer rate constant.[657,901,1015,1016] Thomas obtained solutions for Langmuir and linear systems. The linear solution is considerably simpler, and for sufficiently long columns reduces to complimentary error functions. The predicted concentration profiles for breakthrough are very similar to those shown in Figure 2-19. Superposition can be used for the linear case. For a differential pulse of feed, Equation 2-106 is valid but X is now the linear Thomas solution for breakthrough. The predicted concentration profiles from Thomas' analysis are very similar to those predicted by Equations 2-107 and 2-111. Since Thomas obtained solutions for both breakthrough and elution for Langmuir isotherms under nonequilibrium conditions, his solutions are often used as the basis for design of adsorption and ion exchange columns. The Thomas solutions are the basis for effective equilibrium pathway solutions[110-112,114] which appear to be useful for the practical design of some commercial adsorption systems.

C. Height of a Theoretical Plate

In their classic paper, Van Deemter et al.[1008] compared the continuous flow staged model results to their approximate solution, Equation 2-111, of Lapidus and Amundsen's[630] mass transfer case. They found that the height of an equilibrium stage was

$$H = \frac{2D_T}{v} + 2\left(\frac{(1 - \epsilon')k_i}{\epsilon' + (1 - \epsilon')k_i}\right)^2 \frac{\epsilon'\, v}{k_T\, a_P} \tag{2-114}$$

Thus for linear isotherms the zone spreading caused by diffusion and mass transfer are linearly additive. The total effective diffusivity can be expanded as the sum of molecular diffusion and eddy dispersion,

$$D_T = \gamma D_M + E_D \tag{2-115}$$

where γ is a labyrinth or tortuosity factor. Since the eddy diffusivity is proportional to velocity and particle diameter, Equation 2-115 can be written:

$$D_T = \gamma D_M + \frac{\lambda}{2} v\, d_p \tag{2-116}$$

Then Equation 2-114 becomes

$$H = A + \frac{B}{v} + C\, v \tag{2-117}$$

This equation is known as the Van Deemter equation although that seems slightly unfair to Zuiderweg and Klinkenberg. This equation is strictly applicable to linear isotherms only.

In the Van Deemter equation the A term is the eddy diffusion or flow contribution,

$$A = \lambda\, d_p \tag{2-118}$$

This velocity-independent term can be greatly decreased by careful design and packing of the column. Use of narrow particle size distributions also decreases the A term.[605] The B term is caused by molecular diffusion and is

$$B = 2\gamma D_M \tag{2-119}$$

In Equation 2-117 the molecular diffusion contribution is inversely proportional to the velocity. Thus this contribution becomes important at low velocities, particularly for gas systems where D_M is relatively large. The C term is due to mass transfer resistance. Since the overall mass transfer resistance can be represented as the sum of film resistances in the mobile and stationary phases

$$\frac{1}{k_T\, a_p} = \frac{1}{k_M\, a_p} + \frac{1}{k_i\, k_s\, a_p} \tag{2-120}$$

The C term can be expanded to

$$C = 2\left[\frac{(1 - \epsilon')k_i}{\epsilon' + (1 - \epsilon')k_i}\right]^2 \left[\frac{\epsilon'}{k_M\, a_p} + \frac{\epsilon'}{k_i\, k_s\, a_p}\right] \tag{2-121}$$

Since the mass transfer occurs by diffusion through the stagnant mobile phase in the pores and into a film of bound stationery phase, $k_M a_p$ is inversely proportional to d_p^2 and directly proportional to molecular diffusivity while k_s is inversely proportional to d_f^2; d_p is the particle diameter and d_f is the film thickness of the film coating the stationary phase. For adsorbents d_f is replaced by d_p.

A plot of H vs. velocity according to the Van Deemter equation is shown in Figure 2-20. The total H depends on the sum of the three contributions. In a well-designed and well-packed system the flow term, A, is usually quite small. Note that there is an optimum velocity for minimum H. The absolute minimum value of H is about 2 d_p. Un-

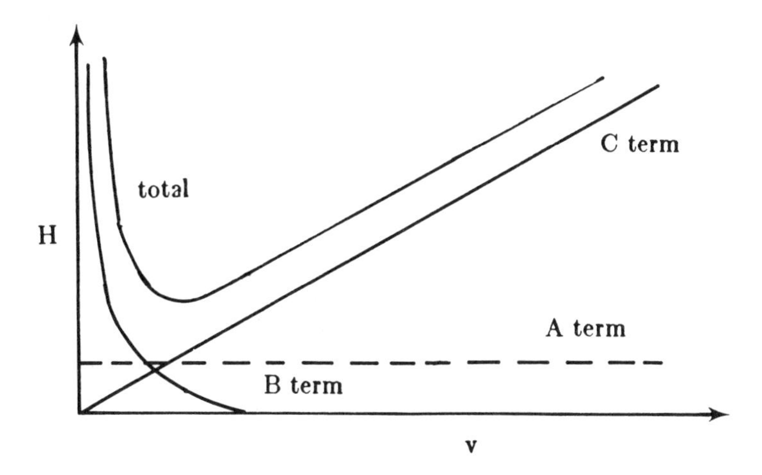

FIGURE 2-20. Variation of H with velocity according to Van Deemter equation (Equation 2-117).

fortunately, this velocity is too low for practical operation. Thus the mass transfer, C terms, usually dominate, and usually the $k_M a_p$ resistance is dominant. Thus H is essentially proportional to d_p^2 and inversely proportional to the molecular diffusivity. For very small particles (less than 10 μm) kinetics can become important and the C terms may not dominate.[517]

Many studies of zone spreading have followed the original publication of the Van Deemter equation.[354,416,517,605,870,930,1109] Using random walk and stochastic analysis, Giddings[416] was able to refine the Van Deemter equation. Several modified forms of the Van Deemter equation are in use. The classical expanded form of the Van Deemter equation is

$$H = A + \frac{B}{v} + C_M v + C_{SM} v + C_s v \qquad (2\text{-}122)$$

where the mass transfer terms have been expanded. C_M is due to extraparticle mass transfer.

$$C_M = c_M d_p^2/D_M \qquad (2\text{-}123a)$$

C_{SM} is due to diffusion in the stagnant mobile phase

$$C_{SM} = c_{SM} d_p^2/D_{SM} \qquad (2\text{-}123b)$$

and C_s is due to diffusion in the stationary liquid phase coating the solid or to diffusion in the solid.

$$C_s = \frac{c_s d_f^2}{D_s} \qquad (2\text{-}123c)$$

This last term is often very small in liquid systems. Giddings[416] assumed that lateral dispersion could move solute from slow to fast streams in the column. This flow-diffusion coupling theory gives:

$$H = \frac{B}{v} + C_{SM} v + C_s v + \frac{1}{\dfrac{1}{A} + \dfrac{1}{C_M v}} \qquad (2\text{-}124)$$

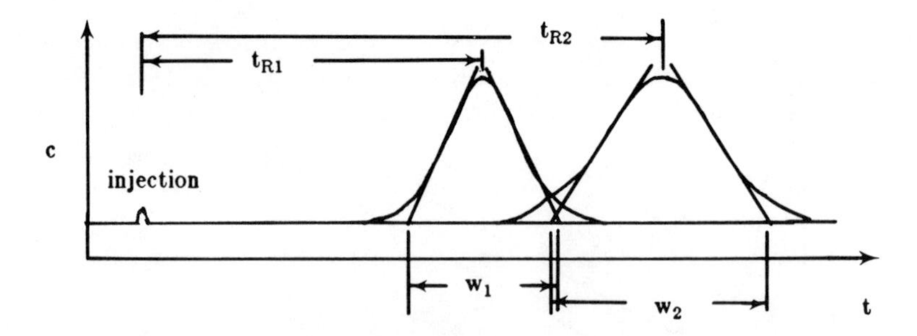

FIGURE 2-21. Separation of two Gaussian peaks. See Equation 2-125.

where the A, B, and C definitions are unchanged. Other forms of the modified Van Deemter equation have been developed[517,605,606] and the basic predictions have been experimentally verified.[354,517,605,606,930,1109]

Since at practical operating velocities H is essentially proportional to d_p^2 and inversely proportional to D_M, zone spreading can be reduced by decreasing the particle diameter and by increasing the molecular diffusity. Decreasing d_p is discussed further in Chapter 3 (Section II.D). For liquid systems D_M can be increased by using a low-viscosity solvent or raising the temperature. However, raising the temperature also usually decreases the selectivity of the packing. The diffusivity in the pores, D_{SM}, can be increased by using large-pore packings. This is particularly important when large molecules are being separated.

Note that column diameter does not enter into any of the expressions for zone spreading. *If* the system is well designed and well packed, there is theoretically no reason why H obtained with small pulses should increase. The trick is proper design and packing of the column. Extracolumn zone broadening effects are not included in the Van Deemter equation, and can destroy the separation. Any dead zones or mixing zones in the feed system, distributors, piping, valves, and detection system can greatly increase H. All external volumes should be minimized by careful design and by placing equipment close together. Since solutes that are less strongly adsorbed (low k_i) will be more affected by extracolumn effects, the system can be tested with pulses of slow and fast solutes.[354] Skewed profiles or larger H values for the faster solutes are an indication that extracolumn zone broadening is important. Most commercial analytical chromatographs have been designed to minimize extracolumn effects. Since it always seemed less important, most large-scale adsorption and ion exchange systems have not been designed to minimize extracolumn zone broadening. This is a mistake when difficult separations are required.

Equations 2-87 and 2-111 were derived for differential pulses for systems with linear equilibrium. Thus the various forms of the Van Deemter expression, Equations 2-117, 2-122, and 2-124 are strictly valid only for this situation.

D. Resolution of Two Components by Linear Chromatography

One of the very useful things which can be done with the linear theory of chromatography is to predict the separation of two components. For Gaussian peaks the resolution between two solutes is defined as[354,416,870,930,1109]

$$R = \frac{2(t_{R_2} - t_{R_1})}{w_1 + w_2} \qquad (2\text{-}125)$$

where the terms are illustrated in Figure 2-21. An R value of 1.5 represents complete separations while with an R of 1.0 the two peak maxima are separated by 4σ which is about 2% overlap. For large-scale systems with nonlinear isotherms and considerable overlap of the peaks, Equation 2-125 is not a good measure of the separation.

For differential pulse inputs in linear systems w_1 and w_2 are easily determined from the Gaussian distribution while the retention times can be found from Equation 2-94. The result is the "fundamental equation" of linear chromatography,

$$R = \frac{1}{2}\left(\frac{\alpha_{21} - 1}{1 + \alpha_{21}}\right)\frac{\bar{k}'}{1 + \bar{k}'}N^{1/2} \tag{2-126}$$

In this equation k'_i is the relative retention of solute i,

$$k_i' = k_i\ V_s/V_M \tag{2-127}$$

$\bar{k}'_i$ is the average relative retention of the two solutes, and α_{21} is the selectivity

$$\alpha_{21} = k_2'/k_1' = \frac{k_2}{k_1} > 1.0 \tag{2-128}$$

Thus α_{21} is the ratio of the distribution coefficients of the two solutes, and is the same as Equation 2-10.

Equation 2-126 shows that the resolution can be increased by increasing α_{21}, $\bar{k}'$, or N. Changes in α_{21} have by far the most effect. For α_{21} near 1.0 a very large number of stages are required. For α_{21} above 1.15 the number of stages becomes reasonable. This is just another way of saying that the solutes must have different distribution coefficients (different solute velocities) to separate. In large-scale systems an α_{21} above 1.50 or 2.0 is desirable. Changing the selectivity requires changing the chemistry of the system. This can be done by changing or modifying either the stationary phase or the mobile phase. When α_{21} is large, the stage requirement will be low. Thus H can be fairly large and, according to the Van Deemter equation, high velocities can be used (see Figure 2-20). Thus, increasing selectivity allows both shorter columns and high flow rates.

Another way to increase resolution is to increase $\bar{k}'$. However, as the relative retentions increase the solute velocities become slower and operating time increases. A value of $\bar{k}'$ between 4 to 6 seems to be a reasonable compromise.

Finally, resolution can be increased by increasing N. This can be done by decreasing H or increasing L. The plate height can be decreased by decreasing the particle diameter. Decreasing d_p and increasing L both increase the pressure drop so neither can be done indefinitely. If only small changes in resolution are required, increasing the column length is often the easiest thing to do. If large changes are needed, adjusting the system to change α_{21} is usually a good idea, particularly for large-scale systems where throughput is important.

In large-scale, nonlinear systems, Equation 2-126 is no longer valid. However, the qualitative conclusions drawn from this equation are valid.

E. Combining Zone Spreading Predictions with Solute Movement Theory

Exact superposition of the zone spreading predictions onto the solute movement theory is valid only for linear systems. However, we can approximately combine the results to make qualitative prediction. By adjusting the predictions with experimental data, we can develop a semiempirical theory.

The solute movement theory predicted three cases: linear systems, diffuse waves for nonlinear systems, and shock waves for nonlinear systems. For linear systems, superposition of zone spreading is valid. For diffuse waves the solute movement theory predicts that the zone spreading or band width is directly proportional to the distance traveled (see Figure 2-6B). Since linear chromatography theory predicts that zone spreading is proportional to $\sqrt{L}$, the diffuse wave effect is dominant. Thus the diffuse wave predictions shown in Figure 2-

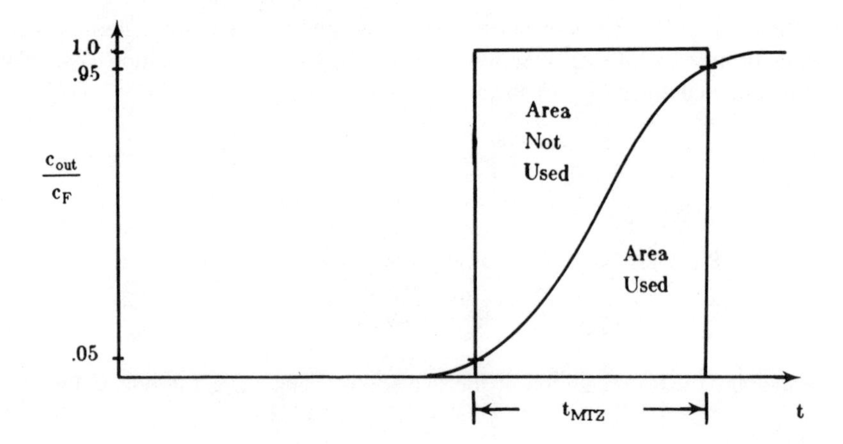

FIGURE 2-22. Breakthrough curve for step input. LUB approach.

6C are quite accurate except at the corners. At very low concentrations the Langmuir isotherm becomes linear and thus the solute movement theory underpredicts zone spreading. At very high concentrations the Langmuir isotherm approaches an asymptote and thus also underpredicts zone spreading.

For nonlinear systems where the solute movement theory predicts shock waves, a constant pattern is generated. There is zone spreading around the shock wave (see Figure 2-9), but once developed it does not change shape. Linear chromatography results are qualitatively useful in this case. Although the zone has a constant shape for a given system, the amount of zone spreading depends on the system properties. The qualitative predictions of the Van Deemter equation will be correct. Thus if we increase the velocity in the region of Figure 2-20 where H increases then the width of the constant pattern will also increase.

The approximate combination approach has been used to model large-scale, nonlinear, coupled GLC.[463,536,864] First, the nonlinear isotherm effect, coupling effects of the solutes, and sorption effects were modeled with the local equilibrium theory. Then dispersion effects were added as if they were independent. This is an approximate procedure, but it appeared to work rather well.

VII. SIMPLE DESIGN PROCEDURES FOR NONLINEAR SYSTEMS

The solute movement theory is very useful for developing and understanding operating methods, but it is not a design method. The linear theories are useful for dilute systems, but not for most commercial processes where the isotherms are nonlinear. More complex theories have been and are constantly being developed, but have limitations for design. In this section we will outline a simple, semiempirical approach for design.

A. Mass Transfer Zone (MTZ) or Length of Unused Bed (LUB) Theories

In actual practice, shock waves will spread due to dispersion and finite rates of mass transfer. However, this spread is opposed by the isotherm and sorption effects which tend to form shock waves. The net result of these opposing forces is to form a ''constant pattern'' wave which does not change shape as it moves in the column. An example was shown in Figure 2-9C. Once formed the wave does not change shape as it moves down the column. This behavior will occur whenever the solute movement theory predicts a shock wave.

A common industrial design procedure for isothermal or approximately isothermal systems uses constant pattern analysis based on experimental data. What is actually measured is the ''breakthrough curve'' as shown in Figure 2-22. This is the outlet concentration which results when a step input in concentration is used. The "mass transfer zone" (MTZ) is the region where concentration is changing and thus mass transfer is occurring. The length of the MTZ in time units, t_{MTZ}, is easily measured from the breakthrough curve (Figure 2-22). Usually the MTZ is arbitrarily measured from a concentration of 0.05 c_F to 0.95 c_F, since it is hard to tell exactly where the S-shaped pattern starts and ends. This problem is exacerberated by the noise typical in commercial adsorption.

We wish to use the breakthrough data to determine the MTZ length inside the column. The MTZ must move at the shock wave velocity u_{sh}. Then,

$$L_{MTZ} = u_{sh} t_{MTZ} \tag{2-129}$$

The MTZ inside the column is the region where concentrations are changing and thus mass transfer is occurring. Upstream of the MTZ the bed is saturated at the feed concentration while downstream of the MTZ the bed contains no solute. Typically, adsorption is stopped when breakthrough just starts. Since the concentration in the MTZ is varying, the sorbent is not saturated through the entire length of the bed. Thus part of the bed is unused. The MTZ data can be used to design the column using the LUB method.

The LUB approach is based on the original work of Michaels[726] and has been applied to ion exchange and adsorption.[273,510,639,645,741,865,990] The basic idea is to determine from the breakthrough curve the fraction of the bed not used. Many different approaches which do this have been developed. We will use Figure 2-22. Inside the column the fractional bed utilization can be determined as

$$\text{Fraction of bed utilized} = \frac{L - L_{MTZ}\left(\dfrac{\text{area not used}}{\text{total area in MTZ}}\right)}{L} \tag{2-130}$$

The ratio of area not used to total area in MTZ is the fraction of the MTZ which is not useful for adsorbing solute at the saturation concentration. This ratio can be determined from Figure 2-22. If the breakthrough curve is symmetric the ratio is 1:2. For symmetric breakthrough curves Equation 2-130 simplifies to:

$$\text{Fraction of bed utilized} = (L - 0.5L_{MTZ})/L \tag{2-131}$$

The value of L_{MTZ} can be obtained from Figure 2-22 by use of Equation 2-129. Note that if u_{sh} is very low L_{MTZ} may be very short even though it appears long in Figure 2-22.

The amount of solute which can be held by the bed is

$$\text{Capacity} = (\text{Fraction of beds utilized}) m q_{sat} \tag{2-132}$$

where q_{sat} is the saturation capacity of the adsorbent for the given feed concentration and m is the mass of adsorbent in the column. Obviously, the capacity can be increased by increasing any of the three variables in Equation 2-132. The fractional bed utilization can be increased by increasing L or decreasing the MTZ. The MTZ will be shorter for rapid mass transfer rates or small-size adsorbent particles. The MTZ tends to be quite large for large molecules such as proteins which have low rates of mass transfer. The saturation capacity, q_{sat}, depends on the adsorbent used and the temperature. The effect of increasing L/L_{MTZ} is discussed in Chapter 3 (Section II.C).

The LUB approach has several limitations. The most obvious limitation is it is only applicable to constant pattern systems. For a favorable isotherm this would be the adsorption step. Some other approach will be required for the desorption step. The LUB approach is strictly applicable for isothermal systems,[273] although it is often applied for nonisothermal systems with modest temperature fluctuations. Large changes in temperature can destabilize the MTZ and give nonconstant patterns.[639] The LUB approach has been adapted to nonisothermal systems,[639] but it becomes more complex. Scale-up also becomes more difficult since small columns are usually not adiabatic while large columns usually are adiabatic.

The LUB approach requires experimental data. This is really less of a disadvantage than it may seem at first since more fundamental theories also require data. Obtaining data on the chemical system which will actually be separated is very helpful if any trace components are present since they may affect the adsorbent markedly. The same particle size should be used in the laboratory column. To avoid wall effects, values of the ratio diameter/$d_p > 30$ should be used for Newtonian fluids and greater than 50 for non-Newtonian fluids.[271]

Constant pattern systems are amenable to significant mathematical simplification. The constant pattern will always form when conditions are right.[283] Because the shape of the pattern is invariant, the partial differential equations can be transformed into ordinary differential equations based on the distance from the stoichiometric front (the shock wave location). This simplification has led to a deluge of theories for constant pattern systems. These theories are reviewed in detail elsewhere.[165,510,522,657,865,901,912,990,1015-1017,1101] In the remainder of this section a simple isothermal model will be developed.[901]

The usual starting point for a theoretical analysis is Equation 2-73 plus some mass balance expression such as Equation 2-74. The constant pattern will be a function of τ,

$$\tau = t - z/u_{sh} \tag{2-133}$$

since τ is a measure of the time or distance from the center of the pattern. Neglecting axial dispersion, we can reduce Equation 2-73 to an ordinary differential equation by introducing τ. The result is

$$\epsilon'\left(1 - \frac{v}{u_{sh}}\right)\frac{dc}{d\tau} + \rho_B(1 - \epsilon')\frac{dq}{d\tau} = 0 \tag{2-134}$$

This equation has the obvious solution:

$$\left(\frac{v - u_{sh}}{u_{sh}}\right)\epsilon' c - \rho_B(1 - \epsilon')\, q = \text{constant} \tag{2-135}$$

If the bed is initially clean, $c = q = 0$ and the constant must be zero. Equation 2-135 can be written for arbitrary conditions in the wave, c and q, and for final conditions after the wave has passed, c_{final} and q_{final}. Dividing one equation by the other,

$$\frac{c}{c_{final}} = \frac{q}{q_{final}} \tag{2-136}$$

Substituting Equations 2-133 and 2-136 into Equation 2-74 we obtain:

$$(1 - \epsilon')\rho_B \frac{q_{final}}{c_{final}}\frac{dc}{d\tau} = -k_M a_p(c^* - c) \tag{2-137}$$

It is usually assumed that the stagnant fluid c^* is in equilibrium with the solid, but $c^* \neq c$. The variable c^* can then be substituted for from the isotherm and q can be removed with Equation 2-136. For example, with the general Langmuir expression,

$$c^* = \frac{q}{a - bq} = \frac{c(q_{final}/c_{final})}{a - bc(q_{final}/c_{final})} \tag{2-138}$$

Substitution of this equation into Equation 2-137 gives the simple ODE,

$$\rho_B(1 - \epsilon')\left(\frac{q_{final}}{c_{final}}\right)\frac{dc}{d\tau} = k_M a_p\left[c - \frac{c(q_{final}/c_{final})}{a - bc(q_{final}/c_{final})}\right] \tag{2-139}$$

which is easily integrated.

The MTZ can be obtained by rearranging Equation 2-139 and integrating from $c = 0.05c_F$ to $0.95c_F$. The result is

$$\Delta\tau = \frac{\rho_B(1 - \epsilon')\left(\frac{q_F}{c_F}\right)}{k_M a_p}\left\{\frac{-a}{a - \frac{q_F}{c_F}} \ln\left[\frac{0.05}{0.95}\,\frac{a - \frac{q_F}{c_F} - 0.95bq_F}{a - \frac{q_F}{c_F} - 0.05bq_F}\right]\right\}$$

$$+ \ln\left[\frac{a - \frac{q_F}{c_F} - 0.95bq_F}{a - \frac{q_F}{c_F} - 0.05bq_F}\right] \tag{2-140}$$

If $z = L$, $\Delta\tau = t/_{c=0.95c_F} - t/_{c=0.05c_F}$ which is the t_{MTZ}. If $t = t_{breakthrough}$, $\Delta\tau = L_{MTZ}/u_{sh}$. Equation 2-140 will be useful later on.

B. Design of Total Cycle

Someday, computer codes will exist to allow design of the total adsorption/desorption cycle for arbitrary systems. Although progress is being made and restricted but important problems can be solved quite rigorously, the day of a complete solution for any problem is not yet here. In the meantime, relatively simple procedures can be used for preliminary designs of many systems.

For the adsorption of one solute with favorable isotherms and modest heat effects the LUB approach is satisfactory. This requires column experiments as shown in Figure 2-22. During desorption a proportional pattern usually develops. The local equilibrium (solute movement) theory usually gives quite reasonable predictions for proportional pattern behavior. Thus, if the equilibrium isotherm under the conditions of desorption is determined, the solute movement theory can predict the desorption part of the cycle. This combined approach is "quick-and-dirty" but should be acceptable in most cases.

For linear systems the theories of Section VI. work very well. The major applications of these theories has been in analytical chromatography. They are useful in a limited number of large-scale chromatography applications (see Chapter 5). Nonlinear gas chromatography has been designed by superposition of local equilibrium models and dispersion. More complicated situations will usually require numerical solution of at least part of the cycle. More detailed models used are reviewed elsewhere.[59,112,165,513a,536,657,686,832,865,901,912,1006,1015-1017,1068,1073,1101] Ruthven's book[865] is a detailed, up-to-date resource for modeling.

VIII. SUMMARY

This chapter is long because it serves as a basis for all of the other chapters. Since in most adsorption and chromatography techniques equilibrium or deviations from equilibrium are the basis for separation, good equilibrium data is necessary. Although a number of equilibrium isotherms were given, the list was not exhaustive. The solute movement theory is a method for visualizing what is happening in the column. It provides a means for thinking about how equilibrium and operating methods are combining to cause the separation. Since this theory is simple compared to other theories of adsorption, it can be used as a tool to think about, explore, compare, and try on paper new operating schemes. We will be using the solute movement theory to do this in the remainder of this book.

Mass transfer and dispersion work to limit or destroy the separation. The staged models and rate theories quantify these effects. For linear systems the zone spreading effects are conveniently included in the Van Deemter equation. Thus the Van Deemter equation is a very useful tool to explain and then reduce zone spreading. The zone spreading for linear systems is proportional to the square root of the distance traveled.

For nonlinear systems with favorable isotherms a constant pattern is formed when a dilute solution is displaced by a concentrated solution. The zone spreading is now independent of distance traveled and the LUB approach can be used for design. For the same chemical system a proportional pattern (diffuse wave) will form when a concentrated solution is displaced by a dilute solution. Now the zone spreading is proportional to the distance traveled. Since the isotherm curvature controls zone spreading for proportional pattern systems, the equilibrium theories often given excellent predictions.

Chapter 3

PACKED BED ADSORPTION OPERATIONS

I. INTRODUCTION

In this chapter we will consider several common operating methods for recovery of a single solute or for removal of all solutes as a waste stream. We will also consider processes where the adsorption and desorption steps are often conceived of as two separate parts. In Chapter 4 the adsorption and desorption steps tend to be more tightly coupled. Chapters 3 and 4 complement each other, and the division between chapters is somewhat arbitrary. Chapter 5, which also covers packed beds, considers chromatographic separations where fractionation is done.

Many common industrial processes are included in this chapter. These include trace contaminant removal, drying of gases and liquids, solvent recovery with activated carbon, and wastewater treatment with activated carbon. Many of these processes can be done by several different operating methods, and thus the process will appear in several chapters. For example, solvent recovery with activated carbon is done in packed beds (Chapter 3), in countercurrent moving beds (Chapter 6), and in rotating beds (Chapter 8).

This chapter is structured to first look at the operation of a packed bed in general. Methods for desorption and for increasing the efficiency of the process are discussed. Then the specific adsorption processes are briefly covered with an emphasis on thermal desorption processes. Desorption using solvents is covered in a separate section at the end of this chapter (and in Chapter 6 [Section V.B]) and pressure swing adsorption is covered in Chapter 4.

II. OPERATION OF PACKED BEDS

A variety of operating cycles have been developed for packed bed adsorbers. In this section we will look at some of the alternatives in a general sense, and develop a background which will be useful when we look at individual processes.

A. Adsorption-Desorption Methods

Essentially, four approaches have been developed for adsorption-desorption cycles. The first is a nonregenerative design where the charge of adsorbent lasts the life of the system without being regenerated.[17,253,282] This type of system is used in sealed dual-pane windows, in some food products, and as an antidote in poisonings. This method will not be considered further here.

The second approach is to adsorb until the adsorbent is saturated and then replace the adsorbent. Totally new adsorbent may be used as in the canister of a gas mask.[666,687] Gas mask canisters are often complex mixtures of several sorbents in series, and they would be difficult to regenerate. Also, incomplete regeneration could be serious. Thus total replacement of the canister is the usual procedure. Total replacement of adsorbent in the form of canisters is also used occasionally in air purification systems for odor removal.[608,687] The adsorbent is usually activated carbon. Since the concentration of the odor-causing chemicals is usually very low, the adsorbers can be kept on stream for very long periods (often years) before the canisters have to be replaced. (Some canister systems are regenerated in place and fit into the fourth category of operating methods.) These systems will not be discussed further. Canister systems are also used in treating drinking water with activated carbon (see Section IV.B).

A similar approach is commonly used for wastewater treatment with activated carbon and for treating molasses with activated carbon, except regeneration of the carbon in a separate furnace is planned for. The regeneration may be done on or off site. The adsorption and desorption cycles are effectively decoupled. The adsorption cycle consists of loading the adsorbent into the column, adding a charge if liquids are being processed (the sweetening-on step), processing the fluid during the feed step, and then removing the used adsorbent from the column. This type of cycle is common when trace amounts of strongly adsorbed materials are removed, and the feed step can be quite long.

The third type of operation consists of loading regenerated sorbent into the column, sweetening on, doing the adsorption step, and emptying the column. The sorbent is sent to a separate vessel where it is regenerated and then immediately reused. Regeneration can be done by any of the means used for the fourth type of operation (see below). This type of operation is often used in the food and pharmaceutical industries since it clearly keeps the regeneration materials away from the process fluids. This type of system differs from the second type of operation since the adsorption and regeneration steps are fairly closely coupled, and the adsorption step cannot be designed completely independently of the desorption step.

In the fourth type of operation the adsorption and regeneration steps are done in the same vessel and the adsorbent is not transferred from one vessel to another. A typical cycle consists of adsorption, regeneration, and preparation for the next cycle.

A variety of regeneration methods are used in the third and fourth type of operations. These include:

- Noncondensing hot gas
- External heating
- Steam or other condensing hot gas
- Hot liquid
- Drain column and then use hot gas
- A gas or liquid desorbent (see Chapters 4 and 6 also)
- Brine (fluid of high ionic strength)
- Pressure swing (see Chapter 4)
- Vacuum (see Chapter 4 also)
- A combination of the above

Details of these regeneration steps will be explored when the individual processes are considered.

The steps required to prepare the column for the next adsorption step depend on the regeneration step. In some cases where hot gas is used for regeneration no preparation step is required, while in other cases the bed may have to be cooled. When steam is used for regeneration, the bed may need to be dried and/or cooled. When desorbent or brine are used, they need to be washed out from the column. With pressure swings and vacuum regeneration the pressure needs to be increased to the operating pressure. In addition to these steps, a backwash may be used to remove suspended solids which have been trapped on the bed. Also, a sweetening-on step may be needed to add a concentrated viscous material to the column, and a sweetening-off step may be used before the regeneration step to recover valuable feed material in the column.

B. Desorption Cycles

When the adsorbent is to be regenerated in the same vessel or in a separate vessel which is closely coupled to the adsorption step, the desorption or regeneration step is crucial to the commercial success of the process. Despite the importance of desorption, many articles in the literature look at only the adsorption step instead of the entire cycle. If only the

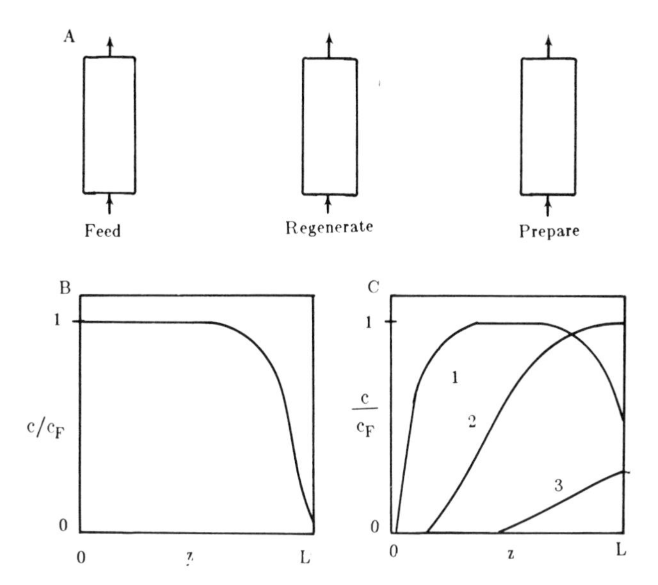

FIGURE 3-1. Co-flow regeneration. (A) Steps in cycle; (B) concentration profile in column at end of feed step; (C) concentration profiles in column during regeneration. 1, Early in regeneration step; 3, end of regeneration.

adsorption step is looked at, incorrect conclusions may be drawn about desired operating conditions. It is important to always consider the entire cycle.

Regeneration can be done with either co-flow (in the same direction as the feed in the adsorption step) or counter-flow (in the opposite direction to the feed). Co-flow regeneration is shown schematically in Figure 3-1A. Either up-flow or down-flow may be used. The feed step is continued until breakthrough occurs as shown in Figure 3-1B. With co-flow regeneration the regeneration fluid flows in the same direction and pushes the concentration profile through the previously clean part of the bed. This is illustrated in Figure 3-1C where the concentration profiles are shown for three times during the regeneration step. Co-flow desorption thus contaminates the clean part of the bed, which is undesirable.

Counter-flow regeneration is shown schematically in Figure 3-2A. The feed step is essentially the same as in co-flow systems, and the concentration profile at breakthrough is illustrated in Figure 3-2B. Regeneration is now done in the opposite direction as shown in Figure 3-2C. With counter-flow operation the clean end of the bed stays clean. Regeneration is seldom run long enough to completely clean the column. Thus a residual amount of solute (the ''heel'') is often left in the column. With co-flow (Figure 3-1C) this residual will appear in the product during the next feed step. With counter-flow regeneration the residual is at the feed end of the column, and the solute can be readsorbed.

The solute movement theory developed in Chapter 2 can be used to compare co- and counter-flow regeneration. The advantages of counter-flow are most evident when the isotherms are nonlinear and have a favorable shape (Langmuir or Freundlich). Consider a case where a hot fluid is used for regeneration. (The solute movement theory can also be applied to other regeneration methods.) Heating decreases adsorption and makes the solute wave velocities higher (Equations 2-28, 2-32, or 2-35). The resulting solute movement diagrams for co- and counter-flow are shown in Figures 3-3A and B, assuming the thermal wave is faster than the solute waves. Co-flow does not produce a pure final product unless the column is completely regenerated. A pure product can be produced with counter-flow even though the column is not completely regenerated. In addition, the diffuse wave which is generated during the regeneration step can be partially recompressed (or swallowed by the shock wave)

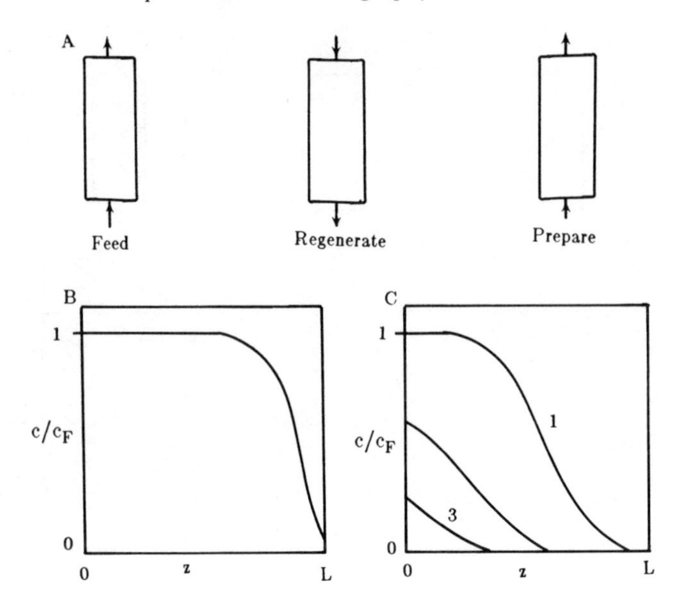

FIGURE 3-2. Counter-flow regeneration. (A) Steps in cycle; (B) concentration profile in column at end of feed step; (C) concentration profiles in column during regeneration. 1. Early in regeneration; 3, end of regeneration.

as shown in Figure 3-3B. The step increase in concentration when feed is added causes a shock wave which decreases in slope as less concentrated material is intersected, and the diffuse wave disappears. Counter-flow regeneration is usually used for adsorption systems because of these advantages. The adsorbate will leave the column as a quite concentrated peak which then tails when the diffuse wave exits. Thus, adsorption columns can serve as concentrators for dilute streams, and may be the cheapest way to concentrate. Co-flow has advantages for fractionation systems where a heel can often *not* be left in the column and is used in chromatography (Chapter 5) and in simulated countercurrent systems (Chapter 6).

In some cases the cooling period shown in Figures 3-3 is not required.[109] This is the case when the thermal wave set up by the cool feed solution moves faster than the concentration wave (see Section III.B). Even when a cooling step is desirable it can usually be much shorter than shown in Figures 3-3A and B. A short cooling step followed by feed will push the hot fluid out of the column before the solute wave breaks through.

How much regeneration should be done? Partial regeneration keeps the MTZ inside the column, and makes the regeneration product much more concentrated in solute. This procedure also minimizes the amount of desorbent or purge fluid required. (Of course, there are situations where it is desirable to completely regenerate the bed, but they are unusual.) A loading diagram can be constructed using the equilibrium curve and the solute movement diagram shown in Figure 3-3B. Figure 3-4A illustrates the loading in the bed when partial regeneration is used. Note that the useful capacity is considerably less than the total capacity, but regeneration is stopped at a convenient time. Typically, this would be some multiple of the adsorption time so that several columns can be used to allow continuous processing of the feed. Complete regeneration is shown in Figure 3-4B. For systems with nonlinear "favorable" isotherms desorption is slower than adsorption. This occurs because the MTZ during the adsorption step approaches a constant shape while the diffuse wave during desorption grows proportionally to the distance traveled. With longer desorption periods, the useful capacity increases, but the capacity/time and the outlet concentration of the purge stream decrease. Thus, partial regeneration is usually favored. This is particularly true when the isotherm is very nonlinear and the diffuse wave would show considerable tailing. This

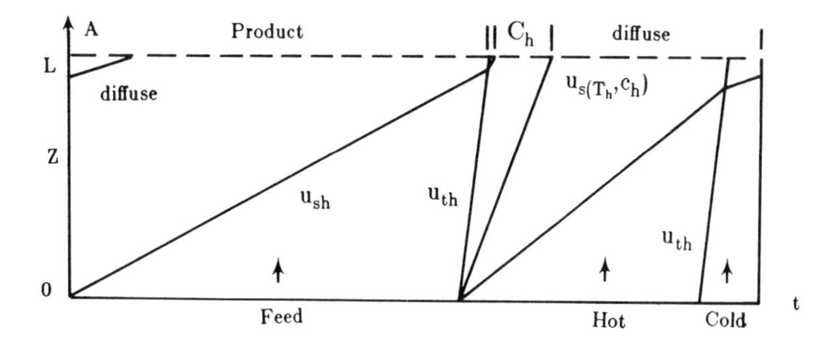

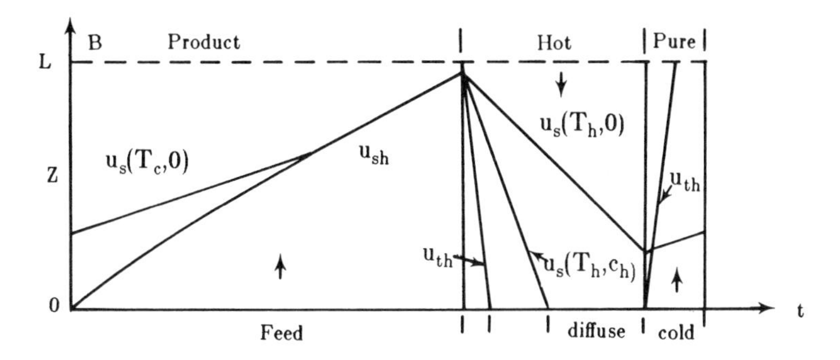

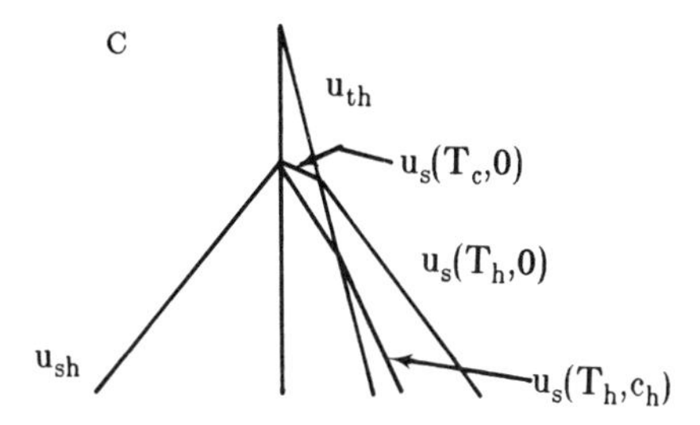

FIGURE 3-3. Solute movement diagram for system with nonlinear isotherms. (A) Co-flow; (B) counter-flow; (C) blown-up diagram for intersection of thermal waves and solute waves for counter-flow.

situation occurs in drying of gases and in solvent recovery with activated carbon. As a general rule it is best to keep the system small and to cycle fairly rapidly.

C. Increasing Fractional Use of the Bed

The fractional use of the bed for the situation shown in Figure 3-4A can be very low. This results in a very low adsorbent productivity (kilogram of adsorbate per kilogram adsorbent-hour) and thus may require rather large beds. In addition, the concentration of adsorbate in the purge product may be quite low. Several methods can be used to increase the fractional bed use and increase the productivity.

For the usual situation for adsorption where the isotherm is nonlinear and the shape is "favorable", the fractional bed use can be increased by increasing the bed length. With favorable isotherms a constant pattern wave is formed and the MTZ will reach a constant length. Then as the bed length, L, increases the ratio L_{MTZ}/L decreases and the relative length of the bed which can be fully loaded increases. This effect is shown in Figure 3-5[676] for the adsorption part of the cycle only. There is a decreasing rate of improvement as the

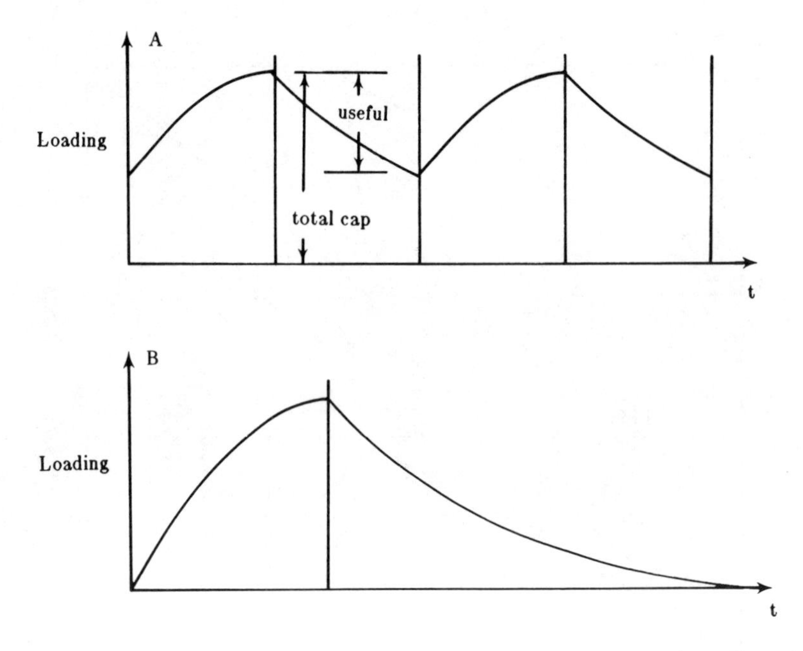

FIGURE 3-4. Total column loading during cycle. (A) Partial regeneration; (B) total regeneration.

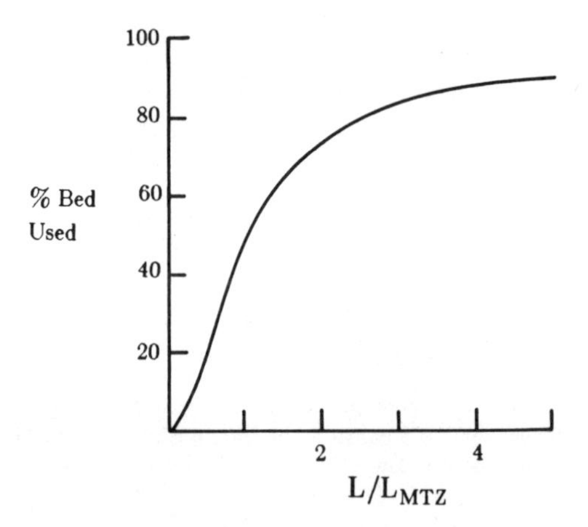

FIGURE 3-5. Fraction of bed used as function of L/L_{MTZ}. (From Lukchis, G. M., *Chem. Eng.*, 80(13), 111, 1973. With permission.)

bed length increases past two to three times the L_{MTZ}. The appropriate bed length thus depends very heavily on the length of the MTZ, and thus on the mass transfer rate and the equilibium isotherm. Increasing the bed length will also increase the pressure drop. The fractional use of the bed and the sorbent productivity can be increased by decreasing the particle diameter. This is discussed in detail in Section II.D.

The bed utilization can also be increased by using layered beds. One way of doing this is to use one layer of large-diameter particle followed by a layer of the same adsorbent of a considerably smaller diameter.[382b] In the first layer the pressure drop will be low, but the MTZ will be quite long. If the isotherm is favorable the MTZ will be sharpened in the second layer since there is less resistance to mass transfer in the smaller particles. The layered bed can then have a small L_{MTZ} at the outlet which means a small ratio of L_{MTZ}/L

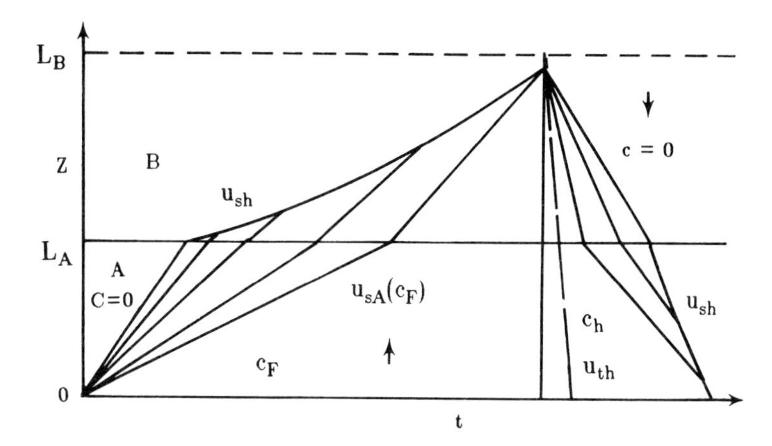

FIGURE 3-6. Solute movement diagram for layered beds. Adsorbent A has unfavorable isotherm while B has a favorable isotherm.

and hence a large fractional bed use. At the same time the pressure drop will be considerably less than a bed which was entirely small particles. Use of this procedure has been reported with two sizes of zeolites.[112]

A second way to use layered beds is to use two different adsorbents.[12,112,268,1016] If a cheap adsorbent with a modestly favorable isotherm is covered with an expensive adsorbent with a very favorable isotherm, the shock wave will be tightened up when it reaches the second layer. Thus, bed utilization will be higher or a purer product will be produced but at lower cost. This procedure is done with mixtures of silica gel and zeolite molecular sieve, and other combinations of adsorbents. Layered beds of different adsorbents are also used to remove several adsorbates from the feed stream. A unique application of this concept is on spacecraft.[308,309]

Another use of layers is to use a guard bed which protects the main bed from contaminants which might polymerize or irreversibly adsorb. The guard bed can use sorbent which has lost some of its capacity, but is still useful.

The two layers could be two different sorbents which have different isotherm curvatures. If a layer of adsorbent with an unfavorable isotherm is covered with a layer with a favorable isotherm, the column will have particularly advantageous properties. A diffuse wave will form in the adsorbent with the unfavorable isotherm during the adsorption step. This diffuse wave will be refocused in the second layer of adsorbent. The result at the outlet of the column will be a shock (constant pattern) wave. Then the bed utilization during the adsorption step can be quite high. During counter-flow desorption a diffuse wave will form in the bed with the favorable isotherm, but this wave will be refocused in the layer of adsorbent with an unfavorable isotherm. The result is a shock wave at the feed end of the column. This type of system can then have tight constant patterns during both adsorption and desorption. This situation is illustrated using a solute movement diagram in Figure 3-6. Note that when dispersion is ignored the diffuse waves can be recompressed. In Figure 3-6 sorbent B has a favorable isotherm while sorbent A has an unfavorable isotherm. This could be achieved by using an adsorbent such as activated carbon for sobent B, and a reverse-phase chromatographic packing for sorbent A. The wave slopes vary when the boundary between sorbents is passed. Since the isotherms have different curvatures, the waves can either speed up or slow down at the transition point. The exact shape of the solute movement diagram will depend on the sorbents and their isotherm shapes. With only one adsorbent a diffuse wave must form during either the adsorption or regeneration steps. This two-layer procedure has been studied for ion exchange systems.[595]

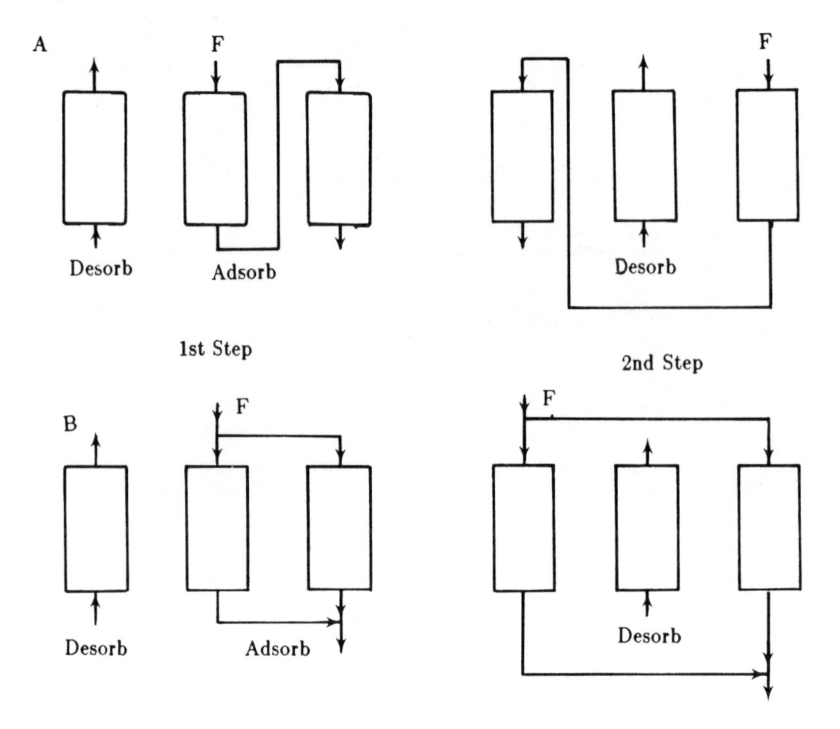

FIGURE 3-7. Increasing bed utilization. (A) Series arrangement; (B) parallel arrangement.

For removal of a divalent ion by ion exchange the column can be operated so that equilibrium is favorable for sorption of the divalent ion during the loading step and unfavorable for the divalent ion during the regeneration step. This is possible since the equilibrium expression, Equation 2-23, depends on the total equivalents per liter of ions in the solution. The result will be shock waves during both loading and regeneration steps, and a minimum of dispersion. This approach is commonly used for water softening where concentrated sodium chloride solution is used for regeneration. This approach does not appear to be generally applicable to adsorption.

Another procedure which can be used to increase fractional bed use is to use multiple beds either in series (Figure 3-7A) or in parallel (Figure 3-7B). In the series case, two or more columns are connected in series while the other column(s) is regenerated. The first column in the series can be totally saturated while the MTZ is held in the remaining columns. The freshly regenerated column is put on at the tail end of the series. If counter-flow regeneration is used, the column does not have to be completely regenerated. This procedure is commonly used,[247,327,380,382b,469,522,527,557,591,655,677,687,785, 946,954,1016] particularly for systems with long MTZs such as activated carbon wastewater treatment.[327,380,522,527,557,591,785]

Many theoretical studies of the series systems have been reported[247,327,591,655,946,954] using numerical methods to solve the equations. An alternate way to use columns in series, but with only two columns, is discussed in Section III.A. and shown in Table 3-1. When there are several columns in the series at one time the method is often called a ''merry-go round'', and is related to simulated moving bed techniques (see Chapter 6, Section V.A.).

In the parallel system shown in Figure 3-7B, the bed which will be desorbed next is run past the allowable breakthrough concentration, but the product concentration is kept below acceptable levels.[380,527,946] The parallel arrangement has the advantage of a lower pressure drop, but always has a lower bed utilization and a less concentrated effluent than the series case.[946] The solute movement theory for nonlinear isotherms can easily be used to analyze both the series and the parallel cases.

Table 3-1
NO_x CYCLE[556]

Step 1.	Col. 1 adsorbing Col. 2 desorbing	Step 2.	Col. 1 adsorbing Col. 2 cooling
Step 3.	Col. 1 adsorbing Col. 2 adsorbing in series behind Col. 1	Step 4.	Col. 1 desorbing Col. 2 adsorbing
Step 5.	Col. 1 cooling Col. 2 adsorbing	Step 6.	Col. 1 adsorbing in series behind Col. 2 Col. 2 adsorbing

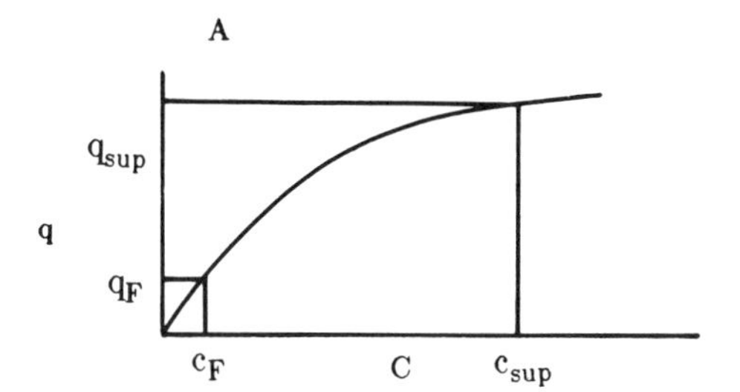

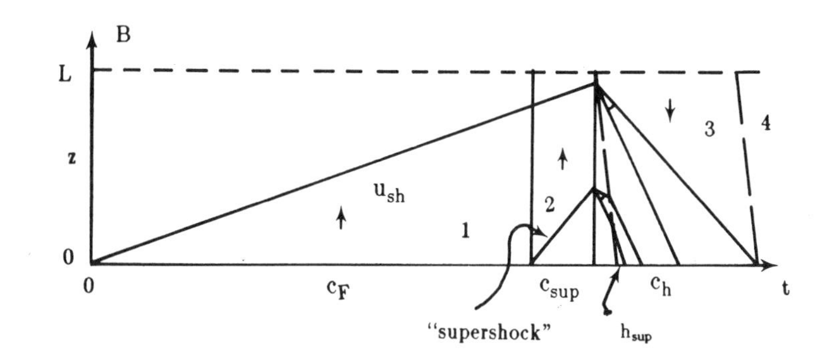

FIGURE 3-8. "Superloading" procedure. (A) Illustration on equilibrium diagram; (B) solute movement diagram. 1, Feed step; 2, "superloaded"; 3, solvent rinse; 4, water rinse.

A final method for increasing the bed utilization is the "superloading" or two-feed adsorption concept.[381,382,382a,1026] An adsorbent bed was used to remove phenol from water. The bed was regenerated with acetone followed by a water rinse. The eluant from regeneration was distilled and acetone was the distillate product. The bottoms product separated into two layers: a phenol-rich layer which was recovered and an aqueous layer with 10% phenol (a process flow sheet is shown later in Figure 3-20). This layer was sent to the adsorption column at the end of the usual loading step. Even though the bed was saturated with phenol at the feed concentration, the bed still had considerable capacity for phenol at the very high concentrations in the waste aqueous phase. This is shown schematically for a favorable isotherm in Figure 3-8A. This superloading procedure increases the concentration of phenol in the bed and thus gives a higher concentration feed to the distillation column. This procedure should be adaptable to other adsorption problems. It would be generally applicable whenever two feeds of different concentration need to be processed. The more concentrated feed enters after the dilute feed. This procedure is roughly analogus to a distillation column with two feeds. The solute movement diagram for superloading is shown in Figure 3-8B. The superloading produces a shock wave, the velocity of which can be calculated from Equation 2-28. During regeneration an acetone wave and two diffuse waves will be formed. Complete regeneration of the column is not required, and a heel can be left in the column. The solute movement diagram can look different depending on the isotherm shape and the periods for each step.

D. Intensification of Adsorption: Particle Diameter Effects

Particle diameter has a major effect on the behavior of the packed bed. In this section we will explore this effect for packed bed adsorption systems. The changes in analytical liquid chromatographic columns over the last 25 years have been tremendous. By decreasing the particle size and increasing the pressure drop, column lengths and analysis times have decreased by almost two orders of magnitude. Unfortunately, very little of this process intensification has appeared in large-scale adsorption or chromatography. In this section we will show that many packed bed commercial adsorption systems can operate with shorter columns and faster cycles to increase the productivity of the adsorbent by at least an order of magnitude. This requires decreasing the particle diameter, but can be done with *no* increase in total pressure drop. Continuous systems have different requirements (see Chapter 6, Sections II and III) and this section does not apply to moving bed systems.

In Chapter 2, Section VI.C we saw that under most circumstances the plate height for linear systems is directly proportional to d_p^2. If pore diffusion controls this is absolutely true. Thus, N is proportional to (L/d_p^2). By decreasing the particle diameter we can increase N and thus shorten the MTZ, or we can decrease both d_p and L while keeping N constant.

Decreasing the particle diameter also affects the pressure drop. For laminar flow, which is almost always the case in chromatography and adsorption, the pressure drop equation is[138,549]

$$\Delta p = \frac{v\mu L}{K d_p^2} \tag{3-1}$$

For rigid particles the permeability K can be calculated from[138]

$$K = \frac{\alpha^3}{150(1 - \alpha)^2} \tag{3-2}$$

while for deformable gel particles[549]

$$K = K_o \exp\left(-\frac{a\,\Delta p}{L}\right) \tag{3-3}$$

In this section we will restrict ourselves to rigid particles. Equation 3-1 shows that the total pressure drop is proportional to (L/d_p^2) which is exactly the same functional dependence as N.

For process intensification we can decrease both the particle diameter and L, keeping (L/d_p^2) constant. Now Δp and N are unchanged. The zone of changing concentrations (MTZ) requires the same number of stages in the column, N_{MTZ}, and the fractional length of the MTZ is the same

$$\frac{L_{MTZ}}{L} = \frac{N_{MTZ}}{N} = \text{constant} \tag{3-4}$$

Thus the fraction of the column which can be used per cycle is unchanged. Since the column is shorter it will saturate faster. Thus, the cycle period is decreased by the same factor as

L. Over many cycles the throughput is unchanged but the adsorbent productivity [moles throughput]/[(kg adsorbent) (hour)], increases by the same factor as L decreased. The faster cycling and the shorter column length cause this increase in productivity. The smaller particle diameters allows one to shorten the bed and cycle faster.

An alternate solution is used commercially by the Graver Company.[623] They attach very small particles to fibers to give a system with very rapid mass transfer and low pressure drop. However, currently these systems are not regenerated.

The productivity argument was based on the Van Deemter equation which is valid for linear systems. How about nonlinear systems? The distance that shock and diffuse waves travel is proportional to time. Thus if the cycle time and L are decreased by the same factor, the fraction of the column length traveled by the wave is unchanged. This can be seen by changing both L and t in Figures 2-6 and 2-9. Since the zone spreading in a diffuse wave is proportional to L, the fraction of the column occupied by the diffuse wave will be unchanged when L and t are both decreased. For constant pattern waves the width of the MTZ depends on a dynamic balance between dispersion and the compressive effects of the shock wave. Equation 2-140 shows that L_{MTZ} is inversely proportional to $K_m a_p$. If pore diffusion controls, $K_M a_p$ is inversely proportional to d_p^2 and thus L_{MTZ} is proportional to d_p^2. If we scale by L/d_p^2, L/L_{MTZ} is constant and the fractional bed use remains constant.

The question of whether the column will scale L/d_p^2 can also be asked using the Thomas solution for Langmuir isotherms (e.g., see References 901 or 1016). When this is done it is easy to show that if pore diffusion controls, the number of transfer units is proportional to L/d_p^2 and Equation 3-4 is valid if (L/d_p^2) is constant. A good example of this is to rework example 10.6 in Reference 901 to include this scaling. This is shown in Table 3-2. The problem consists of adsorbing 10 mol % methane from hydrogen at 10 atm and 25°C on activated carbon. Flow rate was 100 ft^3/min and the adsorption time was 10 min. Cases 1 to 3 were previously reported.[901] The authors concluded that there was an advantage in decreasing particle size from 4 to 10 mesh, but not from 10 to 20 mesh. Case 4 is a new calculation which uses 20-mesh particles but reduces the period so that n and nT (terms in the Thomas solution[901]) are kept at the same value as in Case 2. This results in a bed which is roughly four times smaller than Case 2, has a period roughly four times shorter, but has essentially the same number of transfer units and the same pressure drop. Thus, Case 4 has the same relative bed utilization as Case 2 but with an adsorbent productivity which is roughly four times greater. There is an advantage to decreasing the particle size if the cycle period is also decreased.

In this example, scaling was done to keep the number of transfer units n constant and not exactly as $1/d_p^2$. Scaling by $1/d_p^2$ would have reduced the column length by a factor of 4 instead of the 3.92 actually observed. This problem does not scale exactly as $1/d_p^2$ since $k_f a_p$ is not negligible in the determination of $k_T a_p$, and $k_f a_p$ is proportional to $1/d_p^{1.415}$. The difference is not very significant and there is still an obvious advantage in decreasing d_p, L, and the adsorption period.

So far, only the adsorption step has been discussed. Can the desorption step also be done faster? The answer is yes, if certain conditions are met. The time for the desorption step will also scale by the same factor used to decrease the column length. For thermal desorption the time for the thermal wave to break through is proportional to L. Thus the fraction of the desorption period required for breakthrough is constant. The analysis of Van Deemter et al.[1008] could be applied to the energy balance. The result would be a Van Deemter equation for the number of thermal stages in the column. If thermal pore diffusion controls, then N_{th} will be proportional to (L/d_p^2). This means that

$$\frac{\text{Width thermal transfer zone}}{L} = \text{constant} \tag{3-5}$$

Table 3-2
EXPANDED RESULTS OF THOMAS CALCULATION FOR EXAMPLE 10.6 OF SHERWOOD ET AL.[901]

Case	dp (cm)	Period (min)	$k_p(k_M)$ (cm/sec)	$a(a_p)$ (cm^2/cm^3)	Superficial Vel (cm/sec)	k_f (cm/sec)	$\kappa(k_T)$ (cm/sec)	κa ($k_T a_p$)	n	nT	Length (ft)	Adsorbent volume (ft^3)	Δp in, H_2O
1	0.476 (4 mesh)	10	0.0327	8.26	3.0	1.06	0.0364	0.300	30.6	17.65	10.05	185.7	0.42
2	0.168 (10 mesh)	10	0.0927	23.4	3.0	2.02	0.1016	2.38	160.9	143	6.63	122.5	2.24
3	0.0841 (20 mesh)	10	0.0185	46.6	3.0	3.05	0.200	9.33	579.9	562.4	6.12	113.1	8.26
4	0.0841	2.55	0.0185	46.6	3.0	3.05	0.200	9.33	160.9	143	1.69	31.2	2.29

Note: Cases 1 to 3 are from Sherwood et al.[901] with a minor correction in case 1. Case 4 is same particle diameter as case 3, but has period selected so that n and nT are same as in case 2.

as long as (L/d_p^2) is constant. Since there is no change in the fraction of the bed covered by the thermal transfer zone, the period of desorption will scale in the same way as L. Thus, if the column length is decreased by a factor of ten then both the adsorption and desorption periods are decreased by a factor of ten, and the adsorbent productivity is increased by a factor of ten. Obviously, this does not apply for a jacketed, externally heated system since radial heat transfer will not be speeded up. If a large proportion of the energy is used heating the ends of the column, energy use will increase with rapid cycling. This can be controlled with insulation.

If the desorption is done with a solvent or desorbent, similar arguments can be used to show that the desorption time will scale in the same way as L does if (L/d_p^2) is constant.

Another common method is pressure swing adsorption (see Chapter 4). Usually, the time for the pressure wave to pass through the system is not important since this time is much shorter than the cycle time. If the bed is tight and cycle times are short, pressure waves can be very important;[558] however, for most pressure swing adsorption (PSA) systems at least a modest process intensification is possible without any changes in the operating principle.

Decreasing the particle diameter and the column length will affect the design of the adsorber. Suppose we decrease d_p by a factor of $\sqrt{10}$ and to keep (L/d_p^2) constant decrease L by a factor of 10. The cycle time will also be decreased by a factor of 10. The following factors must be included in the new design.

1. The smaller particles must be carefully packed to avoid channeling effects.[653] Large-scale columns have been efficiently packed with small particles.[159,342,558] Thus this can be done with care. For the examples considered here the decreases in d_p are not huge. Use of the same small-diameter particles in both laboratory- and large-scale systems could make scale-up easier since wall effects are less likely to be important in the large-scale system,[271] and mass transfer inside the particles will be the same.
2. A tight fractionation of particle size is required. This is always desirable since the largest particles control H while the smallest particles control the pressure drop. As d_p decreases it is important that $\Delta d_p/d_p$ not increase. Until there is a market for them, the small particles may be more expensive, or may have to be specially ordered. This may make this intensification process uneconomic in a particular case, but tends to be a *Catch 22* situation.
3. The particles should be spherical since this will reduce channeling.[653]
4. The packing should be stable and have a long life. Adsorbents which rapidly foul will not be very useful in a separation which requires rapid cycling.
5. The end fittings need to be redesigned. The column is now short and fat, and end effects could be much more important than in long columns. Proper distribution will probably require a change in end fitting design. Short, fat columns are common in large-scale gel permeation chromatography[544,549,800] because the gels are compressible. Similar designs could be used for rigid particles, and have been used in ion exchange systems.[26,342] In addition to proper distribution the ratio (dead volume to packed volume) must be controlled. Since the packed volume has decreased by a factor of ten, careful column design is required to keep this ratio small. This is particularly important for liquid systems.
6. Smaller particles are easier to fluidize. Thus, a hold-down screen or frit will be required. This is particularly important with rapid back-and-forth cycling. If a significant change in d_p is done (greater than that illustrated in Table 3-2), the support plate and the hold-down plate will have increased pressure drops. If this becomes a problem the column diameter can be increased to lower the pressure drop. A significant increase in productivity can be obtained in many commercial systems before this problem becomes significant.

7. The piping and valves connected to the column must be designed to keep the dead volume small.
8. The control valves and other equipment must be capable of cycling faster. For thermal systems this should be no problem since the cycle is reduced from hours to minutes. For common PSA systems the cycle will be reduced from minutes to 10 to 30 sec. Since faster cycles are operated commercially,[558] this should not be a problem.
9. The feed must be free of suspended solids. This is a most important requirement. Beds packed with small particles are highly efficient filters. A dirty feed will clog such a bed and cause the pressure drop to increase rapidly. Very small amounts of suspended solids can be handled if feed flows downward and an upward flow elution step is used. Fluidized bed equipment (see Sections II and III in Chapter 6) is well suited to handling dirty feeds.

This process intensification will have a minor impact on the operating cost, except for depreciation. Some other tradeoffs are possible to simultaneously intensify the process and decrease the operating cost. Equation 3-1 shows that pressure drop will be constant if $(v\ L/d_p^2)$ is constant. The velocity is easily decreased without changing the throughput by increasing the column diameter. Suppose we decrease d_p by a factor of $2\sqrt{10}$, decrease L by a factor of 10, and double the column diameter. For constant throughput Δp will be unchanged, the amount of adsorbent will be decresed by a factor of 2.5, and the number of stages will increase. The number of stages increases since in the usual range of operating conditions where pore diffusion controls H is proportional to $v\ d_p^2$ (see Equations 2-122, 2-123a, and 2-123b and Figure 2-21) and thus N is proportional to $(L/v\ d_p^2)$. With larger N a higher fraction of the bed can be utilized and desorption will be more efficient. This means less energy in a thermal desorption, less desorbent in a chemical desorption, or less product loss in a pressure swing desorption. Thus both capital and operating costs will decrease. As Figure 3-5 shows, this method is useful when L/MTZ is less than 2 to 3. Similar tradeoffs can be made if modest increases in pressure drop are allowable.

There are situations where this process intensification may not work. With many ion exchange systems swelling of the resins occurs and space must be available for this swelling. Because of this extra dead volume it may be impossible to keep the ratio (dead volume/column volume) small if d_p and L are both decreased. However, ion exchange systems with small resin particles, rapid cycling, high resin productivity, and very small resin inventories are available commercially.[26,342] In activated carbon systems where the carbon is regenerated in a furnace, very small particles may cause problems during the regeneration. First, the furnace may have to be redesigned; second, the loss in weight and hence diameter may be too large for small-diameter particles. In some designs where the packing is moved periodically it may be difficult to properly move small particles and there will be a limit to how much d_p can be reduced. There are many other adsorption systems where process intensification is feasible and probably is economic.

How much can the particle diameter be decreased? This question cannot be answered without careful design and economics studies. However, some of the technical limits can be delineated. As the particle diameter decreases, film diffusion will become increasingly important. Since $k_f a_p$ is approximately proportional to $(d_p)^{-1.4}$ instead of $(1/d_p)^2$, H will no longer scale as d_p^2. Thus the exact scaling used will no longer be applicable. Although N could still be held constant, the pressure drop would have to be increased. In the range of particle sizes currently used, pore diffusion usually controls and this limit is not important. This is the case in the example shown in Table 3-2. For particles less than 10 μm in diameter, kinetic effects can be important, and the scaling may break down.[517] The use of alternate procedures such as attaching small particles to a fiber[623] or making the adsorbent as fibers[28,725] probably allows further reductions in the particle diameter.

More important current limitations are the ability to pack small particles in short, fat columns, and the ability to design short, fat columns with small dead volumes. Perhaps the ultimate is to dispense with particles entirely and use a homogeneous sheet of packing with uniform pores. This could be made from fused particles or by polymerizing a gel as is currently done in gel electrophoresis. This ultimate step will require a total rethinking of the design of adsorption and chromatography systems.

III. ADSORPTION OF GASES WITH THERMAL REGENERATION

Gas adsorption processes with thermal regeneration are very common industrially. Thermal regeneration changes the equilibrium parameters and dilutes the adsorbate in the gas phase since a sweep gas is used to remove the adsorbate. These two effects combine to cause desorption. A variety of operating methods have been developed for these processes. In this section we will briefly look at trace contaminant removal, drying and sweetening, and solvent removal with activated carbon. Most of these systems are designed to use particles that are too large, and thus the columns are too big and cycles are too long.

A. Trace Contaminant Removal

One of the simplest and most common applications of adsorption is the removal of trace contaminants. This application takes advantage of the usual favorable shape of adsorption isotherms. At very low adsorbate concentrations in the gas the amount adsorbed can be very high. Since the feed concentration is low, the saturation of the adsorbent at higher concentrations never controls the process. A shock wave with a very low velocity will form during the adsorption step and very long adsorption periods can be used. When very long adsorption periods or intermittent adsorption periods are used, one critical question is how much residual capacity remains.

Regeneration may be done off-line or the adsorbent may be discarded. This is economical because of the very long periods between regeneration. When regeneration is done *in situ,* a counter-flow of hot gas is usually employed. In some cases one column is sufficient since regeneration is so infrequent. With more frequent regeneration two or more columns are used with one or more columns adsorbing while the remainder are being regenerated and then cooled in preparation for the next adsorption step.

Odor control is a special case since parts per million or less of some chemicals can produce an odor problem. Activated carbon often has a high affinity for the odor-causing chemicals and is used as the adsorbent. With low concentrations and a very nonlinear isotherm a sharp shock wave is formed, and the MTZ will be very short. Thus, thin layers of packing can be used. Usually, pressure drop must be minimized since the blower horsepower is the largest operating cost. Designs which maximize the surface area and have a very short bed depth are used. These include annular systems such as canisters and corrugated bed systems.[608,687] Examples of these systems are discussed in detail elsewhere.[608,687]

A considerable amount of work has gone into developing processes for removal of sulfur compounds. Removal of SO_2 is commonly done with a thermal cycle.[17,253,274,588] A zeolite molecular seive can be used to adsorb SO_2 in the tail gas from a sulfuric acid plant. The tail gas typically has between 2000 to 4000 ppm SO_2, which is too low for making sulfuric acid but much too high to exhaust. The molecular seive effectively adsorbs SO_2 in the presence of CO_2, and can be easily regenerated. Desorption is done counter-flow with air heated in a furnace. During desorption the SO_2 exits as a peak starting at around 0.3% and then rising to about 4% before subsiding.[274] This gas is returned to the acid plant where it results in a 2 to 3% increase in acid production. Removal of H_2S (sweetening) is discussed in the next section.

Molecular sieves can also be used for NO_x removal.[17,253,556,588,608] The sieves adsorb NO_2

but not NO; however, in the presence of oxygen the NO will react to form NO_2. The gases need to be dried first since water will be preferentially adsorbed by the sieves, thus deactivating the sieve. With no water present the sieves can be regenerated by counter-flow at 150°C. If water is present 250°C is required. The recovered NO_2 is returned to the acid plant. More efficient bed use and higher NO_2 concentrations in the purge gas can be obtained by using two columns in series.[556] The cycle for doing this is listed in Table 3-1. This cycle could be adapted for other adsorption problems when regeneration can be done faster than the adsorption step.

Another example of trace component adsorption with thermal regeneration of the adsorbent is the removal of traces of mercury vapor from air.[17,253,730] A gas stream containing mercury with 1 to 2 ppm(v) can be reduced to less than 1 ppb(v) with a proprietary adsorbent. Regeneration is done counter-flow with hot air. When the effluent gas is cooled, droplets of liquid mercury condense and are removed in a separator. The gas from the separator is mixed with fresh feed and returned to the adsorber. Since regeneration takes a maximum of 2 days and adsorption can last for 7 days, one adsorber will sit idle for 5 days.[730] Higher bed utilization could be obtained by using the two-column scheme shown in Table 3-1, but at the cost of a higher pressure drop and hence higher blower operating costs. This would be an inexpensive way to increase the capacity of existing systems.

B. Drying and Sweetening of Gases

Adsorptive drying of gases followed by thermal regeneration has been a standard unit operation for many years.[112,206,253,268,301,529,571,575,608,616,645,687,706,731,865,1017,1079] Basmadjian's review[112] is particularly complete and up to date. A large amount of data on properties of adsorbents and equilibrium isotherms are available in Basmadjian's review[112] and in other sources.[12,206,608,616,645,865,1017] Equilibrium isotherms such as Equations 2-4, 2-8, and 2-11 have been used. Pressure swing adsorption (PSA) is also used for drying gases and is covered in Chapter 4 (Section II). Other separation methods, such as condensation and glycol drying,[529,608,706] are also used for drying gases. Adsorption has the advantage of drying to very low final humidities, but may not be the best method for hot saturated gases. In this case a combination of methods such as cooling and condensation followed by adsorption may be the best method. For example, reducing the temperature from 50 to 25°C reduces the water concentration from 12.2 to 3.1 v/v%, and this decrease obviously decreases the load on the adsorption dryer.[731] After partial condensation, the liquid droplets must be removed from the gas stream in an efficient vapor-liquid separator.

The basic adsorption system will use two beds if the MTZ is short, and three beds with the option of running two in series if the MTZ is long. Silica gel and activated alumina can give dew points from −40 to −50°C, while zeolite molecular sieves are capable of giving dew points as low as −100°C. The beds are usually regenerated with a hot gas although open steam or external steam heating are also used. A regeneration temperature of 350 to 400°F is typically used except for zeolite molecular sieves which require 500 to 700°F.[608] Counter-flow regeneration is more energy-efficient than co-flow regeneration, but the latter is simpler and has lower capital costs. Thus both counter- and co-flow regeneration schemes are used. The bed is often cooled co-flow to the feed direction before reintroducing the feed. One desires to operate with high adsorbent productivity, low product losses, and low energy use. Fast cycles tend to increase adsorbent productivity, particularly if the bed utilization is fairly high. Short beds with fast cycles have been recommended for drying in LNG plants.[418] Bed utilization can be increased by using beds in series, short beds with small-diameter particles, or an adsorbent such as molecular sieves which have a short HETP. However, excessive pressure drops must be avoided. Using a layer of silica gel or activated alumina followed by a layer of molecular sieve will keep pressure drops low, minimize costs, and keep bed usage high[112,268] (see Section II.C).

Product losses occur at the beginning of the regeneration step. Gases in the void spaces and in the head spaces will be swept out and lost. It may be possible to recycle these gases to the on-line column to avoid loss. Minimizing the head spaces will help minimize these losses and will help reduce energy requirements. Minimizing the HETP increases the bed utilization and helps to reduce product losses. Several methods for reducing the energy use have been developed:

1. Counter-flow regeneration will reduce energy usage. Incomplete regeneration as illustrated in Figure 3-2C will reduce energy requirements per kilogram of product.
2. Internal insulation can be helpful to decrease the amount of energy used to heat the shell.
3. When compressed air is dried the heat of compression can be used to regenerate the bed. This is done by feeding the hot compressed air directly to the regenerator, and then to a cooler and liquid-vapor separator before the adsorption dryer.[106] When this is done no supplemental heating is required. This approach could probably be applied to other gases which need to be compressed and dried.
4. When the thermal wave velocity is greater than the solute wave velocity, it may be possible to operate without a cooling step.[109,112,783] The cold feed gas then serves as the coolant. To prevent premature breakthrough of the adsorbate, a very short cooling step which cools only the feed end of the column may be used, but commercial operation without any cooling operated without problems for more than 1 year.[783]
5. Operating so that the thermal wave stays inside the bed during the feed step will reduce the energy requirements. This requires that the cross-over ratio, Equation 2-65, be near 1. This type of operation keeps the heat of adsorption inside the bed where it is available for desorption. Pressure swing adsorption (see Chapter 4, Section II) uses this principle, and recently[399,403,404,625] this method has been used commercially to thermally regenerate dryers removing water from the ethanol-water azeotrope. With large amounts of water the temperature increase in the bed can be 15°C or higher. A cross-over ratio, R_T of about 1.0 is required, which usually means fairly high water content in the feed gas. This operation might be aided by adding a small amount of inert material with a high heat capacity with the adsorbent. Layers of inert material in a heat regenerator at the end of the bed can also help conserve energy.
6. When hot gas desorption is done at elevated pressures, the system will often be mass transfer-controlled. Roughly half of the energy in the desorbent gas will be used for heating the adsorbent and for desorption.[252] Less desorbent and less energy would be used if desorbent were fed until thermal breakthrough occurred, and then the gas were shut off. After a delay period for mass transfer, desorbent gas would be started again to sweep out the interstitial gas. Another delay period would be followed by another pulse of desorbent gas. This procedure would be continued until sufficient desorption has occurred. This delay procedure will save energy, will probably take longer, will be more difficult to operate, and will result in increased dispersion because of diffusion during the delays. Naturally, an alternative is to reduce the pressure during desorption.

These methods are also applicable to the processes discussed in Sections II.A and C.

One major problem in gas drying is the slow loss in adsorbent capacity with time.[112,206,253,529,571,575,616,645] An adsorbent lifetime of from 1 to 4 years is common when natural gases are dryed, although much shorter lifetimes can occur. This drop in activity is usually due to the irreversible adsorption of heavy trace components. Guard beds using partially deactivated adsorbent can help protect the active dryer. Another solution is to use 3 Å zeolite molecular sieves. The large molecules which irreversibly adsorb cannot fit into the very small pores of the 3 Å sieve. The price for using 3 Å sieves is somewhat lower

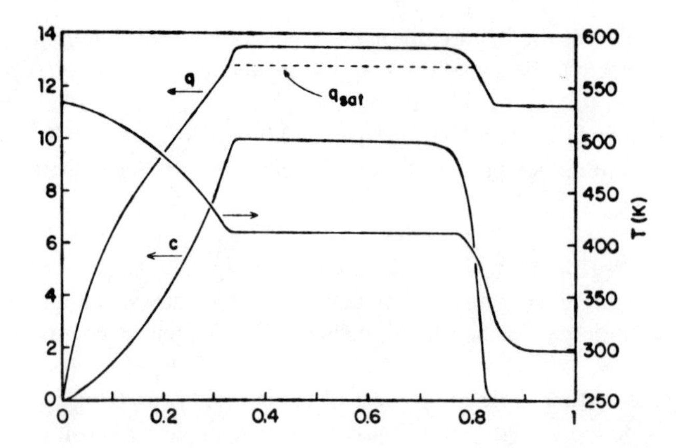

FIGURE 3-9. Concentration and temperature profiles predicted by staged model for regeneration of molecular sieves adsorbing water. (From Friday, D. K. and LeVan, M. D., *AIChE J.*, 28, 86, 1982. With permission.)

mass transfer rates and thus somewhat longer MTZs. When the adsorbent does deteriorate with time, it is common to reduce the cycle time as aging occurs. When the necessary cycle time becomes so short that the column can no longer be regenerated in the available time, the adsorbent must be replaced.

Another cause of deterioration of the adsorbent is the presence of liquid water in the bed. This is particularly detrimental for silica gel beds. Separators are commonly placed upstream of the beds to remove any entrained water. The separator will also help to remove suspended sulfur particles in natural gas streams.[783] A guard bed of liquid water-resistant adsorbent before the high-capacity gel bed can also help prevent problems. However, Friday and Levan[389,391] used a staged model to show that condensation of liquid can occur in the bed during a co-flow regeneration step. This was previously unexpected and has since been collaborated with the local equilibrium theory[390] and by experiments.[391] The concentrations and temperatures predicted inside the bed by the staged model are shown in Figure 3-9.[389] When the solid phase concentration exceeds the saturation concentration, condensation will occur. Physically this high concentration occurs because of ''roll-up'' of the adsorbate. The first part of the column becomes hot and removes the adsorbate. This adsorbate catches up to the colder part of the bed where the highly concentrated gas is adsorbed (in essentially the same way as ''superloading'') and the concentration shoots up. When the saturation concentration is exceeded, condensate forms in the pores. In small capillaries the saturation concentration may be well below the saturation limit determined from vapor pressures. The condensate will revaporize when the thermal wave arrives.

Many other theories have been applied to gas drying and are reviewed in detail by Basmadjian[112] and Ruthven.[865] In the simplest case where very little water is present, isothermal theories can be used. However, since heat effects are usually important in drying, the theories usually need to be nonisothermal. The simplest nonisothermal theories which include heat of adsorption effects are the coupled equilibrium theories discussed in Section III.D.2. An alternative is to use the local equilibrium theory with ''effective equilibrium curves''.[110-112,865] Complete numerical solutions of the partial differential equations have also been done, but many of these solutions are flawed since the mass balances are not satisfied.[112] Carefully done numerical analyses can fit the experimental results for adiabatic drying.[818,865] Industrially, modifications of the MTZ or LUB approach are often used for design,[112,252,399,608,642,645,734] including the results from a large number of experimental runs in both laboratory- and large-scale systems.

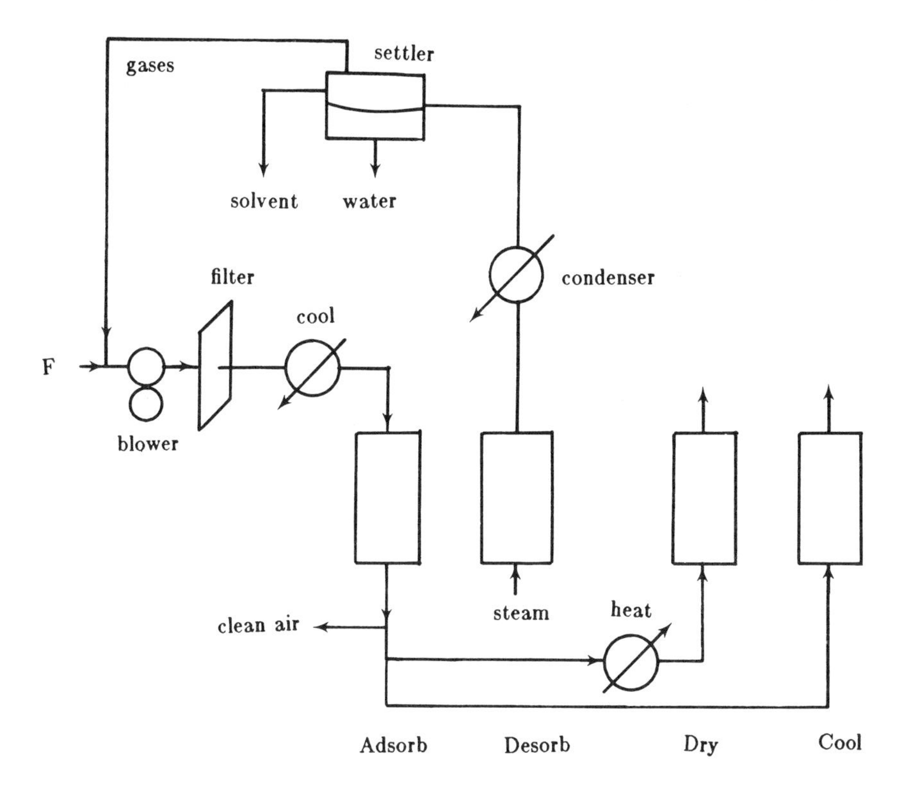

FIGURE 3-10. Typical activated carbon solvent recovery system.

Gas sweetening operations are similar to drying except H_2S, CO_2, and H_2O are all removed.[112,206,253,261,608,616,645] These gases can lead to several problems. First, the acids can attack molecular sieves and special acid-resistant grades must be used. Second, carbonyl sulfide, COS, is formed by the reaction of H_2S and CO_2 which is catalyzed by molecular sieves.[261] Unfortunately, the COS is not as strongly adsorbed as the H_2S, and will break through early. The use of 5 Å sieve appears to minimize the catalysis. Removal of the acid gas prior to dehydration may be necessary to prevent this problem. A third point is that carbon dioxide is less strongly adsorbed than either hydrogen sulfide or water. Thus the system can be operated to remove all three gases or just water and hydrogen sulfide. If carbon dioxide is allowed to break through it will peak at a concentration above its feed concentration because it will be desorbed by the H_2S. Simultaneous removal of acid gases and dehydration with molecular sieves is practiced in small peak shaving LNG plants.[418]

C. Activated Carbon Solvent Recovery

An extremely common but somewhat specialized process is the removal and recovery of solvent from air using activated carbon. This process has been used commercially since the 1920s. A glance at some older references[123,672,687,688] shows that surprisingly, the basic system is almost unchanged, except for more sophisticated controls, in almost 60 years of use. This may mean that major innovations are overdue.

1. Basic System

The basic system for solvent recovery with activated carbon in fixed beds is shown in Figure 3-10. The air stream is usually collected at one quarter to one half of the lower explosion limit when the gas stream is air. Any increases in the solvent concentration will decrease the capital and operating costs. This gas stream is forced through a filter to remove any particulates which might clog the packed beds, and is often cooled. The operation is usually at 80 to 90°F. Cooling is both a safety precaution and a method for making adsorption more efficient. The gas is then sent to the carbon adsorber and the clean gas is either reused or exhausted. The most common desorption method is to use steam counter-flow to the gas flow. Steam is very convenient because of its large latent heat. The steam-solvent mixture is condensed and sent to a settler if water and the solvent are immiscible. The noncondensible gases are recycled through the adsorber. After desorption is complete, the carbon may be dried and then cooled. These steps are optional and it may be possible to return the hot, wet carbon back into adsorption service.

A large variety of solvents can be recovered with activated carbon systems. Solvents with molecular weights from about 45 to 200 can usually be recovered. Lower molecular compounds tend to be too weakly adsorbed while the higher molecular weight compounds are difficult to desorb.

With minor modifications, the system shown in Figure 3-10 is a standard unit operation for solvent recovery.[123,165,297,395,560,571,608,616,618,638,667,672,687,688,690,786,990,996,1016] A variety of packaged units are available.[13,27,28,251,302,305,355,560,689,696,710,725,789,797,798,927,1027,1095] The usual configuration for the adsorber is to lay the column horizontal and support a horizontal layer of carbon. This gives a maximum surface area and minimizes the pressure drop in the bed. Since power for the blower is a major cost, minimizing Δp is a major consideration. The pressure drop is readily determined from charts[157,687,996] or from general equations for flow in packed beds.[138,157] The usual range is from 0.5 to 30 in. of water pressure drop. Relatively shallow beds with gas velocities from 60 to 120 ft/min are commonly used. Either downward or upward flow of the gas can be used. Downward flow is often used since there is much less bed churning and attrition of the carbon. When upward flow is used, the gas velocity should satisfy.[157,640]

$$\frac{G^2}{\rho_G \, \rho_B \, d_p \, g} < 0.0167 \qquad (3\text{-}6)$$

to prevent lifting the bed. Equation 3-6, G is the mass velocity of the gas and g is the acceleration due to gravity. When adsorption is strong, annular columns can be used to maximize area and minimize pressure drop.[28,165,725] In small units vertical columns are often used. Other configurations for activated carbon adsorption are discussed in Chapter 6 (Section II.A.2 and II.A.3) and Chapter 8 (Section II.C).

Steam is most commonly used for desorption, since it often has numerous advantages. A lot of energy can be put into the column very quickly to heat the shell, the carbon, and desorb the adsorbate. The steam should be at such a pressure that it will condense at least 30°C above the boiling point of the solvent. The steam purge also serves to dilute the adsorbate and sweep it from the column. Water can often be left on the column and its evaporation during the adsorption step keeps the temperature from rising substantially. This water can easily be exhausted to the atmosphere and the water does not have to be recovered. Many solvents are immiscible with water and the solvent can be recovered by condensing the vapor and separating the two phases. The solvent can often be reused without further processing. When the solvent is miscible with water, the solvent can be recovered by distillation. Unfortunately, this greatly increases the cost of the system. If the solvent forms an azeotrope with water, separation becomes even more expensive, and alternate desorption steps may be preferable.

Complete desorption is usually not employed since an excessive amount of steam would be required. Thus the column loading during the cycle looks like Figure 3-4A. Design is usually for a working capacity of 25 to 30% of the capacity at saturation.[1027] This often leads to a working capacity which is about 7.5 wt % of the weight of the carbon. The usual

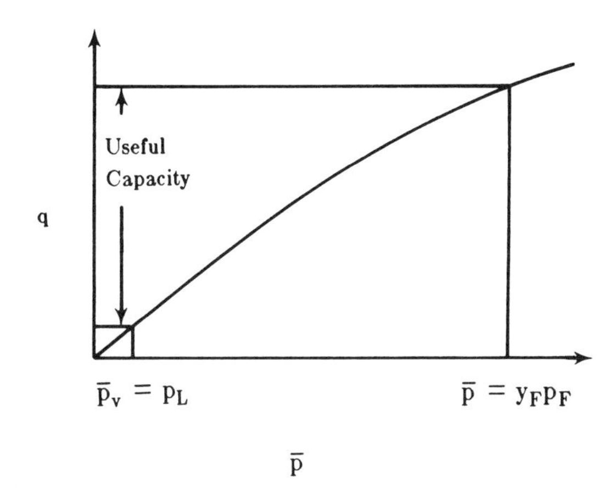

FIGURE 3-11. Steam usage during desorption. (From Danielson, J. A., Ed., *Air Pollution Engineering Manual,* Publ. #AP-40, EPA, Research Triangle Park, N.C., May 1973.)

range for steam use is from 1 to 5 lb of steam per pound of adsorbate recovered. The effect of time of desorption on the amount of steam required is shown in Figure 3-11.[297] More solvent can be recovered by desorbing for longer periods, but at the price of increased steam use. Only about 10% of the energy in the steam is used for desorption. The rest either exits the column or is used to heat the shell and the carbon. Much of this energy can be recovered (up to 70%[797]) and used elsewhere in the plant.

If the solvent is miscible with water the bed is usually dried after desorption to reduce the amount of water fed to the distillation columns downstream. Pebble bed regenerators can often be used to heat the air used for drying and keep heating costs down. If the bed does not have to be dried, the evaporation of water left on the bed will keep the bed temperature from rising excessively. However, excessive humidity (>50% relative humidity) is detrimental to adsorption since water condenses in the pores because of capillary condensation.[618,851,996] The adsorbed water blocks the organics from some of the pores. This reduces the column capacity and results in quicker breakthrough of the organics.[442] Thus, a short drying period may be desirable. Drying can also be done with a cooling step.[701] If the inlet gas is above approximately 50 to 60% relative humidity the condensation of water will interfere with solvent adsorption. To solve this problem it may be necessary to preheat the gas to decrease the relative humidity. The cost of doing this is that operation will be at a higher temperature where adsorption is weaker. An alternative is to dry either the feed gas or the product.

The cooling step shown in Figure 3-10 can often be skipped since the thermal wave usually moves faster than the solute wave. However, in some cases hot activated carbon is a catalyst for undesirable side reactions and the cooling step is required to limit reaction. The cooling step also removes about 30% of the water and can increase solvent recovery by approximately 4%.[701,786] Thus, in many situations cooling will be economical.

The entire process shown in Figure 3-10 should be automated. For safety reasons (carbon beds can burn) a maximum temperature of about 150°C should be set. Cooling the air before it enters the adsorber also serves as a flame arrester and makes the process safer. If trace quantities of materials which can polymerize are present, guard beds can be used to protect the main bed.

After the adsorber, the vapors are cooled and condensed. In an integrated plant some of this heat can be recovered. The noncondensible gases should be returned to the adsorber since they will be saturated with the solvent.

Costs for carbon adsorption systems have been published.[104,1013] Small-scale systems (less than 10,000 lb of carbon) are usually purchased as packaged units. The cost data shows an economy of scale since the exponent in the cost equation based on the weight of carbon is 0.481.[1013] Larger systems are usually custom designed and here the exponent, 1.20, is greater than one and there is no economy of scale. When distillation is required, the distillation system may control the costs.

2. Alternatives

A variety of alternate operating procedures have been developed. First, the basic column shape can be changed. Canisters[1027] or annular shapes[28,616,725] can be used to either provide easy replacement of the carbon or large cross-sectional area for flow. Continuous flow (see Chapter 6, Section II.A.2 or II.A.3) and rotating beds (see Chapter 8, Section II.C) have also been used in some applications, particularly for larger-scale systems.

Carbon forms which are different from the usual granular activated carbon have been recently developed. Carbon fibers[28,583a,682,725] reportedly make the adsorption sites more accessible to the gas phase and reduce mass transfer resistance. The fibers are supposed to be easier to regenerate, but adsorptive forces are weaker and some low molecular weight solvents are not strongly retained. Nonwoven felt-like structures with high carbon contents have recently been developed.[29] but are not yet commercial. Monolithic blocks of fused granular activated carbon have also been developed. These systems should have very short MTZs, but will have to be used in thin layers since pressure drop per centimeter will be high.

A variety of alternate desorption techniques have been developed for special cases. Steam has the disadvantage of adding water to the system and the alternatives try to avoid this. The alternatives may also have an advantage at remote locations where steam is not available.

1. Hot gas desorption has been mentioned as an alternative for many years,[395,667] but only recently has it become fairly common commercially.[13,210,824] The advantage of using a hot inert gas for desorption is that solvent recovery can be done by condensation (perhaps after passing through a desicant column) and distillation columns are not required for solvents which are miscible with water. The disadvantage is large volumes of hot gas are required to transfer the required amount of heat into the column. This can conveniently be done by recycling the same gas through the column, the condenser, and the heater. Heat pumps are useful for reducing energy requirements.[210,824] The outlet concentration of the desorbed solvent and the amount of hot gas used can be minimized by stopping the desorption cycle shortly after the peak of solvent concentration leaves the column. This will follow thermal breakthrough. This leaves a large heel in the column which will be resharpened by the shock wave during adsorption. Rapid cycles minimize energy requirements. The addition of heating coils, which are steam heated, inside the column can also help desorption. Unfortunately, the heat transfer coefficients are quite low inside the bed. The theories used for drying of gases with thermal regeneration (see Section III.B.) can be adapted to hot gas desorption of solvents.
2. A second commercial approach is to use vacuum desorption.[30,667,687,711,830] Vacuum systems again have the major advantage of avoiding contamination with water. The total pressure is reduced to below the partial pressure of the adsorbate and desorption occurs. Unfortunately, if concentrations are low the required pressure can be very low. With low feed concentrations the useful feed capacity of the bed can be quite low. This is illustrated with the equilibrium isotherm shown in Figure 3-12. The partial pressure of the feed gas is $\bar{p} = y_F p_F$. During evacuation all nonadsorbed gases are removed and $y \to 1.0$. Thus the partial pressure during the lowest part of the vacuum desorption is $\bar{p}_v = p_L$. If y_F is low, $y_F p_F$ will be near $\bar{p}_v$, resulting in a small swing

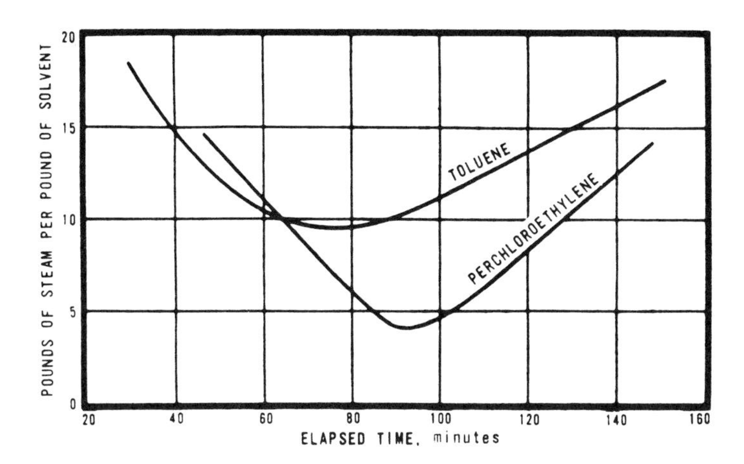

FIGURE 3-12. Isotherm for vacuum desorption system, amount adsorbed vs. partial pressure.

in partial pressures and the resulting useful bed capacity will be small (see Figure 3-12). Vacuum operation has proven to be economical only when the concentration of recovered solvents is high (3 to 50 vol %[30]). In these cases the vapor produced during regeneration may be as high as 95 vol % solvent. This is safely above the upper explosion limit. A common application of vacuum regeneration has been for recovering gasoline vapors in gasoline storage or loading facilities. In this application the adsorbers will rapidly pay for themselves.

3. A third type of desorption process which is still under development is the use of a solvent for the desorption. This can be done by a liquid-liquid extraction-type process or with a supercritical fluid.[312] An alternative is to use a vapor desorbent which is then condensed and separated from the recovered solvent.[1062] The hot desorbent is removed from the bed with steam or with a hot inert gas. Unfortunately, this approach requires complete desorption of both solvent and desorption agent to avoid contamination of the purified air and to avoid cross-contamination of solvent and steam systems.
4. A fourth alternative is to not recover the solvent and to regenerate the carbon in a kiln. This is particularly useful when only traces of a very high molecular weight solvent must be removed.
5. The fifth alternative is really a variant of steam desorption. The solvent is adsorbed on the carbon and then desorbed. Then, instead of condensing and trying to recover the solvent, the steam, solvent, air mixture is burned with additional air and any additional fuel required.[1028] The advantage of this approach over straight incineration of the dilute gas stream is that much less gas needs to be heated, and much less or no additional fuel is required. The incinerator can be used as the boiler to heat the steam needed for desorption. Often the energy requirements for the steam can be met by the incineration. Of course, this approach does not produce as valuable a product as the processes which recover solvent.

Some of these alternative desorption techniques are also applicable to the other adsorption systems discussed in this chapter.

3. Theories

Until recently, very few theoretical studies of solvent recovery with activated carbon had been published in the open literature. Most design seems to have been done on the basis of experience and pilot plant measurements.[395,638,996] Recently, some theoretical results have

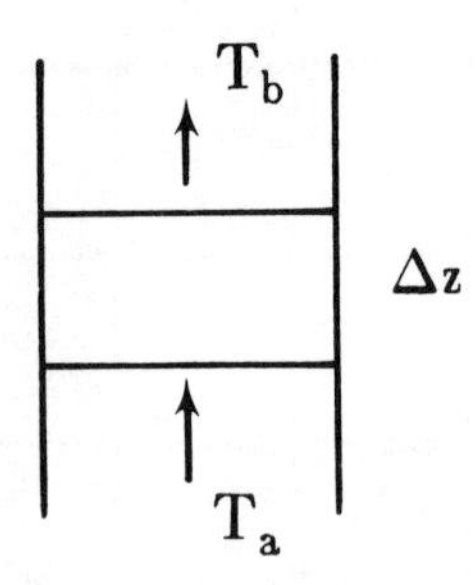

FIGURE 3-13. Segment of column for thermal shock wave when steam is added to the column.

been published, and with slight modification the solute movement theory can be applied to this system.

The solute movement theory is directly applicable to the adsorption or feed step. Since equilibrium isotherms of most solvents on activated carbon are nonlinear and favorable, a shock wave usually forms [$(1 + Kc_F)$ is fairly large]. The thermal wave requires some modification to include the effect of condensation of steam. Because of the condensation of steam a thermal shock wave will result. The energy balance can be done over a segment of the column of length Δz as shown in Figure 3-13. In order for the thermal wave to traverse the entire segment the time interval, Δt, must be related to Δz by

$$\Delta t = \Delta z / u_{steam} \tag{3-7}$$

If we assume a constant vapor velocity, v, before and after the thermal shock wave, and constant heat capacities a simple expression for the thermal shock wave velocity can be derived. Since saturated steam is usually used, we will choose T_a as the reference temperature so that the condensed liquid water has an enthalpy of zero. The energy balance over segment Δz and time interval Δt is

$$\text{In} - \text{Out} - \text{Accumulation} = 0 \tag{3-8}$$

This becomes

$$\begin{aligned} &\alpha\, v(\Delta t)\, \rho_f[C_{Pf}(T_a - T_{ref}) + y_{STM}\lambda_w] - \alpha\, v(\Delta t)\, \rho_f[C_{Pf}(T_b - T_{ref}) + y_{w_b}\lambda_w] \\ &- \Delta z\Big\{(\alpha + \epsilon(1 - \alpha))\, \rho_f[C_{Pf}(T_a - T_{ref}) + y_{STM}\lambda_w - C_{Pf}(T_b - T_{ref}) - y_{w_b}\lambda_w] \\ &- (1 - \alpha)(1 - \epsilon)\rho_s C_{ps}(T_{sa} - T_{sb}) - \frac{W}{A_c} C_{pw}(T_{wa} - T_{wb})\Big\} = 0 \end{aligned} \tag{3-9}$$

where λ_w is the latent heat of water, and y_{STM} and y_{w_b} are the mole fractions of water in the steam and in the bed before regeneration. Assuming thermal equilibrium so that $T_f = T_s = T_W = T$, substituting in Equation 3-7, and solving for u_{steam} we obtain:

$$u_{steam} = \frac{[C_{Pf}(T_a - T_b) + \lambda_w(y_{STM} - y_{w_b})]v}{\left[1 + \frac{\epsilon(1 - \alpha)}{\alpha}\right][C_{Pf}(T_a - T_b) + \lambda_w(y_{STM}\ y_{w_b}] + \left[\frac{1 - \alpha}{\alpha}\right](1 - \epsilon)\frac{C_{Ps}\rho_s}{\rho_f}(T_a - T_b) + \frac{W}{A_c}\frac{C_{pw}}{\rho_f\alpha}(T_a - T_b)} \tag{3-10}$$

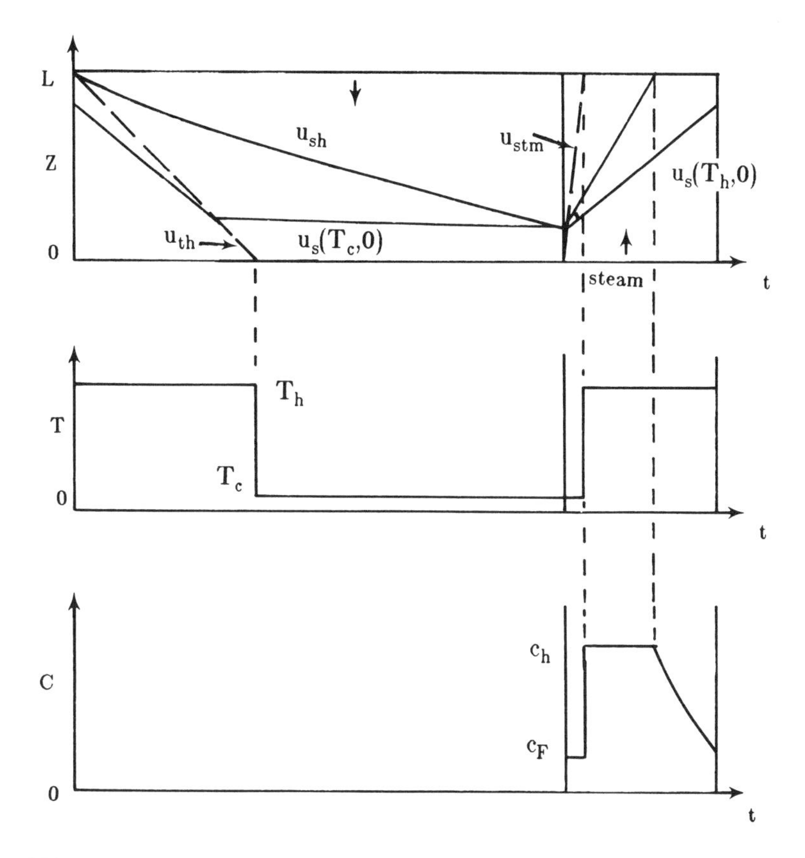

FIGURE 3-14. Solvent recovery with activated carbon and steam desorption. (A) Solute movement theory for solvent recovery with activated carbon. Nonlinear isotherms, no cooling or drying steps; (B) predicted temperature profile; (C) predicted solute concentration profiles.

This equation should be compared to Equation 2-41, which gives the thermal wave velocity when there is no condensing steam. If a noncondensible gas is used $(y_{STM} - y_{w_b})$ is approximately 0 in Equation 3-10, and the terms $(T_a - T_b)$ will cancel out. The result is then Equation 2-41. Since the latent heat effect is quite large compared to the specific heat effects, Equation 3-10 predicts a steam wave velocity which is close to v. As was mentioned earlier, this is the major advantage of using a condensible vapor; the column can be rapidly heated.

Following the steam wave will be a diffuse wave as the solvent desorbs. The entire operation is illustrated in Figure 3-14A. The predicted temperature and composition profiles are shown in Figure 3-14B and C. Note that delays in appearance of the temperature and concentration changes are predicted. A separate cooling step is not shown in this figure and cooling results from the addition of the cool feed gas. Note that the diffuse wave which results during steam regeneration is not completely removed from the column, but is recaptured during the next adsorption step by the shock wave. Figure 3-14 is an oversimplification since the following are ignored:

- Velocity changes when the gas is heated.
- Spreading of the shock wave.
- Spreading of the thermal and steam waves.
- The effects of the heat of adsorption and desorption. These effects can be quite large when fairly concentrated streams are adsorbed.

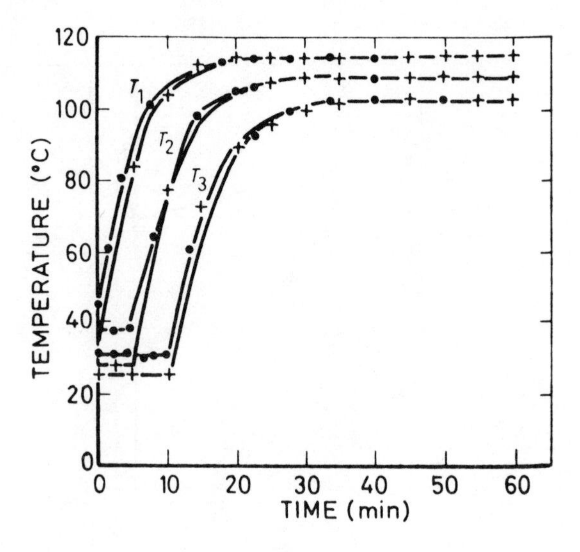

FIGURE 3-15. Outlet temperature profiles during steam desorption. T_1 is near steam inlet to column while T_3 is near outlet; + and ● are data points for two different runs. (From El-Rifai, M. A. et al., *Chem. Eng.*, 269, 36, 1973. With permission.)

- Condensation of solvent in the column during desorption. In drying operations adsorbate can condense during the desorption step.[389,390] This might also occur with steam desorption.

Even though this model is quite simple it does a reasonable job of predicting the qualitative behavior. Other theories for solvent recovery with activated carbon have been developed.[214,350,434,701,875] The one closest to that presented here is a local equilibrium model which includes coupling of solute and energy effects.[214] This model is similar to those used for other coupled systems (see Chapter 2, Section IV.D.2).

The adsorption step can be modeled as a constant pattern solution since the equilibrium isotherms are favorable. In many cases a first-order irreversible kinetics model[434] can be used. This is valid since in many cases the isotherm is so steep that a good fit is obtained. The advantage of this approach is a very simple solution can be obtained[434]

$$\ln \frac{c}{c_F} = \frac{k_A\, c_F}{\rho_B\, q_{sat}}\, t - \frac{k_A\, m}{Q\, \rho_B} \tag{3-11}$$

where m is the weight of carbon used, Q is the volumetric flow rate, and the first-order rate constant k_A can be fit to the data. A plot of log c/c_F vs. t gives k_A from the slope. This irreversible equation was an excellent fit up to 50% of breakthrough.[434] Since the adsorption step is rarely run past the start of breakthrough, this is a good semiempirical model. Note that temperature effects are included in the empirical rate constant. The sharp breakthrough curves predicted by this model have been observed experimentally.[434,875] However, when very large temperature increases are observed a semiempirical model such as this would have to be treated with caution for scale-up since energy losses from the bed are diameter dependent.

The steaming step has been studied experimentally.[350,875] Typical experimental results are shown in Figure 3-15.[350] The lag or dead time shown in Figure 3-15 is predicted by the solute movement theory in Figure 3-14. The zone spreading is what we would expect for a constant pattern-type system with finite rates of heat transfer. The data of Figure 3-15 could be empirically fit to a lumped parameter-type model.[350]

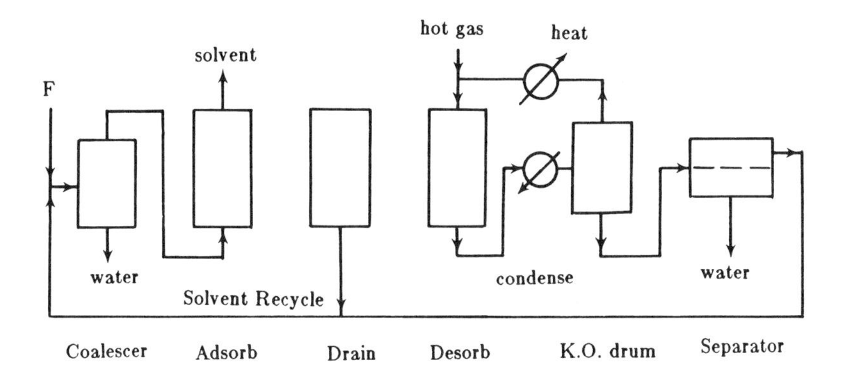

FIGURE 3-16. System for adsorptive drying of liquids.

A detailed numerical model has been used to explore the behavior of carbon adsorbers for acetone recovery with and without cooling steps.[701] The cooling step reduced the temperature and removed about 30% of the water from the bed. Both of these effects increase adsorption of acetone during the feed step. The net result was a 4% increase in acetone recovery. This increase was sufficient to make addition of a cooling step economical.

Obviously, solvent recovery with activated carbon followed by steam desorption is a problem which has not been completely solved theoretically. Numerical models will be required to solve the complete problem.

IV. ADSORPTION OF LIQUIDS WITH THERMAL REGENERATION

In contrast to gas systems, where *in situ* thermal regeneration is very common, in liquid systems it is less common. Thermal regeneration of activated carbon is separate kilns or furnaces is quite common, and is discussed in Section IV.B.

A. Drying Liquids

The drying of liquids by adsorption is practiced commercially.[112,206,301,467,615,745,1115] A schematic diagram of the usual method is shown in Figure 3-16. The liquid feed should first be sent to a coalescer or knock-out drum to remove water droplets. The feed is then sent to the adsorption column which can be packed with activated alumina, silica gel, zeolite molecular sieves, or other desicants. Since the water content in the organic liquid is often very low, the adsorption period may be quite long.

Once breakthrough starts, the column is drained to remove unadsorbed solvent contained in the void spaces. This recovered solvent is recycled to the adsorber during the adsorption step. Draining also reduces the amount of energy required for the desorption step since the drained solvent does not have to be vaporized. If the solvent fits into the pores, it may be difficult to drain this liquid from the pores. A vacuum may be used to help remove this solvent.[615] Another possibility would be to pack the adsorbent in a centrifuge (see Chapter 8, Section III). An alternate solution is to use an adsorbent such as a 3 Å or possibly 4 Å zeolite molecular sieve which has pores which are too small for the solvent to penetrate.

After drainage, the column is usually desorbed with an inert hot gas. Nitrogen, carbon dioxide, butanediol, air, butane, and other gases have been used. The higher the heat capacity of the gas the more rapid heating and desorption will be. After condensation to remove water and residual solvent, the gas can be reheated and recycled. Burying heating coils in the bed allows the use of steam condensation to aid the inert gas purge.[205] The temperature of the regeneration must be high enough to desorb the water, but not so high as to cause excessive coking. For gasoline drying,[745] 400 to 450°F was suggested. Coking problems can also be controlled by using a zeolite molecular sieve with pores small enough that the solvent will not fit into pores[301,467] or by using vacuum to remove all the solvent before thermal regeneration.[615]

Theories for liquid drying have not been extensively reported in the literature and more research would be quite helpful. Available equilibrium theories to predict isotherms are excellent when the solvent adsorption is negligible, but are inadequate when there is competitive adsorption of the solvent and water.[112] More theoretical and experimental work is needed for the competitive adsorption case. Very little modeling of the column behavior has been done. The solute movement theory could easily be applied to the adsorption step when the solvent does not adsorb. This theory will be similar to other applications of the theory. This theory can also be applied to the regeneration step if it can be assumed that all of the liquid has drained out. The result will be similar to Figure 3-3. Note that the feed step in the solute movement diagram is in the liquid phase while the regeneration step is in the vapor phase. Thus different physical properties need to be used to calculate the wave velocities in the different steps. Unfortunately, so many assumptions are involved in this theory that it is not useful for design purposes. The length of unused bed (LUB) or MTZ approach has been applied to the adsorption step,[745] but this was done empirically. Apparently, no models for the drainage or regeneration steps have been published. Obviously, this should be a fruitful area for future research.

B. Activated Carbon Water Treatment

Water treatment with activated carbon is commonly used and has developed almost as a completely independent technique. In this section we will discuss the use of packed columns which are then unloaded so that the carbon can be thermally regenerated in a separate furnace or kiln. For home treatment of relatively clean water the entire canister is usually replaced, and the carbon is discarded. Many reviews on water treatment with activated carbon have appeared[4,251,291,474,526,679,687,714,793,926,944,1069] which are more detailed than this section. Stirred tanks (see Chapter 6, Section I.A.4), pulsed moving beds (see Chapter 6, Section III.A), and fluidized or expanded beds (see Chapter 6, Section II.A.1) are also commonly used. We will first discuss the treatment of drinking water with activated carbon and then wastewater treatment.

Activated carbon has been used for treating drinking water at least since Biblical times.[582,687,811] The systems are commonly used in bottling plants to remove trace amounts of impurities such as chlorine.[582] The chlorine is removed by the reaction

$$2Cl_2 + 2H_2O + C = CO_2 + 4HCl$$

The carbon will last for 6 months to 2 years before being exhausted. Either fresh carbon may be used or the carbon may be regenerated in a kiln.

Activated carbon is commonly used in homes to remove traces of obnoxious chemicals.[32] Schematic diagrams of two typical under-the-sink units are shown in Figure 3-17.[32] The column type shown in Figure 3-17A has a lot longer effective length and worked better to prevent breakthrough. The annular systems had the tendency to have premature breakthrough, and could concentrate less strongly held chemicals and produce a concentrated pulse of these chemicals during extended operation.[32] These adsorbers are made so that some suspended matter will settle out in the housing. Water with a lot of suspended matter must be prefiltered or the adsorber will clog. The beds are usually replaced instead of being regenerated. A major disadvantage of these systems is that the typical homeowner has no reliable way, other than taste, to determine when breakthrough has occurred. Thus the systems are either discarded too soon or left on line too long. Because of the long life of these systems, the homeowner is likely to forget the unit and replace the cartridge only after it is obviously

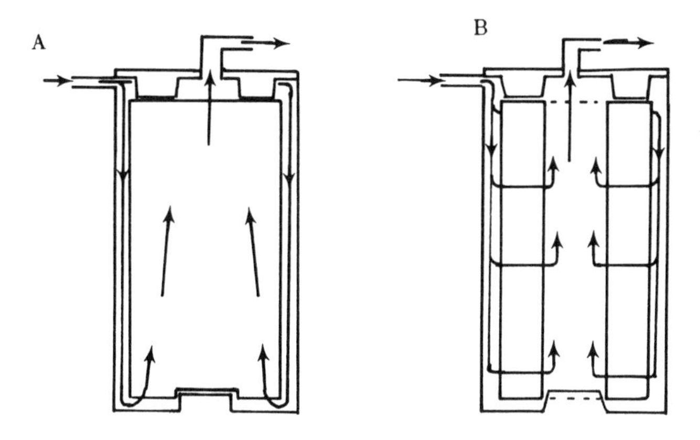

FIGURE 3-17. Household activated carbon adsorber for treating tap water. (A) Column type; (B) annular type. (From *Consumer Reports*, 48(2), 68, 1983. With permission.)

not working. Development of an easy way to determine when to replace the cartridge would be a major advance.

Activated carbon treatment of wastewaters is often done with packed columns which are periodically unloaded so that the carbon can be separately regenerated. Since the molecules are large and the pores are small, mass transfer zones (MTZs) tend to be quite long, and it is common to use columns in series. The adsorption cycle is often very long. During this long cycle some apparent regeneration of the carbon often occurs. This regeneration may be due to bioregeneration where organisms growing on the surface of the carbon break down adsorbed species.[21,222,327,368,714,715,945,1034] When biodegradation occurs, the column can be left on-line indefinitely and some BOD and COD reduction will occur. This continuous operation without regeneration can be used for the first column in a series of columns. An alternate reason for the partial regeneration is very slow mass transfer into micropores.[715,792,1073] Transfer into the macropores is quite rapid and breakthrough occurs quickly. If given enough time, some of the solutes will diffuse into the micropores and adsorb there. This opens up the sites in the macropores and appears as a partial regeneration. Support for this view of partial regeneration comes from equilibrium studies which show that equilibrium of large solutes on activated carbon can take 30 days.[791] Probably, both slow adsorption and biodegradation occur in commercial systems. There does not appear to be a great synergism between the bacteria and carbon, but this question has not been completely resolved. As Benedek[715] stated: "there is a bottomless amount of literature", and no attempt to review this literature will be made here.

Wastewater treatment differs from many other purification problems since the feed is usually not well characterized and the components in the feed can vary drastically over time. Laboratory studies are commonly used for design.[8,130a,213,291,327,679,772,1075] Unfortunately, advances in modeling techniques are not very helpful if the nature of the contaminated wastewater is not known. Designing the large-scale system from small-scale column studies is usually cheaper than completely characterizing the system and then obtaining the required equilibrium data. Equilibrium equations such as Equation 2-16 or more complex forms need to be fit to the data. Wastewaters also often contain suspended solids which will clog a packed bed. Expanded or semifluidized beds (see Chapter 6, Section II.A.1) and stirred tanks (see Chapter 6, Section II.A.4) are commonly used when suspended solids would be a problem. The breakthrough curve for an expanded bed can be very close to that obtained with a packed bed;[702] thus, there may be little penalty involved in using an expanded bed.

Regeneration of the activated carbon usually involves sending it to a separate furnace or kiln. This may be done on-site or by off-site contractors. Since some of the carbon capacity is usually lost, the regenerated carbon is usually mixed with virgin make-up carbon. The subject of thermal regeneration of activated carbon in separate furnaces or fluidized beds is beyond the scope of this book. Interested readers are referred to the extensive literature on the subject.[141,251,255,256,259,270,314,473,489,557,594,607,664,665,679,704,713,811,815,826,880,899,905,1001,1002,1005,1024] Typical losses of carbon range from 4.5 to 8% per regeneration.[905] When on-site regeneration is used, the carbon adsorbers usually work quite well, but the regeneration furnace can cause considerable operating problems.[314]

Alternatives to thermal regeneration include biological regeneration which is usually slow and incomplete, wet-air oxidation,[227,421] solvent regeneration,[251,284,507,664,952,1110] and super-critical fluid regeneration.[312,530,739] In solvent regeneration the organics are desorbed with a solvent which must then be steamed off the carbon. This method often gives incomplete desorption, but has the major advantage of recovering the organics. When the adsorbate is valuable solvent regeneration can be economical. The supercritical fluid regeneration schemes are still experimental at this time. See Section V.B. and C for more details of both these methods.

The solute movement theory can be applied to wastewater treatment with activated carbon if the solutes are assumed to be independent. The results will be similar to the other solute movement diagrams shown in Chapter 2 and in this chapter. Unfortunately, this simple model is not adequate for design since it does not include mass transfer effects which are *very* important in these systems. The LUB or MTZ approach (see Chapter 2, Section VII.A) can also be used when experimental data is available. This approach is commonly used for industrial design.[522,1069]

Theories which try to predict column behavior from first principles are much more difficult since the feed is often not well characterized, the feed is variable, and activated carbon is a very complex material. Most theories ignore the first two problems and try to model adsorption for a well-defined, constant concentration feed. Single pore models[166,223,327,654,686,714,722,981,1033] can give good agreement between theory and experiment, but cannot show the relaxation effect predicted by two pore models.[791,1073,1075] Multicomponent Langmuir isotherms are popular for theoretical studies, but cannot give a good fit to the experimental data. Multicomponent Freundlich-Langmuir isotherms (Equation 2-16) are more versatile[686] and give better fits; however, with a lot of constants the better fits do not have theoretical significance. Ideal Adsorbed Solution (IAS) theory was capable of giving excellent fits to the isotherms,[981,1033] and could be used to accurately predict breakthrough curves for well-characterized laboratory systems. Solutes can be grouped together if their equilibrium data and mass transfer coefficients are within a factor of two.[719] A model including bacterial growth has recently been developed.[1034] All of these more detailed models require numerical solutions.

V. GAS AND LIQUID ADSORPTION WITH DESORBENT REGENERATION

An alternative way to desorb is to use a desorbent or solvent to remove the adsorbates. The desorbent may be removed before the next feed step, or the feed may be started immediately following the desorption. In any case, the recovered adsorbate will be contaminated with the desorbent, and equipment to recover the desorbent is required. Both gas and liquid systems are done commercially, but data in the literature are relatively scarce. Reducing particle diameter and designing shorter columns with faster cycling would generally be advantageous.

A. Gas Systems

Gas systems are regenerated with desorbent when thermal or pressure swing desorption cycles are not adequate to remove the adsorbate. Desorption is done both with less strongly

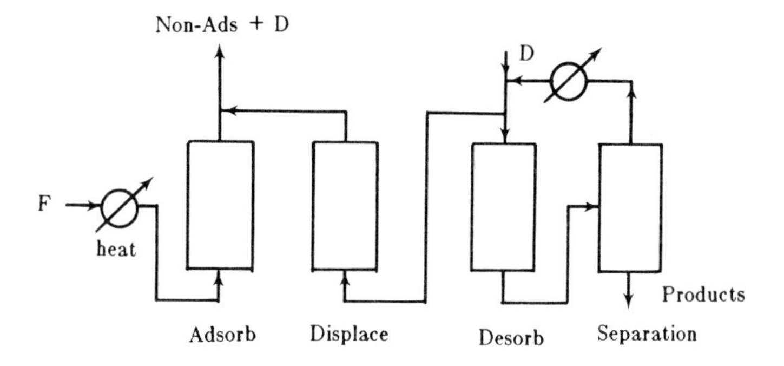

FIGURE 3-18. General flow sheet for separation of *n*-paraffins using a desorbent for desorption.

adsorbed gases[17,62,107,219,253,254,578,627,641,645,678,852,865] and with more strongly adsorbed gases.[107,678] The use of less strongly adsorbed gases is much more common. When the desorbent gas is not adsorbed, the method is often known as purge gas stripping. These systems can be considered a type of gas chromatography with backflushing where one of the components is not adsorbed and the other is quite strongly adsorbed (see Chapter 5, Section IV.C). When the purge gas is adsorbed but less strongly than the adsorbate, the operation is often known as displacement desorption.* In these systems the desorbing gas competes for sites with the adsorbent and helps to remove the adsorbent from the system.

The major commercial application is separating medium molecular weight (C_{10} up to C_{20}) straight chain hydrocarbons from branched chain and cyclic hydrocarbons.[17,62,219,253,254,578,627,641,645,865] There are a variety of commercial processes which differ mainly in the details of the desorption method used. A general process flow sheet is shown in Figure 3-18. The commercial gas-phase processes would typically use counter-flow desorption and then some separation method to recover the product from the desorbent. Fairly rapid cycles with incomplete desorption are usually used. When the more strongly adsorbed material is fed to the column, a shock wave forms which recompresses the diffuse wave formed during desorption. The bed loading will be similar to Figure 3-4. The second step in Figure 3-18 displaces nonadsorbed fluid in the external pores, and is quite short. All of the processes use 5 Å zeolite molecular sieves as the sorbent. The straight chain hydrocarbons can diffuse into the pores and thus be adsorbed while the branched and cyclic molecules are too large to fit into the pores. The linear paraffins are strongly adsorbed even at 350°C. Since thermal decomposition becomes excessive above this temperature, thermal swings cannot be used. The paraffins can be desorbed with a high vacuum,[62] but it is then difficult to provide the high heat needed for desorption. Pressure swing systems are commonly used for separation of lower molecular weight straight chain hydrocarbons (see Chapter 4, Section II).

The ideal desorbent would rapidly remove the adsorbate, be easy to displace from the bed, be easy to separate from the products, and satisfy the usual desirable characteristics of being nontoxic, nonflammable, readily available, and inexpensive. So far, this magic material has not materialized. Purge gas stripping operations usually use gases such as nitrogen or hydrogen, which are easy to recover from the adsorbate, have high diffusivities, are relatively safe, and are inexpensive. Unfortunately, they do not do a good job of pushing strongly adsorbed materials from the bed. These purge gas systems are used for octane improvement at from 100 to 300 psia and 600 to 800°F.[219] They are also used in chromatographic-type systems (see Chapter 5). Applications of the solute movement theory and other theories to these chromatographic-type systems are considered in Chapter 5.

For the harder to desorb *n*-paraffins, a desorbent which adsorbs helps to push the materials

* Note that there can be some confusion in nomenclature. In displacement chromatography (see Chapter 2, Section IV.D.1 and Chapter 5, Section IV.B) the desorber is more strongly adsorbed.

from the bed. Among the desorbents which have been used are *n*-pentane, *n*-hexane, ammonia, and alkylamines. The higher the molecular weights of the *n*-paraffins being purified the stronger the desorbent should be adsorbed. Processes have been developed by BP, Exxon, VBB Leuna Werke, Shell, Texaco, and Union Carbide. The general flow sheet will be similar to Figure 3-18 and details of the processes are given elsewhere.[62,219,254,641,645,865] Since an exchange adsorption is occurring, the processes are almost isothermal. A qualitative understanding of the operation can be obtained with the local equilibrium model with coupled solutes. Quantitative predictions for these isothermal systems can be obtained using the effective equilibrium pathway method.[114]

B. Liquid Systems

Solvent extraction has been studied for desorption of both activated carbon[227,284,507,952,1110] and resin[381,382,382a,382b,1026] systems. A flow sheet for recovery of a solute from water using activated carbon is shown in Figure 3-19A.[952] The heavily loaded wastewater is fed to the adsorber. After breakthrough occurs, the carbon is regenerated with a solvent. The effluent from the column is sent to a separation scheme such as distillation where both the solvent and the organics are recovered. The column is then drained and steamed to remove the solvent. After condensing the steam, the solvent-water mixture must be separated by suitable means. If water and the solvent are immiscible, it may be possible to send the water layer from the liquid-liquid separator to the adsorber at the end of the adsorption step. This is then a form of two-feed adsorption (see Section II.C.). When the water is highly loaded with organics, this method is reported to be cheaper than thermal regeneration since the organics are recovered.[952,1110] One disadvantage of this approach is that quite complete steam regeneration is required or the water may become contaminated with solvent. (An alternative is the waterwash shown in Figure 3-20.) Economics will also depend heavily on the cost of the solvent-organics and solvent-water separation steps. Thus solvent choice will be very important.

The solute movement theory can be applied to part of this cycle if the velocity of the solvent wave is known. Experimentally,[952] the solvent (acetone) seemed to displace water by plug flow. Then the solvent wave velocity will be

$$u_{solvent} = \frac{v}{1 + \left(\dfrac{1-\alpha}{\alpha}\right)\epsilon K_d} \tag{3-12}$$

The solute movement diagram for the adsorption and desorption steps is shown in Figure 3-19B. During desorption the organic will exit at its feed concentration, and then a large peak will follow breakthrough of the solvent wave. This peak is followed by a diffuse wave. These predictions are qualitatively correct as illustrated by the experimental results shown in Figure 3-19C.[952] A more detailed model, including mass transfer, was solved by finite difference methods.[952] These theoretical results are shown in Figure 3-19C as a solid line and agree well with the experimental results. The effective equilibrium pathway approach[114] can also be applied to these systems.

Solvent desorption of organics adsorbed from water onto polymeric resin systems has been used commercially for several pollution problems. A variety of flow sheets are used depending on the organic being recovered. The method used for recovery of phenol is shown in Figure 3-20.[367,381,382,382a] Before breakthrough of the phenol the wastewater flow to the adsorption column is halted and the column is "superloaded" with 10% phenol from the separator (see Section II.C.). The column is then regenerated with the solvent (acetone or methanol) and then rinsed with water to remove the solvent. The solute movement diagram was shown in Figure 3-8 when the superloading[367,381,382,1026] procedure was discussed. Resin life was over 2 1/2 years and thus is not a problem.[382]

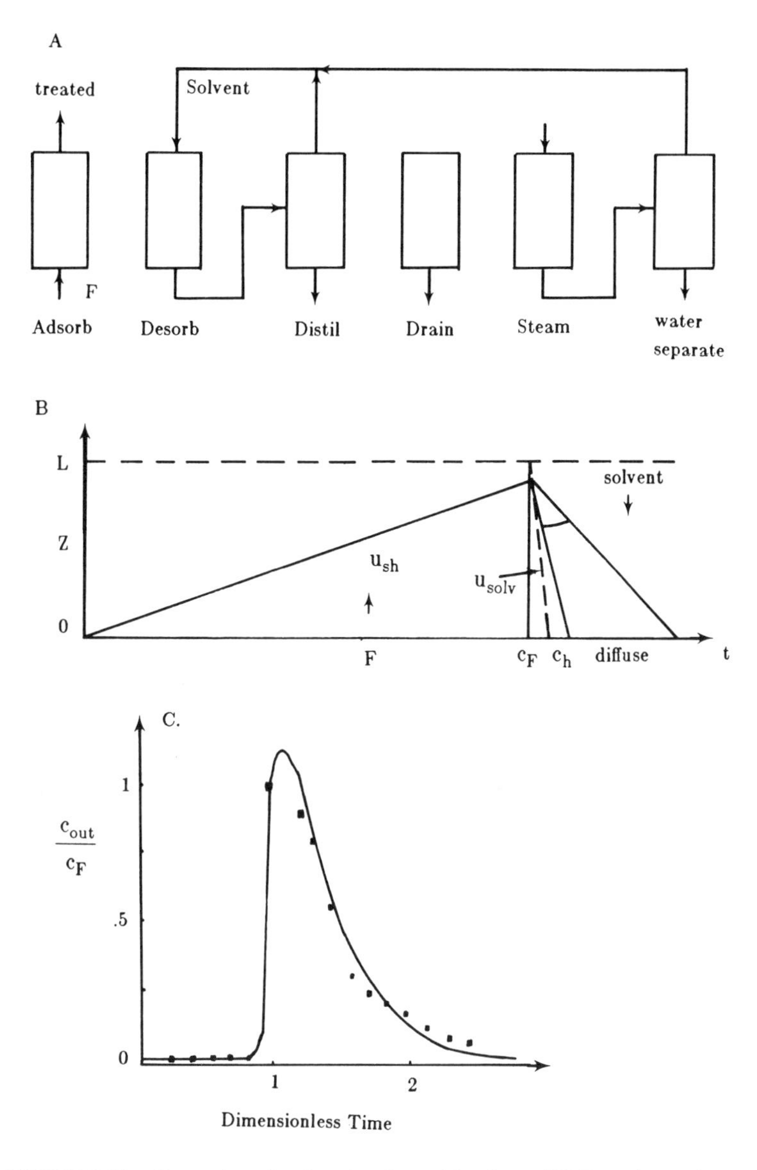

FIGURE 3-19. Desorption of activated carbon with solvent. (A) Operating cycle; (B) solute movement diagram; (C) experimental results for removal of phenol using acetone as solvent.[952]

The resin systems can also be thermally regenerated with steam. This is done commercially for recovery of chlorinated and aromatic hydrocarbons.[381] The cycle consists of adsorption of the organic from the wastewater, draining, and desorption with steam. Usually, the organic can be recovered by condensing the vapors and using a liquid-liquid separator.

C. Supercritical Fluids

Supercritical fluids have recently been introduced as a method for desorption.[312,530,739,804] Typically, solute would be adsorbed from the liquid phase and then the column would be drained. Desorption would be done with a supercritical fluid with carbon dioxide and water being likely candidates. The adsorbate would be recovered from the supercritical fluid by reducing the pressure and/or varying the temperature. One attraction of the method is the tremendous changes in solubility of compounds in supercritical fluids when the fluid is near

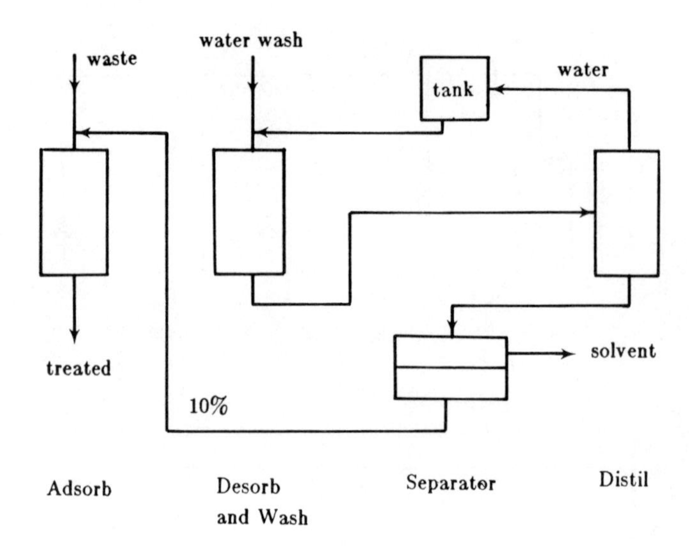

FIGURE 3-20. Phenolic removal and recovery system using polymeric adsorbent and "superloading".

the supercritical temperature and pressure. Thus desorption of many adsorbates will be rapid and the recovery of the adsorbate is simple, but may be costly. Operation is at mild temperatures and thus will be advantageous for thermally sensitive compounds. Also, supercritical CO_2 is nontoxic, has about the same solubility and density as liquids, but the diffusivity is about an order of magnitude higher, and the viscosity an order of magnitude lower. Thus the desorption step should be rapid with a minimum of extra zone broadening caused by mass transfer rate limitations. Pumping the CO_2 is also inexpensive.

Experimental data[804] for the desorption of acetic acid, phenol, and alachlor from both activated carbon and polymeric resinous adsorbents shows that supercritical CO_2 is an effective desorbent. The weaker and less nonlinear the adsorption the more effective the supercritical CO_2 was for removing the adsorbate.

Supercritical fluid desorption does not look like a simple or cheap process. The process requires high-pressure vessels and compressors. Pressure swing versions require compression of the supercritical fluid to high pressures after decompression to recover the adsorbate. Thus operating costs will not be low.[804] A method for recovering energy could greatly decrease operation costs. If used, I expect that the process will be used only for the recovery of expensive products such as flavor ingredients and pharmaceuticals. Even for these compounds the method will have to compete with other separation methods.

Supercritical fluid extraction of caffeine from coffee has been practiced commercially in Germany for many years, but commercial applications of supercritical fluids for desorption have not been reported. Pilot plants have been constructed and are available for lease.[31,739]

D. Ion Exchange Systems (References)

Ion exchange systems are usually operated in a fashion similar to solvent-desorbed adsorption systems. In ion exchange systems the regenerant is usually a solution which is concentrated in the regenerating ion. Because of the special nature of ion exchange resins (e.g., swelling of the resin) and because of the special nature of some cycles (such as using mixed beds for demineralization), the operation of ion exchange systems may be somewhat different than adsorption systems. The underlying chemistry of ion exchange is different and is more closely related to extraction.

A detailed coverage of ion exchange systems is beyond the scope of this book. Ion exchange systems are covered where their operation is quite similar to that of adsorption systems (see

Chapter 4 [Sections III and IV], Chapter 5 [Section VIII.A], Chapter 6 [Sections II.A, III.A, and III.B], and Chapter 8 [Section II]. For packed bed applications the reader is referred to any of the large number of excellent review papers and books.[16,18,56,201,322,334,431,484,493,621,656,753,854,933,943,942,1016,1017]

VI. THE FUTURE OF PACKED BED OPERATIONS

I feel somewhat sheepish making guesses about the future of these packed bed operations since most of them are already used commercially. However, there are a few trends which are fairly evident.

First, efforts at reducing energy use in adsorption operations will continue. This will result in more extensive use of beds in series, of two or more layers of adsorbent in a bed, of short cycles which leave a considerable heel in the column, and of methods such as ''superloading'', which increase the concentration of effluent. Energy reduction efforts will also focus on better ways to regenerate columns. This will include more use of vacuum, solvents, or desorbents, and other ways to replace thermal methods. This also will include an increased rate of replacement of traditional adsorbents with improved adsorbents. An example of this is the use of polymeric resins instead of activated carbon. Efforts to improve thermal desorption techniques will continue. There will probably be more widespread use of cycles without a cooling step, of systems where the thermal wave remains in the column, and more coupling of the adsorption process with other parts of the plant to use waste heat for desorption.

I expect to see a vast increase in efforts to intensify the use of adsorption systems. More systems will be designed with smaller-diameter particles. These systems will be considerably shorter than existing systems, and will operate on much shorter cycles. Thus they will have higher adsorbent productivities. The use of special forms of adsorbents with low mass transfer resistances will also allow rapid cycles and higher productivities. In addition, spherical adsorbents which can be packed to give more efficient columns will be used more widely.

Detailed theoretical understanding of many of the processes lags our ability to apply these processes. Advances in understanding activated carbon systems in both gas and liquid applications should continue; however, they will be slow particularly for wastewater treatment. Understanding of some of the less common methods such as purge gas stripping or purge gas desorption will come when the theoretical experts focus their attention on these areas.

Overall, there will be many small changes which will result in the increased application of packed bed adsorption techniques for solving a wider variety of separation problems.

Chapter 4

CYCLIC OPERATIONS: PRESSURE SWING ADSORPTION, PARAMETRIC PUMPING, AND CYCLING ZONE ADSORPTION

I. INTRODUCTION

In this chapter we consider three cyclic adsorption processes. Essentially, pressure swing adsorption (PSA) adsorbs material at high pressure and desorbs it at low pressure using part of the pure product as a countercurrent reflux. Parametric pumping is similar except a thermodynamic quantity other than pressure is varied. In cycling zone adsorption a thermodynamic variable is changed to force the separation, but flow is unidirectional. The division between Chapters 3 and 4 is somewhat arbitrary. The major difference is that the processes discussed in this chapter were developed with the entire cycle in mind. The processes in Chapter 3 tend to be a separate desorption step attached to an adsorption step.

II. PRESSURE SWING ADSORPTION (PSA) AND VACUUM SWING ADSORPTION (VSA)

Pressure swing adsorption (PSA) was first developed by Skarstrom,[917,918] who called the process ''heatless adsorption''. Since the process was first announced in 1959, there have been many important modifications to the original process. PSA has attracted considerable interest because of its simplicity and low capital investment. PSA is now really a class of processes which are used commercially for drying air, recovering hydrogen, producing oxygen or nitrogen from air, separating hydrocarbon gases, and several specialty applications. Several reviews of the PSA process have appeared.[248,583,627,643,723,825,865,919,988,1051] Vacuum swing adsorption (VSA) is somewhat similar but desorption is done by drawing a vacuum. This process is also used commercially for air separations.

A. Basic PSA Method

One logical place to start to examine PSA is the basic two-column Skarstrom-type system[917,919] shown in Figure 4-1. The high-pressure feed enters the right column where the solute is adsorbed under high pressure p_H. The product gas exits at high pressure and is essentially free of solute. A portion of this pure high-pressure gas is expanded and used as a countercurrent purge in the left column which is at low pressure. After a half cycle period of a few minutes, the columns switch. The feed repressurizes the left column and the left column produces high-pressure product. The right column is depressurized to p_L and then is purged counter-flow to the feed with a fraction of the high-pressure product. After a start-up period, a cyclic steady state is reached where a pure high-pressure product and a somewhat concentrated, low-pressure waste gas are produced. At the cyclic steady state each cycle is time dependent and outlet concentrations vary with time, but the same cycle repeats over and over. The low pressure (p_L) is usually slightly above atmospheric while the high pressure (p_H) depends either on the gas supply pressure or the desired purity and recovery of the product.

This method was first developed for drying compressed air with a desiccant as adsorbent;[917] 3 days were required for start-up with an initially saturated desiccant to dry a 4000 ppm feed gas to 2 ppm. Eventually the product contained less than 1 ppm moisture. This start-up is shown in Figure 4-2.[919] If a dry desiccant were used initially a dry product gas would be produced immediately.[600] The major cost of operation is the loss of product (in this case compressed air) in the waste gas. This loss results in a relatively low recovery of product and has been one of the major problems with PSA systems.

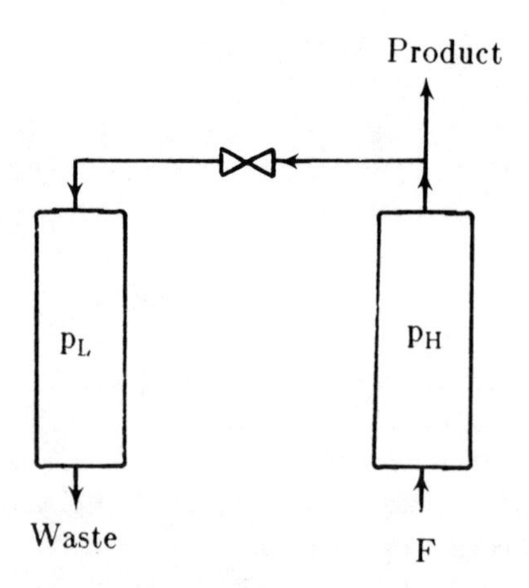

FIGURE 4-1. Skarstrom-type two-column pressure swing adsorber.

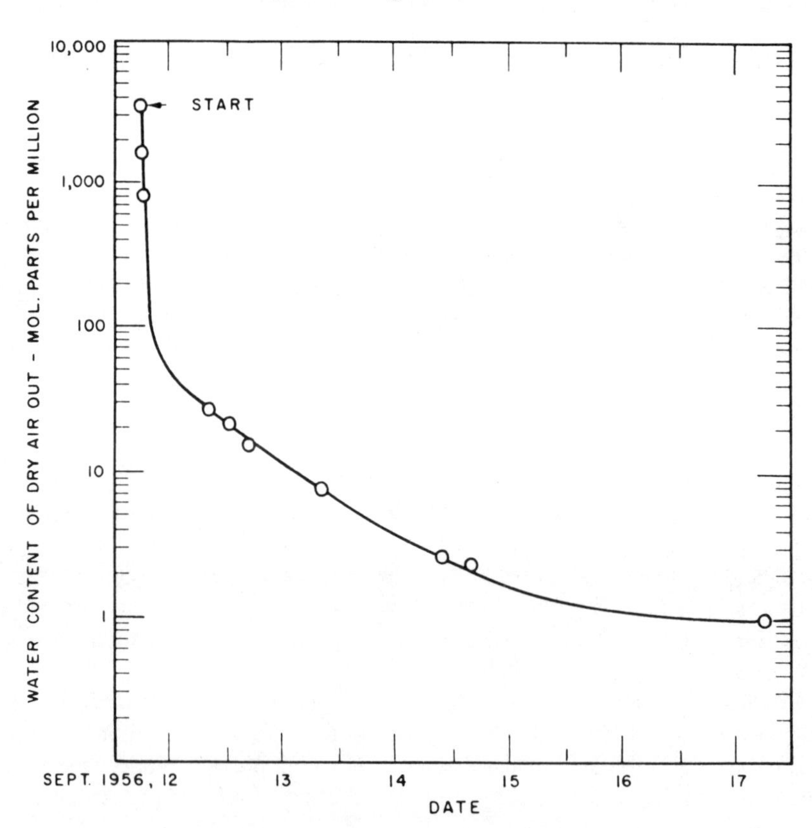

Drying out of adsorption drying system using heatless fractionation.
Flows: wet air in: 1.0 scfm, 40 psig, 3,800 ppm H_2O
dry air out: 0.5 scfm, 40 psig, as shown
regeneration air out: 0.5 scfm, 0 psig
Adsorbent: Mobilbeads, 1 lb each chamber
Cycle time: 3 min dry, 3 min regenerate

FIGURE 4-2. Start-up of PSA air dryer.[919]

With a few modifications the local equilibrium theory developed in Chapter 2 can be used to explain the primary aspects of PSA. This has been done for the two-column system shown in Figure 4-1[224,900] and for other PSA configurations (see Section V). In addition to the assumptions in Table 2-1 (some of which have been relaxed in some analyses), the gas is usually assumed to be ideal and pressure drops are neglected. The solution of the mass balances for solute A for isobaric periods is[224]

$$u_s = \frac{v}{1 + \left(\frac{1-\alpha}{\alpha}\right)\epsilon + \frac{1-\alpha}{\alpha}(1-\epsilon)\,\rho_s\,k_i} \quad \text{Isobaric} \tag{4-1}$$

which is the same as Equation 2-30 with $K_d = 1.0$. The fluid velocity v is usually different during the adsorption and purge steps.

During the blowdown (depressurization) step solute is desorbed and carried towards the outlet end of the column. Thus the mole fraction of solute y in the gas increases and the solute wave moves towards the outlet of the column. Since solute is adsorbed during the repressurization step, the solute mole fraction decreases. The change in the positions of the solute waves during the blowdown or repressurization steps is predicted by equilibrium theory[224] to be

$$\frac{z_{after}}{z_{before}} = \left(\frac{p_{after}}{p_{before}}\right)^{-\beta_A} \tag{4-2}$$

where z is measured from the closed end of the column and

$$\beta_A = \frac{\alpha + (1-\alpha)\epsilon + (1-\alpha)(1-\epsilon)\rho_s\,k_{inert}}{\alpha + (1-\alpha)\epsilon + (1-\alpha)(1-\epsilon)\rho_s\,k_A} \tag{4-3}$$

If the inert gas does not adsorb $k_{inert} = 0$. β_A is the ratio of the amount of inert gas which could be stored in the column to the amount of adsorbate which could be stored. The change in mole fraction corresponding to Equation 4-2 is

$$\frac{y_{A,after}}{y_{A,before}} = \left(\frac{p_{after}}{p_{before}}\right)^{\beta_A - 1} \tag{4-4}$$

The simple equilibrium theory assumes that fluid and solid are always in equilibrium. The exact path taken depends on the way pressure is decreased or increased. However, the final state does not depend on the path when equilibrium is assumed. For nonequilibrium processes the details of the blowdown and repressurization steps become important.

Equation 4-2 predicts that the solute waves move towards the open end of the column during depressurization and towards the closed end of the column during repressurization. This happens because gas is flowing in these directions during these two steps. Equation 4-4 predicts that the solute mole fraction in the gas increases during depressurization and decreases during repressurization. During depressurization the total pressure drops and thus the partial pressure of adsorbate will decrease. The adsorbate will then start to desorb. Since adsorbate is desorbing, the mole fraction of adsorbate in the gas y must increase. The opposite behavior occurs during repressurization. Thus the predictions of Equations 4-2 and 4-4 agree with our physical intuition. More detailed analyses of the depressurization step result in much more complex results.[849]

This simple theory can now be applied to the two-column system shown in Figure 4-1.

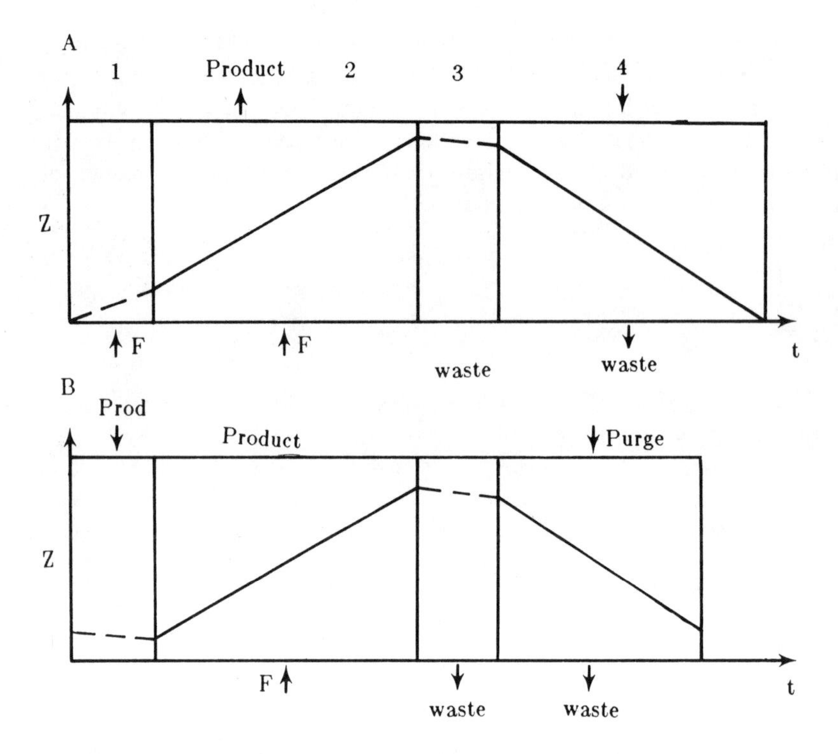

FIGURE 4-3. Solute movement diagrams for PSA, Step 1, repressurization; 2, adsorption at p_H; 3, depressurization (blowdown); 4, purge at p_L. (A) Using feed to repressurize. (B) Using product to repressurize.

Each column undergoes four steps: repressurization with feed, adsorption at p_H, depressurization, and purge at p_L. The solute movement diagram for one column of this system is shown in Figure 4-3A. The dotted lines in the blowdown and repressurization steps indicate that the exact paths are not known during these steps since the pressure history is not known. However, the end points are known from Equation 4-2.

Figure 4-3A is drawn with a higher velocity during the purge step than during adsorption. Typically the volumetric purge to feed ratio, $\gamma = V_{purge}/V_{feed}$, is greater than 1.0. The local equilibrium theory with linear isotherms predicts that any $\gamma \geqslant 1$ will be sufficient.[224,900] In practice, dispersion and finite mass transfer rates cause spreading of the waves and a $\gamma > 1$ is required. A range of 1.1 to 1.5 has been suggested.[919] One major advantage of PSA processes is a large amount of product can be produced even though γ is greater than one. This is possible because of the expansion of the product gas when the pressure is dropped. If the ideal gas law is followed the ratio of moles of purge to moles to feed is

$$\frac{n_{purge}}{n_{feed}} = \frac{p_L}{p_H}\frac{V_{purge}}{V_{feed}} = \gamma\frac{p_L}{p_H} \tag{4-5}$$

The fraction of feed which appears as product will be high if the pressure ratio, p_H/p_L, is high. This expansion effect is one of the major reasons PSA has become a successful commercial process while cycling zone adsorption and parametric pumping have not.

Many variations of the PSA process have been developed. One of the simplest is to repressurize with pure product instead of the feed. This variation is shown in Figure 4-3B. The purge step in Figure 4-3B is drawn with a shorter period and hence a lower γ than in Figure 4-3A. Industrial practice[988,1029] and theoretical calculations[220,599,1106] show that more

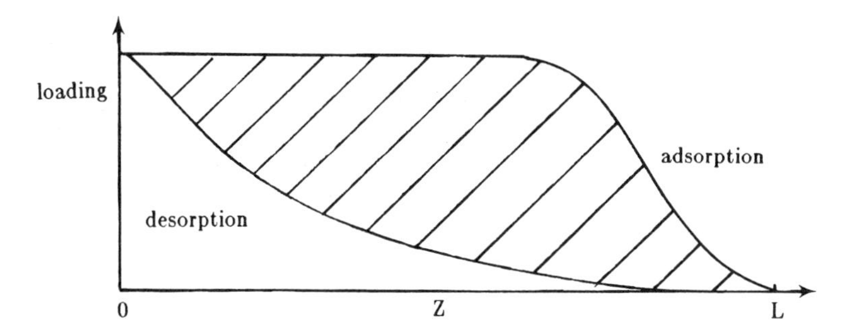

FIGURE 4-4. Loading in a PSA column when removing an adsorbate with nonlinear isotherms.

product is produced when the column is repressurized with product. Since repressurization with product can be considered as more reflux, this does make sense.

In many industrial separations the adsorbates will have nonlinear isotherms. Then a shock wave will be formed during the feed step in Figures 4-3a and 4-3b, and a diffuse wave will occur during the purge step. The solute movement theory is easily adapted to these steps. The blowdown and repressurization steps are more difficult to calculate. The effects of nonlinear isotherms are shown schematically in Figure 4-4 where the loading is plotted vs. the axial distance in the column. A constant pattern wave forms during adsorption and feed is stopped when breakthrough occurs. A proportional pattern (diffuse wave) forms during the purge step. Usually, the column is not completely purged; instead, most of the mass transfer zone (MTZ) remains in the column as a heel. This makes the waste gas more concentrated and uses less purge gas. The working capacity of the bed is represented by the lined region in Figure 4-4. With slow mass transfer the two MTZs will be quite long, and the working capacity can be a very small fraction of the total bed capacity (much less than shown in Figure 4-4). Highly favorable isotherms will give sharp adsorption fronts but very diffuse desorption fronts. The opposite is true for highly unfavorable isotherms. The optimum isotherm shape is a compromise, and is probably close to linear. A higher working capacity can be achieved by using smaller-diameter particles, decreasing L, and increasing the column diameter. This procedure can keep Δp constant while decreasing L_{MTZ}/L and increasing adsorbent productivity (see Chapter 3 [Section II.D]).

Comparison of the solute movement (local equilibrium) theory with experiment[375,900,1099] consistently shows that the local equilibrium theory overpredicts the separation. Considering the very restrictive assumptions made in the derivation, this is not surprising. The agreement between theory and experiment is illustrated in Figure 4-5,[900] which also illustrates the importance of the purge to feed ratio, γ. Figure 4-5 shows the lowest CO_2 concentration observed during a cycle. The outlet gas concentration during a cycle is time dependent, and a series of waves or scallops are actually observed.[900,1099] When mass transfer rates are slow and the separation is kinetically controlled, the observed separation may be the opposite of the local equilibrium theory predictions.[1098,1099] Thus extreme care should be used in employing this simple theory. More complex and accurate theories should be used for design (see Section V).

Adsorber operation is not isothermal. During the adsorption of water vapor from air the adsorber heats up from 2 K[217] to 4 K.[257] This heat is "stored" in the bed and thus is available during the regeneration step. During the regeneration step the bed temperature will drop to 2 to 4 K below the feed temperature. With concentrated systems (50% CH_4) the temperature excursion can exceed 50°C[220,1106] although much smaller temperature increases are seen with faster cycles. In normal operation the relatively fast PSA cycles allow for heat storage since the thermal waves do not have time to break through. If thermal breakthrough did occur

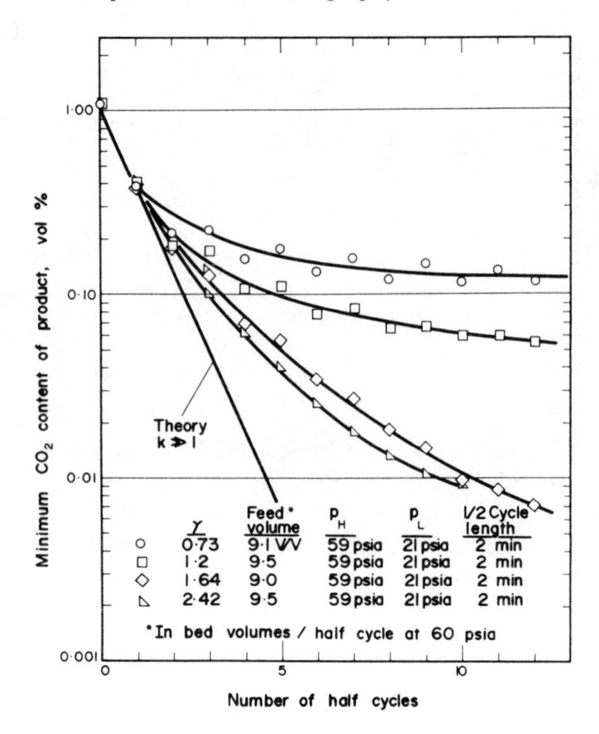

FIGURE 4-5. Removal of CO_2 from He by two-column PSA.[900] (From Shendalman, L. H. and Mitchell, J. E., *Chem. Eng. Sci.*, 27, 1449, 1972. With permission.)

during the cycle, some of the stored heat would be lost. This problem can be solved by adding a nonadsorptive heat storage material to the column.[512,1103] This also decreases the amplitude of the temperature variations, which is desirable. However, the volumetric capacity of the adsorber at the feed temperature will drop since less adsorbent is present. With large temperature variations it may be desirable to couple two columns so that they can exchange heat.[1103] When these columns are operated out of phase, each will be heated or cooled as required.

Most theories have ignored temperature changes and assumed the bed is isothermal. Even if one ignores this temperature variation there will be an optimum temperature for operation. At low temperatures adsorption becomes easier but regeneration becomes more difficult. For hydrogen purification with molecular sieves the optimum range was 10 to 15°C.[1066] Most commercial systems work better when it is cool, and thus a penalty is paid in very hot locations. Operation at higher temperatures may be advantageous if impurities adsorb less. Cyrogenic temperatures are advantageous for separating rare gases.[988] When temperature effects are included in the analysis, adiabatic operation results in lower purities than the ideal isothermal operation.[258]

Gas velocities are typically in the range from 0.01 to 0.5 m/sec. For most systems where internal diffusion controls, velocities in this range will give an average residence time longer than the characteristic time for mass transfer, d_p^2/D_M. Sharp adsorption fronts result, and a clean separation can be obtained.

B. Building Blocks for PSA and VSA Cycles

Many PSA cycles have been developed. One way to look at the multitude of cycles is to look at the individual steps which can be employed. These steps are delineated below:

(1a) Repressurization with feed or (1b) repressurization with high-pressure product or (1c) repressurization by pressure equalization. Steps 1a and 1b were illustrated in Figures 4-3A and B. The repressurization can start from the low pressure p_L and go to p_H, or it can go from an intermediate pressure to p_H or from one intermediate pressure to another. Partial repressurization is used when pressure equalization steps are employed. Repressurization with pure product is in a counter-flow direction to the feed.

(2) Feed (adsorption) at p_H. This step is usually employed to produce the high-pressure product.

(3) Pressure equalization. When one column needs to be repressurized and another depressurized, this can be partially done by equalizing pressure between the two columns.[75,130,218,300,477,1066] This step reduces the amount of product gas lost to waste and thus increases recovery. Pressure equalization may be done several times by matching a column with the column closest to its pressure. Blowdown gas can also be saved in a reservoir for later use in pressure equalization.[643] Commercial systems with ten or more columns have been built to take advantage of pressure equalization.[218,477,1066] Obviously, valving and control systems become more complex as the number of columns increases. Pressure equalization can be done from either end of the column.[477] For instance, at the start of depressurization pure or almost pure product gas can be recovered by depressurizing co-flow with the feed gas direction.[300,477] The material can be recovered as product or be used to repressurize a column counter-flow to the feed direction. During later depressurizations flow should be counter-flow since impurities are being desorbed. Now this gas should be fed cocurrently to the feed direction in the column being partially repressurized. Pressure equalization can also be done by hooking together the two ends of the columns to have both co- and counter-flow depressurization and repressurization simultaneously,[75] or by having feed gas and repressurization gas enter simultaneously.[300] In all cases it is desirable to keep impurities away from the product end of the column.

(4) Blowdown (depressurization). Blowdown may go from p_H to p_L as in Figures 4-1 and 4-3, or partial blowdown to p_L may be used when pressure equalization steps are used. During blowdown some solute desorbs. Thus a certain amount of regeneration occurs during the blowdown step. In some processes this is the only regeneration since separate purge or vacuum regeneration steps are not employed.[33,375,504,558,580,617,999,1098] Blowdown is usually counter-flow, but a period of co-flow blowdown can be useful for removing the less adsorbed gas[220,1106] or producing product.[300] The co-flow blowdown period will usually be short, and is usually followed by counter-flow blowdown.

(5) Purge (regenerate) at p_L. Usually pure high-pressure product gas is expanded and used to purge the column at the low pressure. This is done counter-flow to the feed direction. Sometimes the purge gas is obtained during blowdown of other columns.[300]

(6) Vacuum regeneration. A vacuum can be used to regenerate the column. This may be done instead of the purge step,[34,35,75,190,330,737,969] or in addition to a purge step.[602,603,910,911] Vacuum operation works best when the mole fraction of the strongly adsorbed gas in the feed is fairly high (see Figure 3-11). Vacuum operation minimizes the loss of nonadsorbed product gas, but pulling the vacuum requires a vacuum system and columns usually need to be larger in VSA systems (no purge step). These cycles are also often longer than PSA cycles, and thus adsorbent productivities are lower. Operating costs are often lower since recoveries can be higher and less gas is compressed.[190] Vacuum regeneration can also separate binary gas streams although the more strongly adsorbed gas will not be at extremely high purities.[190,915] These processes are used for simultaneous recovery of oxygen and nitrogen from air. For air separations PSA is favored when energy is cheap and capital is expensive, and VSA is favored in the opposite case.[190]

(7) Purification step (high-pressure rinse or reflux). This procedure is used when there are multiple solutes to be recovered[758,910,911,1064] or removed[915] (see Section II.D) or when a binary mixture is to be completely separated.[737,1064] The less adsorbed solute is swept out at high pressure with either pure carrier gas or a reflux of more strongly adsorbed solute. A compressor may be required to repressurize the reflux gas.[737,911,1064] Flow would usually be cocurrent with the feed gas.

(8) Delay. Some processes[33,558,580,1098] use a delay step where there is no flow in or out of the column. This delay period gives slowly diffusing solute time to adsorb or desorb and gives time for pressure waves to move in the column.[558,580]

These steps can be arranged into a very large number of different processes. The appropriate combination will depend upon the desired objective. As in all classification schemes, some processes cannot be conveniently broken down into the steps listed here.

Additional procedures can be used to develop unique PSA processes. A mid-bed withdrawal line[330,583,643,988] is useful since it gives quite large turndown ratios and can help pull off fairly pure streams of the more strongly adsorbed solute. Magnetically stabilized moving beds (see Chapter 6 [Section II.A.3]) have been adapted to combined pressure and temperature swings.[475] Multiple adsorbents have been used for several purposes. A guard bed packed with a weak or partially deactivated adsorbent is useful to prevent poisoning of the main bed.[7] Packing a column with activated carbon and a desiccant will remove both traces of hydrocarbons and water from compressed air.[587] A layer of desiccant plus zeolite is used commercially to produce dry oxygen from air while a layer of desiccant plus carbon sieve is used commercially to produce dry nitrogen.[35,756] A VSA system with desiccant columns and an adsorbent for nitrogen can dry the air and then separate it into oxygen and nitrogen.[915] In general, two adsorbents packed in series can be used to remove two solutes. Two layers of adsorbent could also be used to sharpen constant pattern (shock) waves. This can be done by placing a layer of small-size particles with a low HETP on top of a layer of larger particles (with a low Δp). Two adsorbents with different selectivities can also be used for multicomponent separations (see Section II.D).

C. PSA and VSA Cycles for Bulk Separations

The usual industrial applications of PSA have been for bulk separations. Impurities are removed and sent to a waste stream. In some cases, e.g., air drying or oxygen purification from air, the waste gas is simply discarded. In other cases, such as hydrogen recovery, the waste gas has value as a fuel. In bulk separations no attempt is made to fractionate the gas into different solutes. In this section some of the basic cycles for bulk separations will be discussed. No attempt will be made to exhaustively cover the many cycles illustrated in the several hundred PSA and VSA patents. Useful reviews cover some of these processes.[219,583,627,643,825,865,919,988,1051]

1. "Slow" Cycles

The "slow" cycles have cycle periods of a few minutes. These processes can be described as combinations of the steps discussed in the previous section. Usually, the solute movement theory gives an adequate qualitative description of the process. (This is not true for carbon sieves which work by a kinetic effect.[37,602,756]) Flow sheets for many processes are in the reviews by Lee and Stahl[643] and by Tondeur and Wankat.[988]

Most one-column systems do not have a source of gas for the purge step. Thus the system either uses only the blowdown step for regeneration or vacuum for regeneration. A typical one-column system[375,504] using only blowdown for regeneration is shown in Figure 4-6A. This cycle is quite simple, but will have difficulty producing as pure a product as systems with purge or vacuum regeneration. Purge can be obtained with reservoirs[64] or dead vol-

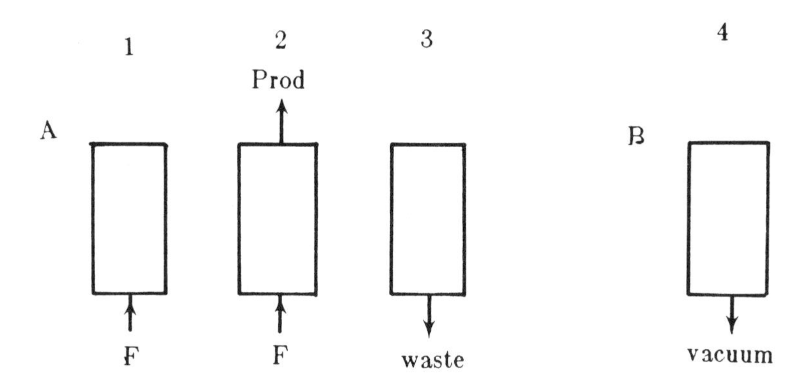

FIGURE 4-6. One-column PSA systems (A) Using only blowdown for regeneration. Step 1, repressurization; step 2, adsorption at p_H; 3, blowdown. (B) Addition of vacuum regeneration step.

ume.[250] A one-column system with vacuum regeneration would either add a vacuum regeneration step as shown in Figure 4-6B, or the feed in Figure 4-6A would be at close to atmospheric pressure and the waste stream would use a vacuum. Vacuum regeneration can do a more thorough job of removing solute and thus produce a purer product gas. However, a vacuum system is required and capital costs tend to be larger. Neither type of one-column system can use pressure equalization steps.

The original two-column configuration[917,919] shown in Figure 4-1 is commonly used for drying compressed air. Small-scale package units are available.[36,310,409,563,782,1079] A variety of desiccants can be used in the column. One-column systems with a reservoir for purge gas storage are also used for drying small quantities of compressed air.[64] Their behavior is similar to two-column systems. The compressed air dryers can be designed to knock out entrained liquids in an action similar to a cyclone before the air enters the dryer. Considerations for design and an empirical design procedure are available.[64] Packaged two-column systems for oxygen and nitrogen recovery from air are also available.[37,756] Molecular sieve zeolites are used when pure oxygen is desired,[300,627,643,919] and carbon sieves are used when pure nitrogen is desired.[37,561,602,603,723] Two-column VSA systems using zeolite molecular sieves have been used to produce both oxygen and nitrogen from air.[190] Packaged units for other separations are also available.[409,519,712] The two-column systems are simple and can produce a very high-purity product, but losses of product gas in the waste stream are often fairly high.

The carbon sieve process is of interest since it is the "only known commercial process using difference in intraparticle diffusivity . . . "[578] as the basis for the separation. The equilibrium adsorption of oxygen and nitrogen are approximately the same. However, the oxygen diffuses into the pores much faster and acts as if it is more strongly adsorbed. This carbon sieve system has been built with both vacuum[561,602,603,723] and atmospheric purge[37] operation, usually using two-column systems. The cycle is very similar to the Skarstrom cycle in Figure 4-1, but the reason for separation is different and the solute movement theory is not applicable. One advantage of the carbon sieve systems is predrying of the gas is usually not required. Capillary condensation of water occurs only if the relative humidity is high. This condensation will decrease the adsorption capacity, but does not affect the selectivities.[953] Carbon sieves will probably be used for other commercial separations since they can be made very selective.[216]

Multiple-column systems allow for pressure equalization steps and thus are capable of higher recoveries. From three[7,300,519,643,988] to ten or more[218,477,643,988,1066] columns have been used. More columns allow for additional pressure equalization steps. This is useful if the high pressure is high enough.[1066] For low-pressure operation additional columns are not

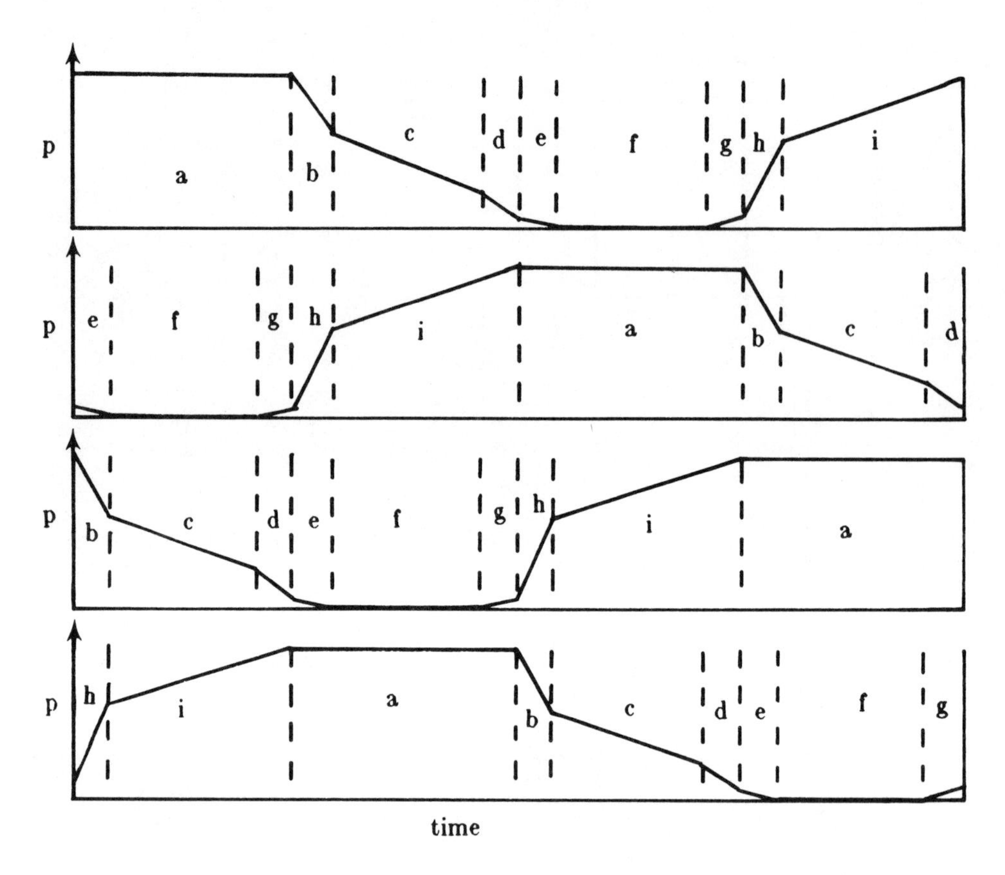

FIGURE 4-7. Pressure-time diagram explaining operation of four-bed PSA system with pressure equalization steps. (From Watson, A. M., *Hydrocarbon Processing,* 62,(3), 91, 1983. With permission.)

useful. The operation of a four-bed system with pressure equalization is illustrated by the pressure vs. time diagram shown in Figure 4-7.[1066] Adsorber 1 first is in the adsorption mode (step a) and then the void space gases, which are mainly hydrogen, are expanded (step b) and used to partially pressurize column 2. The gas from further depressurization of column 1 (step c) is used to purge column 3 while gas from step d is used for the initial repressurization of column 3. The final depressurization (step e) is a blowdown operation. Regeneration (step f) is done with expanded gas from column 4 which is also used for the first repressurization step g. The second repressurization (step h) uses gas from column 2 while the final repressurization (step i) uses pure H_2 product. Timing is set so that the short periods — b, d, e, g, and h — are all of equal duration. Then from Figure 4-7 periods,

$$a = e + f + g \quad , \quad a = h + i \quad , \quad i = c + d$$

This allows easy development of the cycle. Extension to more columns allows for more pressure equalization steps and thus recovery of a higher percent of the hydrogen. Naturally, the valving manifolds and control systems become more complex as more columns are added.

Economics have been presented for four-bed systems.[1097] Multibed PSA systems have been used for a variety of industrial separations. Hydrogen recovery is done with three or more (usually more) beds.[219,287,1097] Over one billion SCFD of H_2 are recovered world-wide.[219] Large-scale oxygen recovery systems use multibed systems. Over 200 PSA systems for oxygen recovery have been installed.[219] Light *n*-paraffins have been commercially recovered by two- or three-bed PSA since 1961.[66,219,723]

One should also remember that it may be advantageous to use PSA or VSA in conjunction with other separation techniques. For example, combining PSA or VSA and cryogenic separation can give both high hydrogen purity and high recovery.[1035,1097]

2. "Rapid" Cycles

Rapid cycle systems operate with considerably shorter periods (usually less than 30 sec), may not satisfy the classification scheme presented in Section II.B, have higher productivities than the classical PSA systems, and can often not be adequately explained by equilibrium models. These methods may represent the wave of the future.

A rapid cycle PSA system which uses one to three columns has been commercialized for removal of nitrogen from a nitrogen-ethylene stream[33,341,558,559,578,580] and for production of medical oxygen.[219] This system is shown schematically in Figure 4-8A with one column. The 1- to 5-ft long column is packed with small particles (40 to 80 mesh) which have a fairly high pressure drop and a low HETP. The cycle is as follows:[33,558,580]

Feed:	0.3 to 1.0 sec
Delay:	0.5 to 3.0 sec
Exhaust:	5 to 2.0 sec

The feed valve is open during the feed step and the exhaust valve is open during the exhaust step. The product valve controls column pressure but it is always open. Nonadsorbed or less adsorbed gas is collected as product while the adsorbed impurities and some carrier gas leave as exhaust. Similar processes but without the delay step were reported earlier.[617,999]

To understand the operation of this process, consider the pressure profiles shown in Figure 4-8B.[580] During the feed step (1) the pressure profile monotonically decreases as expected. The delay step (2) allows the pressure wave to penetrate into the column further. When the exhaust valve is opened the pressure immediately drops at the feed end. Early in the exhaust period (3) there is a pressure maximum in the column and gas flows to both ends from the center of the bed. Late in the exhaust period (4) the column continues to be purged. The feed, delay, and part of the exhaust step produce product while the exhaust step both depressurizes and regenerates the bed. The concentration profiles in the bed are shown in Figure 4-8C.[580] Note that a rather small fraction of the bed (between curves 1 and 3) is available as a useful capacity.

The productivity (kilogram of product per kilogram adsorbent per day) of this one-bed system for recovery of 90% oxygen from air was 2.3 compared to 0.5 for standard three-bed PSA systems.[558] The oxygen recovery was 38% compared to 40% in the PSA system.[b9] Thus a much smaller system can be used at the price of a somewhat lower recovery. Somewhat higher recoveries are achieved with the three-column system.[341,578]

An interesting comparison which was not made would be to use a smaller packing in the standard PSA system, but keep L/d_p^2 constant so that both N and Δp are constant. The cycle period would also be reduced by a factor d_p^2. The amount of feed processed per hour would be essentially unchanged. This should give roughly the same separation as the standard PSA system but the productivity would be considerably higher (see Chapter 3 [Section II.D]). For example, a decrease in d_p by a factor of $\sqrt{5}$ (to about 25 mesh) would reduce L by a factor of 5 (to around 1 to 2 ft). The cycle time would be reduced by a factor of 5 (to about 35 to 60 sec). The amount of product should be about the same but the productivity would increase by approximately a factor of 5. Thus the faster cycling greatly increases productivity. These suggested modifications of standard PSA cycles have not been tried. If they worked they would significantly increase the productivity. One possible problem area which needs to be checked experimentally is whether the pressure waves adversely affect the operation. In large-diameter systems the resulting column would be short and fat, and would not satisfy

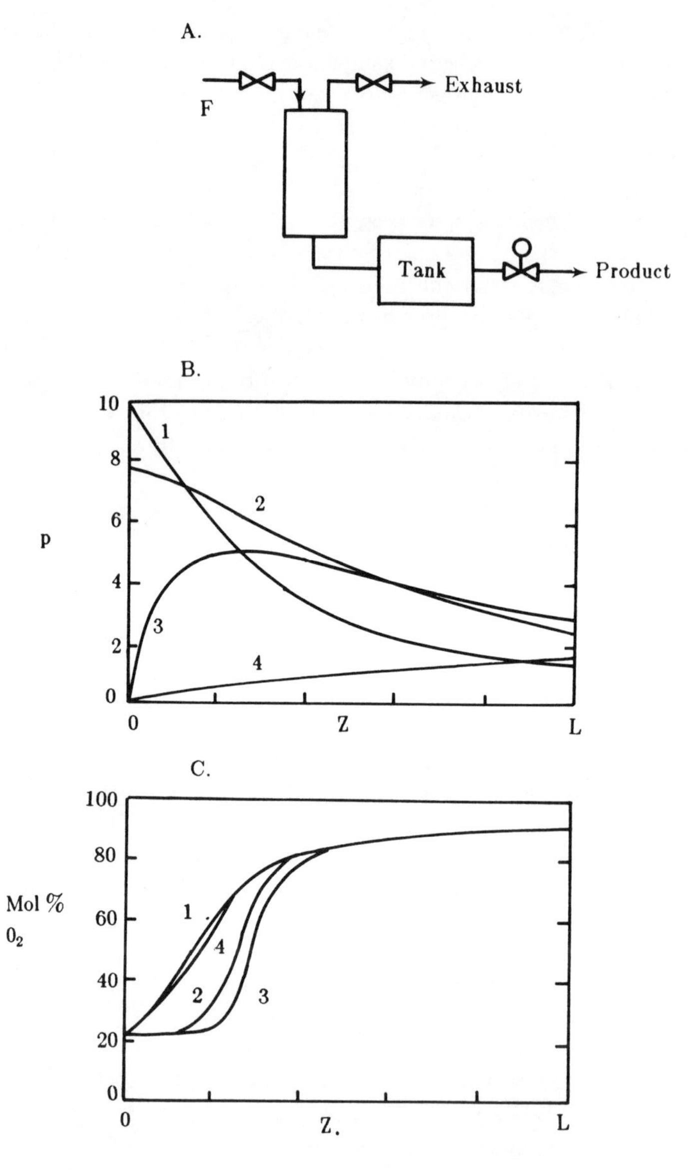

FIGURE 4-8. Rapid cycle PSA process. (A) Single column system. (B) Pressure profiles. 1, Middle of feed; 2, middle of delay; 3, early in exhaust; 4, late in exhaust. (C) Concentration profiles in bed producing 90% oxygen. (From Flank, W. H., Ed., *Adsorption and Ion Exchange with Synthetic Zeolites*, ACS, Washington, D.C., 1980, 275. With permission.)

rules of thumb for L/D. However, short fat columns are successfully operated (see Chapter 3 [Section II.D] and Chapter 5 [Sections VI and VIII]) for other sorption operations. Less waste gas could be achieved by modifying the scaling procedure. If d_p were reduced by a factor of $\sqrt{5}$ and L by the same factor the pressure drop can be kept constant by increasing the cross-sectional area by $\sqrt{5}$ (v decreases by $\sqrt{5}$). The number of stages or the number of transfer units will increase significantly (by a factor of 5 in linear systems with pore diffusion control), resulting in sharper breakthrough curves and a more concentrated waste gas. This can be done with simple cycles and with higher adsorbent productivity.

A second very short cycle system called the "molecular gate" with significantly higher productivities has been developed by Keller and Kuo.[579,581] A schematic diagram of this

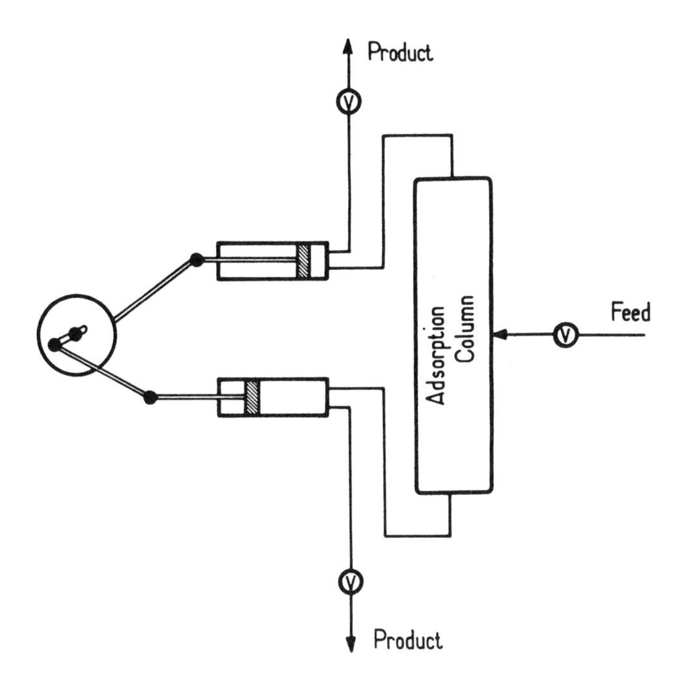

FIGURE 4-9. Center feed, refluxed PSA system.[581]

process is shown in Figure 4-9. The pressurized feed enters the column somewhere between the two ends. The two pistons have different amplitudes and operate out of phase. Thus during parts of the period both push in, during parts one pushes in and the other retracts, and during parts both retract. This results in complex pressure profiles and flow patterns. The process essentially has large reflux ratios at both ends. During part of the cycle the column pressure rises above the feed pressure which effectively shuts off the feed to the column. The more strongly adsorbed solute concentrates in the piston with the larger amplitude while the least strongly held gas concentrates in the piston of smaller amplitude. For separation of 15 psig air the optimium ratio of volume small piston to volume large piston was about 0.25. The optimum phase angle was approximately a 45° lead of the smaller piston. This phase angle caused the flow direction to reverse when the flux from the particles to the gas reverses. Both pistons operated at 30 rpm, although higher frequencies are now used.[579] Productivities for N_2 ranged up to 85.2 and for oxygen up to 15.7. For high-purity N_2 plus argon (99.9%) with 98.5% recovery and simultaneous 95% oxygen with 99.8% recovery the productivities were 4.42 and 1.35. Other gases were also separated, including a 50:50 mixture of H_2 and CH_4. Almost no temperature increase was observed because of the rapid cycling between adsorption and desorption. To some extent this device can be considered a type of pressure swing parametric pump (see Section III).

The local equilibrium model is not satisfactory for explaining these results. The cycle periods (2 sec or less) are too short to approach equilibrium, and kinetic effects are important. Models, including diffusion in and out of a single particle, are qualitatively successful.[579]

So far commercialization has not been announced. Since the pistons are compressing and expanding the gas, energy reuse is critical to keep operating costs low. The major problem with commercialization has been capital expense and the use of rotating equipment.[579] However, the ability to simultaneously produce high-purity oxygen and nitrogen or high-purity hydrogen and methane in a relatively simple device is a breakthrough. Other cycles which can do this are quite complex or require a vacuum. This achievement shows that PSA can produce high-purity products with very low recoveries. This should spur further PSA research.

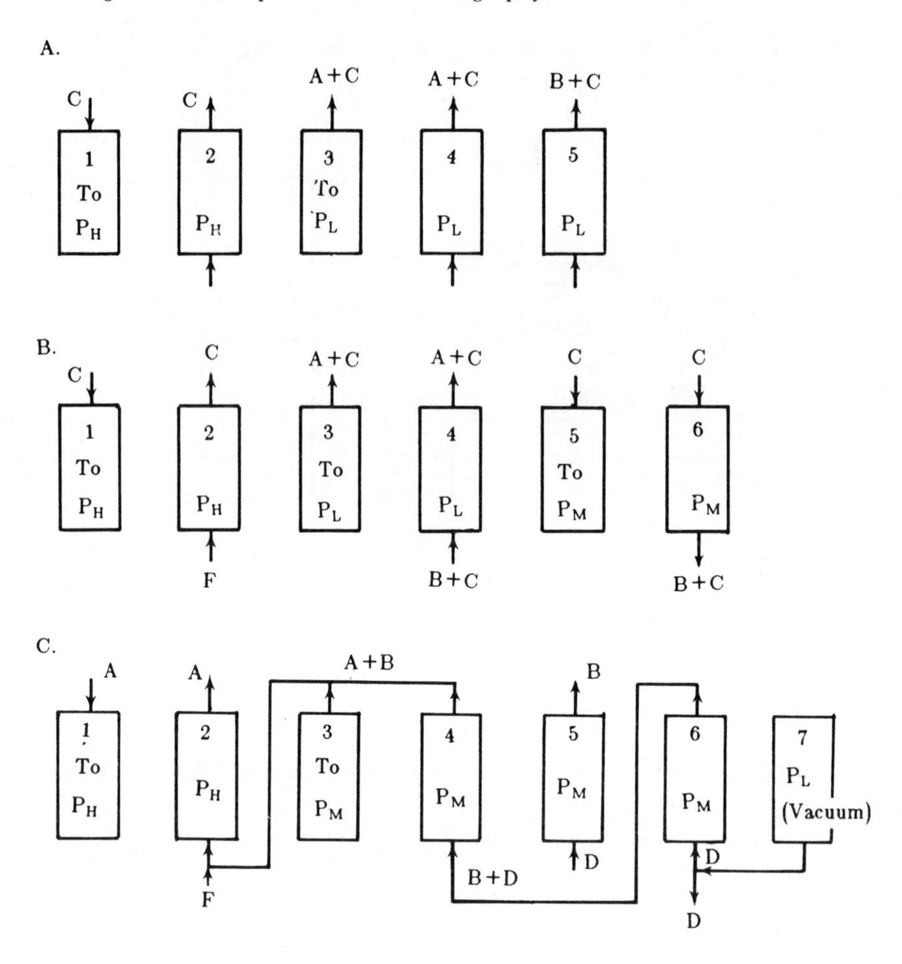

FIGURE 4-10. Multicomponent PSA cycles. A, B, D are solutes and C is carrier gas. (A) No flow reversal. Step 1, repressurization; 2, high-pressure adsorption; 3, blowdown; 4, purification; 5, cocurrent purge. (B) With flow reversal. 1, 2, and 3 same as Part A; 4, purification using expanded B and C from Step 6; 5, partial repressurization; 6, countercurrent purge. (C) With vacuum. 1, Repressurization; 2, feed; 3, blowdown; 4, purification; 5, product; 6, purification; 7, vacuum.

D. PSA Cycles for Fractionation

PSA was first developed as a bulk separator. Recently some attention has been paid to developing PSA and VSA cycles which can fractionate several solutes. Essentially, three approaches for PSA separation have been used. One approach uses a single adsorbent and uses a cycle with a purification step to separate dilute solutes from each other, but the solutes are not separated from the carrier gas.[758] The second type of approach modifies the standard PSA cycles by adding a cocurrent blowdown step.[220,1106] The third type of approach uses two adsorbents with different selectivities.[598,910,911,1064] VSA systems with a single adsorbent are used commercially for simultaneously production of oxygen and nitrogen.[190,915]

Two PSA cycles with a single adsorbent for separating solutes were developed[758] and are shown in Figure 4-10A and B. The cocurrent flow system shown in Figure 4-10A works very much like an elution chromatograph which is producing its own pure carrier in step 2. The faster-moving solute A in carrier gas C is eluted from the column in steps 3 and 4 while solute B remains in the column. Solute B in carrier gas is eluted co-flow during step 5. Additional solutes can be separated by adding more steps. The system shown in Figure 4-10B has similar steps 1, 2, and 3. In purification step 4 a mixture of solute B and carrier gas C is used to remove A from the column. The column is then partially repressurized to medium pressure P_M and the column is purged counter-flow in step 6. Some of the product gas from step 6 is expanded and used as reflux in step 4. An alternate cycle would skip step 5, operate step 6 at pressure P_L, and use a compressor for the reflux gas.

Analysis of these cycles by the solute movement theory is straightforward for linear systems.[758] The cocurrent cycle is useful for separations where the selectivity between solutes A and B is small while the countercurrent cycle works best when the selectivity is large. The same conclusion can be drawn for elution chromatography systems (see Chapter 5). These schemes can both be considered gas chromatographs and offer an alternate way of operating large-scale gas chromatographs. They have the disadvantage of chromatographs that a heel of component A cannot be left in the column in either scheme. A heel of component B can be left in the column in the system shown in Figure 4-10B. This will be an advantage for nonlinear systems.

A VSA system similar to that shown in Figures 4-6A and B can simultaneously produce fairly pure weakly adsorbed gas and strongly adsorbed gas. This is done commercially for production of oxygen and nitrogen from air using zeolite molecular sieves.[190,915] The weakly adsorbed oxygen is produced as product during the adsorption step which is operated near atmospheric pressure. The more strongly adsorbed nitrogen is then desorbed under vacuum. A higher-purity oxygen should be obtained if the column is repressurized counter-flow with oxygen product instead of with feed. A higher-purity nitrogen product is obtained if a co-flow nitrogen flush is added after the feed step to remove oxygen in the voids.[915]

Multicomponent separations similar to the system shown in Figure 4-10B can be done by adding a vacuum desorption step. One possible cycle is shown in Figure 4-10C. All three components adsorb with A adsorbing least and D most. After repressurization with pure A (step 1) and processing of feed plus recycle gas (step 2), a co-flow blowdown (step 3) is used to remove A from the system. Step 4 is a further purification step to remove A if the co-flow blowdown is insufficient. In step 5 pure B is produced by purging co-flow with D. In step 6 the remainder of species B is removed either by co-flow blowdown and/or a purge step. Finally, in step 7 species D is removed by vacuum desorption. Pressure equalization steps can easily be added to the cycle. Required compressors and vacuum pumps are not shown in the figure. The advantage of using a vacuum is all components can be recovered as fairly pure products.

Concentrated binary mixtures can be separated into two products with greater than 90% purity by adding a co-flow blowdown step to the usual four step PSA cycle: repressurize, adsorb, counter-flow blowdown, and purge.[221,1106] The co-flow blowdown has the same function as the purification step shown in Figure 4-10. A heel of the faster-moving component cannot be left in the column, but a heel of slower-moving component can be left in the column. A cycle with co-flow repressurization, adsorption, and co-flow blowdown was capable of partial separation of hydrogen and methane.[1026] Ternary mixtures were separated with a cycle consisting of pressurization, adsorption, co-flow blowdown and counter-flow vacuum evacuation,[220] which is similar to the cycle in Figure 4-10C. The most strongly adsorbed component, H_2S, was concentrated but was not recovered as a pure stream.

Systems with two adsorbents choose adsorbents with different and if possible reversed selectivities for the solutes.[598,1064] With two adsorbents of reversed selectivity for the two components relatively simple schemes can be devised for completely separating a binary mixture.[1064] If one two-column PSA system (Figure 4-1) is packed with one adsorbent and another two-column PSA system is packed with the other adsorbent, the two systems can be operated in parallel. Each system produces a pure high-pressure gas which is the less adsorbed component with that adsorbent. The waste gases from each system are compressed and sent to the complementary system for cleanup. Each of the two-column systems act like normal PSA systems, and thus a heel of the adsorbed component can be left in each system.

Multicomponent systems can be fractionated using more complex two-adsorbent schemes.[619,910,911] One of these cycles[910] has a high-pressure feed step, high-pressure purification step, several pressure equalization steps, blowdown step, purge step, vacuum desorption step and repressurization. Six columns with one adsorbent and three with the other are used. A mixture of hydrogen, carbon dioxide, methane, and carbon monoxide was separated using BPL-activated carbon and 5 Å molecular sieves. A primary product of high-purity hydrogen, a secondary product of high-purity carbon monoxide, and a tertiary product with hydrogen, methane, and carbon monoxide were produced.[910]

If one desires to fractionate all the solutes but not separate them from the carrier gas, simple two-adsorbent systems can be used.[1064] The apparatus for doing this is illustrated in Figure 4-11A. The lower column is packed with adsorbent A which selectively adsorbs solute 1 but very weakly adsorbs solute 2. The second column is packed with adsorbent B which strongly adsorbs solute 2 (the relative adsorption of solute 1 by this adsorbent is irrelevant). During the feed step solute 1 is retained by the first column while the second column retains solute 2. This is illustrated in Figure 4-11B. At the end of the feed step solute 2 is present in the void spaces of the first column. This solute is pushed into the second column during the high-pressure purification step. Now the first column contains only solute 1 plus carrier gas while the second column contains solute 2 plus carrier gas. Each column is separately depressurized and purged with carrier gas. Repressurization can be done either with product or with feed sent to the first column. This cycle is easily extended to systems with more components.[1064] Since solute 1 must pass through column A, a heel of solute 1 cannot be left in column A; however, a heel of solute 2 can be left in column A and a heel of solute 1 can be left in column B. Thus both solutes can be recovered as fairly concentrated mixtures in the carrier.

Other fractionation schemes for ''slow'' cycles are easily developed using different combinations of the basic processing steps. The ''fast'' cycles can also be adapted to binary and perhaps multicomponent separation. There is considerable commercial interest in fractionation of gas mixtures. It remains to be seen if PSA and VSA fractionation schemes become widely used for these separations.

III. PARAMETRIC PUMPING

Parametric pumping (PP) is similar to PSA in that flow reversal (or reflux) is employed after a thermodynamic variable has been changed. However, PP utilizes the shift in the isotherm when temperature or another thermodynamic variable is changed to force the separation, and not the large expansion of purge gas which is used in PSA when pressure is decreased. Thus the molar reflux ratio in PP is usually much higher than in PSA. This, of course, is a disadvantage, but it is hard to see how to avoid this problem when liquids are separated. This high reflux ratio and the resulting marginal economics have kept PP from being commercialized except in one brief instance (see Section III.E). PP was originally developed by Wilhelm and co-workers[856,962,1089-1092] at Princeton University. The process has been extensively studied and reviewed[233,449,450,840,956,957,959,1040,1051] by academics since Wilhelm's original work. Grevillot's[449] review is particularly up to date and thorough.

A. Batch Direct Mode Thermal Parametric Pump

To explain how parametric pumping works we will start with the batch direct mode process shown in Figure 4-12. The column is jacketed and the entire column is heated and cooled externally (direct mode). Fluid flows upwards when the column is hot and is stored in the top reservoir. The column is cooled during a delay period and then fluid flows downward into the bottom reservoir. The column is heated and this cycle repeats. When the column is hot, adsorption of a single solute will be weak and the solute wave velocity will be high. When the column is cooled, the solute is more strongly adsorbed and the solute wave velocity

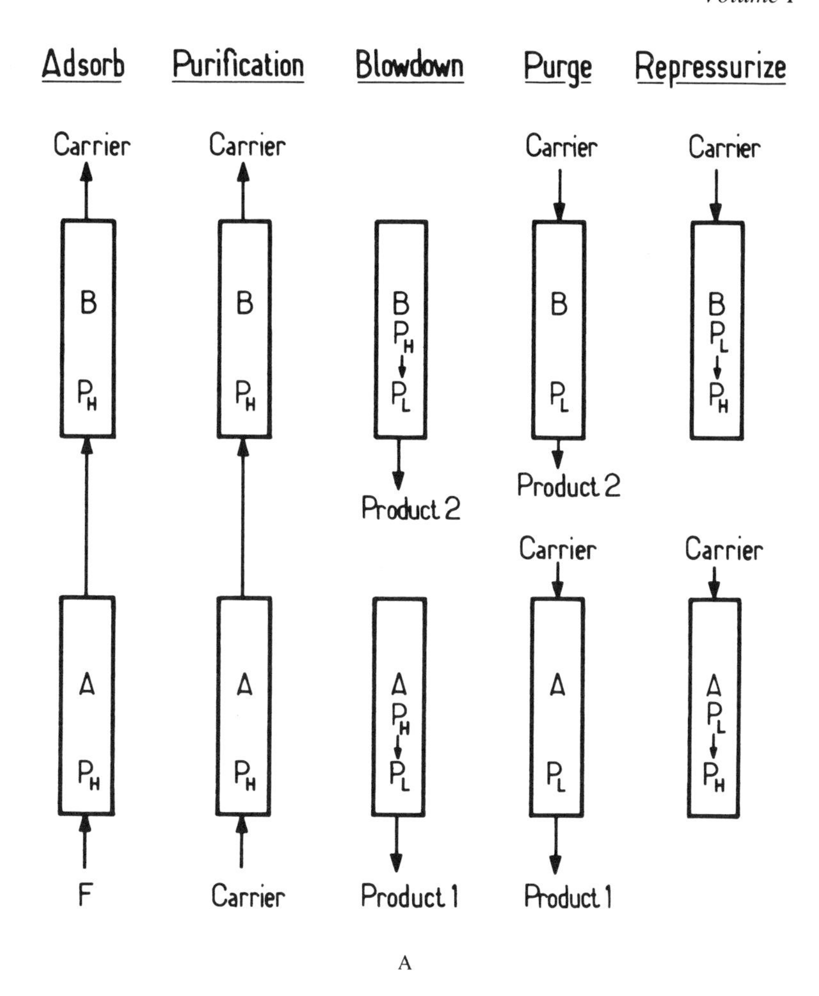

FIGURE 4-11. Multiple adsorbent PSA system for fractionation. (A) Apparatus. (B) Solute movement diagram for linear isotherms. (From Wankat, P. C. and Tondeur, D., *AIChE. Symp. Ser.*, 81(242), 74, 1985. With permission.)

is low. Thus, upward movement of solute will be greater than downward movement and solute will tend to concentrate in the top reservoir.

The solute movement theory is easily applied to batch PP. The solute wave velocities are given by Equation 2-30 for linear systems and Equation 2-28 for nonlinear systems. The parametric pump is often started up with the entire column and the bottom reservoir in equilibrium at the hot temperature. Then solute movement can be followed on a plot of axial distance z vs. time t. Every time the temperature switches the solute will redistribute according to Equations 2-44 or 2-46 for linear isotherms in the direct mode. The solute movement diagram for a linear system is shown in Figure 4-12C. As can be seen, solute eventually is "pumped" to the top reservoir. The concentration at any time and location can be predicted if the type of reservoir is specified so that external mass balances can be written. The usual assumption has been that the reservoirs are completely mixed. Then in Figure 4-12C everything exiting the column from a to b is at c_i (the initial concentration). During the next half cycle this material enters the column from the top reservoir during the period from b to c. The material leaving the column during the cold half cycle has undergone one temperature change from hot to cold. Since more material is adsorbed, this portion will be diluted. From Equation 2-46,

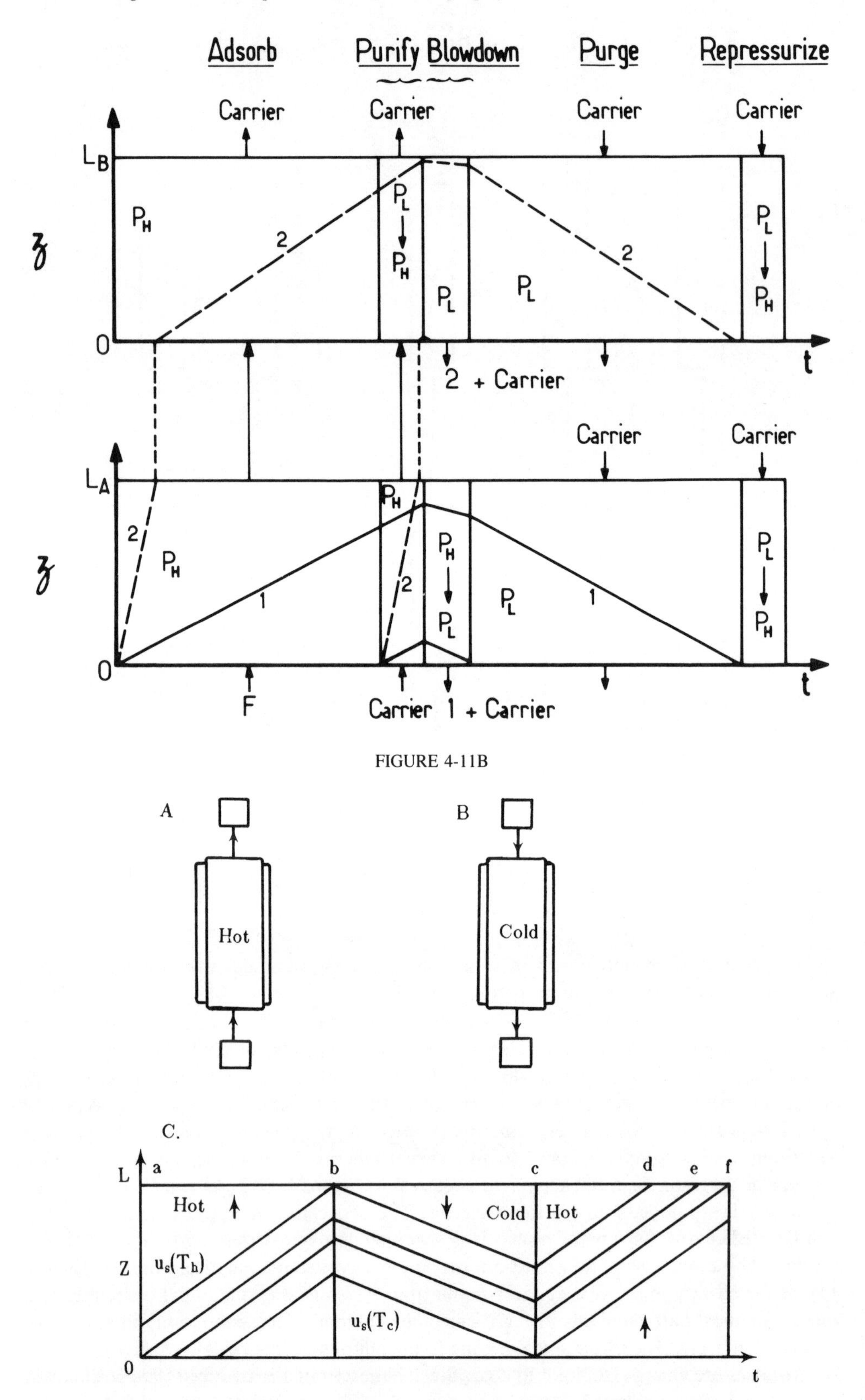

FIGURE 4-11B

FIGURE 4-12. Batch direct mode thermal parametric pump. (A) Hot half cycle. (B) Cold half cycle. (C) Solute movement diagram for system with linear isotherms.

$$c_{bot}(\text{cycle 1, cold}) = c_i \frac{u_s(T_c)}{u_s(T_H)} \tag{4-6}$$

During the next cycle, the solution leaving the top of the column from c to d was at c_i, but it underwent a temperature change from cold to hot and was concentrated.

$$c_{cd} = c_i \left(\frac{u_s(T_H)}{u_s(T_c)}\right) \tag{4-7}$$

The solute exiting from d to e and from e to f both started at c_i and both underwent first a hot-to-cold temperature change and then a cold-to-hot temperature change. These changes cancel and the concentration from d to f remains c_i. The average concentration for the second cycle is

$$c_{top,2} = c_{cd} \left(\frac{\overline{cd}}{\overline{cf}}\right) + c_i \left(\frac{\overline{df}}{\overline{cf}}\right) \tag{4-8}$$

The relative distances are easily determined from the solute wave velocities. This process can be continued for any number of cycles.

For linear isotherms a closed form solution for bottoms concentration can be obtained.[58,805] In terms of the variables defined here this solution is

$$c_{bot,cycle\ n} = c_i \left[\frac{u_s(T_c)}{u_s(T_H)}\right]^n \tag{4-9}$$

The bottoms concentration is predicted to decrease exponentially and the separation factor will increase exponentially as the number of cycles increases. The top reservoir concentration asymptotically approaches a limiting concentration where essentially all of the solute is pumped into the top reservoir. The theory also predicts that each half cycle should be short enough to avoid breakthrough of solute directly from one reservoir to the next.

This simple theoretical description of parametric pumping appeared after the classical experiments of Wilhelm and Sweed[1092] were published. Their results for the separation of toluene from *n*-heptane on silica gel are reprinted in Figure 4-13.[1092] The separation factor $>10^5$ was considered astounding, but it really means that the bottom reservoir concentration was quite low. The separation increased when the temperature difference was greater and when the cycle period was longer (slower flow) since this allowed a closer approach to equilibrium. Note that for the first 15 to 20 cycles the separation factor does increase exponentially. The local equilibrium (solute movement) theory gave a surprisingly good fit to the data.[805] However, this was a semiempirical use of the theory. The equilibrium parameter was fit to the PP experiments and was not obtained from equilibrium data. When equilibrium data is used, the solute movement theory overpredicts the separation. The equilibrium theory cannot explain the differences between curves A and B. With longer cycles and hence slower flow rates operation is closer to equilibrium.

A large number of other mixtures have been separated by batch direct mode PP.[196,233,237,449-453,840,846,879,956,957,959,1040,1051] When ionic mixtures are separated, the physical picture is slightly different from the picture presented earlier. Because of electroneutrality, if one ion is desorbed when the temperature increases, the other ion must be sorbed. Thus one ion must concentrate at the hot end of the column and the other at the cold end. This binary separation of ions is experimentally observed.[196,205,451] Complete separation of the two ions can be obtained by using a center feed system.[205,456] The linear solute movement theory is unable to explain the observed results, but the nonlinear theory with shock and diffuse waves (see Chapter 2

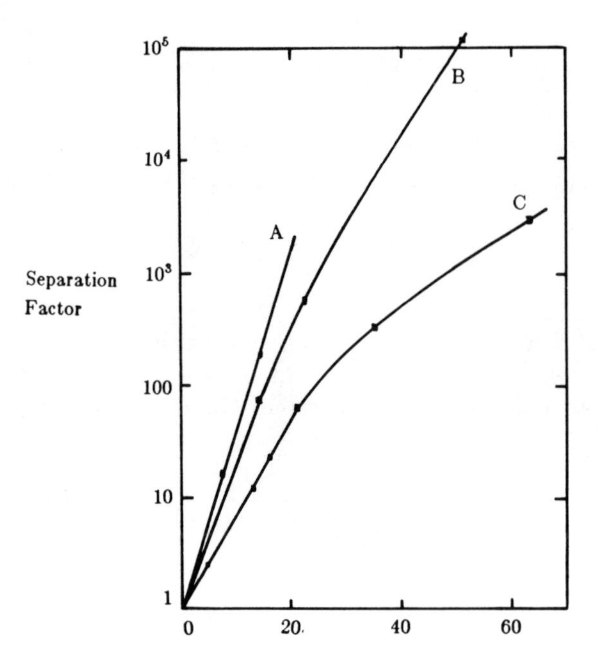

FIGURE 4-13. Batch direct mode thermal parametric pumping separation of toluene from *n*-heptane on silica gel. (A) 140-min cycle, 70 to 4°C, (B) 30-min cycle, 70 to 4°C; (C) 8.5-min cycle, 70 to 15°C. (From Wilhelm, R. H. and Sweed, N. H., *Science*, 159, 522, 1968. With permission.)

[Section IV.A.2]) does qualitatively explain the results.[205] As usual, the model overpredicts the separation. In addition to the separation of ions, the PP experiments show a change in total concentration.[196,451-453] This effect was recently explained with the local equilibrium theory based on the temperature dependence of Donnan electrolyte partition.[453]

There are situations where the local equilibrium theory is qualitatively wrong. Reverse separations have been observed.[237,846] These reverse separations will usually be caused by slow rates of mass transfer so that the process is kinetically controlled. (Of course, this is not necessarily bad. A kinetically controlled separation could be developed.) Other possible causes have been delineated.[846] In most cases the local equilibrium model is qualitatively correct; thus, we will use the local equilibrium model to explain other PP procedures. For design and optimization the local equilibrium model is *not* adequate and more detailed theories are required (see Section V or the reviews by Rice[840] or Grevillot[449]).

B. Batch Recuperative Mode Parametric Pump

In the "recuperative" or "traveling wave" mode the column is adiabatic and the entering fluid is heated or cooled. This process is illustrated in Figure 4-14A. The solute movement diagram for a linear system is shown in Figure 4-14B. Note that the temperature does not change until the thermal wave has passed. Thus, breakthrough of the thermal wave is desired so that concentration changes will be observed at the ends of the column.[961] When the solute waves intersect the thermal wave, the concentrations change according to Equation 2-45 for linear systems. The solute wave velocities change because the distribution coefficient depends on temperature. Analysis of start-up is very similar to the batch direct mode analysis shown in Equations 4-6 to 4-9. When properly operated (i.e., with breakthrough of the thermal waves), large separations are predicted for the recuperative mode. The local equilibrium theory predicts complete removal of solute from the cold reservoir, but the more detailed STOP-GO theory predicts finite separation factors which are somewhat less than the direct

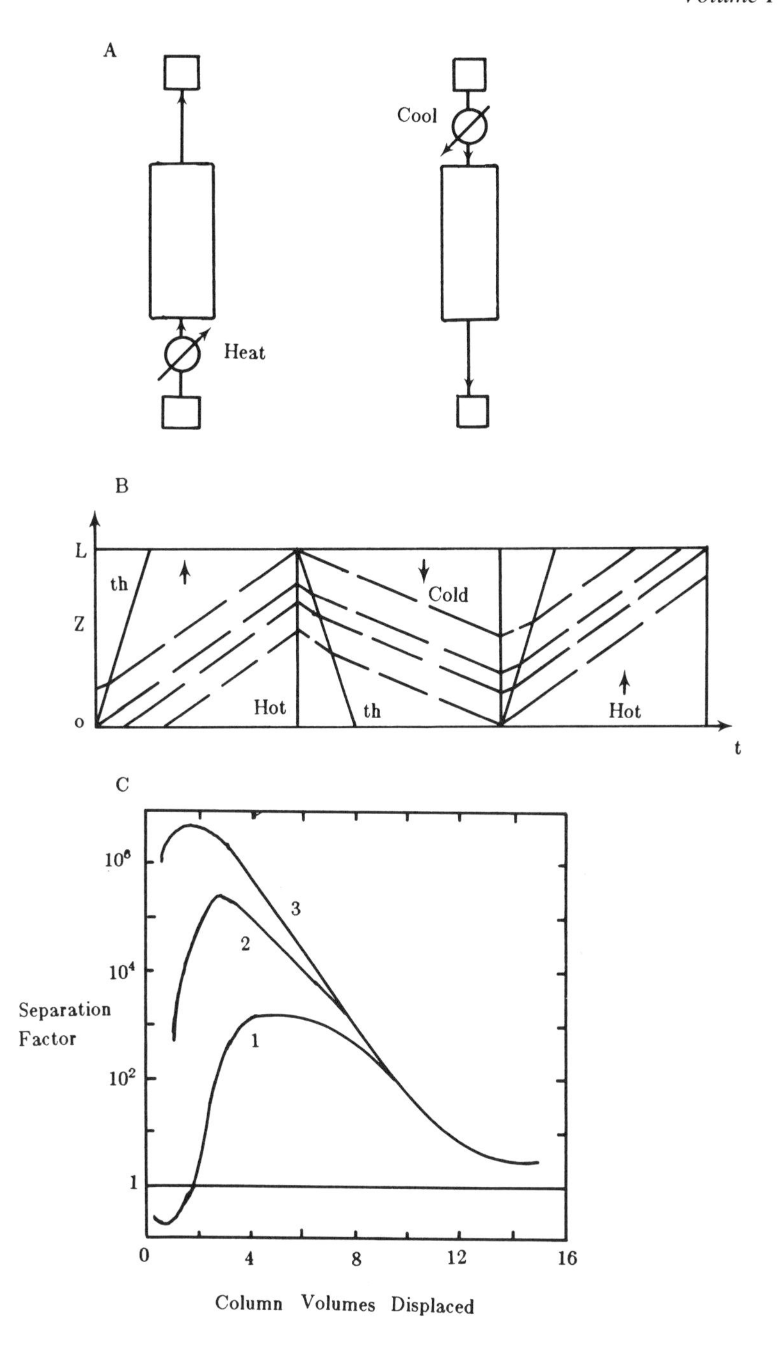

FIGURE 4-14. Batch recuperative mode thermal parametric pumping. (A) Column operation. (B) Solute movement diagram for system with linear isotherms. (C) Simulation using STOP-GO model. 1, recuperative mode; 2, recuperative mode with three intermediate heat exchangers; 3, direct mode.[961] (From Sweed, N. H. and Rigaudeu, J., *AIChE Symp. Ser.*, 71(152), 1, 1975. With permission.)

mode results for slow cycling.[961] This is shown in Figure 4-14C.[961] As shown in this figure, the recuperative mode results can be improved by using intermediate heat exchangers which will reduce the lag time for the temperature to change. Thermal breakthrough is desired to avoid the small reverse separations predicted with low fluid displacements. With fairly rapid cycling or with short delay periods for the direct mode temperature changes, the recuperative mode separation can be greater than the direct mode.[9]

The recuperative mode has two major advantages for large-scale separation. First, in large-diameter systems it is much easier to scale-up than the direct mode. Slow radial heat transfer is a problem in the direct mode and special designs would be required.[940] However, laboratory-scale direct mode systems are simpler than the recuperative mode so most thermal experiments have used the direct mode. Second, in continuous or semicontinuous operation the energy use of the recuperative mode is significantly less than in the direct mode plus it is easier to recover energy. In one calculation the recuperative mode used 10 to 15 times less energy than the direct mode.[1048] In addition, the recuperative mode can be used for thermodynamic variables other than temperature.

Unfortunately, the early recuperative mode experiments operated without breakthrough of the thermal wave and the separations were very small.[856,1089,1091] The reason for these small separations was not understood and at the same time the direct mode was giving very large separation factors.[1091,1092] Research shifted away from the recuperative mode and it has only been quite recently that properly operated thermal, recuperative mode simulations[961] and then experiments[9] have been conducted. The simulations shown in Figure 4-14 show that the early experiments with fluid displacements from one to two column volumes per cycle would be expected to have low separation. Experimental results are discussed in the next section.

The majority of the research reported in the literature refers to the direct mode. However, since the one industrial application of PP used the recuperative mode (see Section III.E) and other large-scale uses would probably also use the recuperative mode, I will emphasize the recuperative mode in the remainder of this chapter. Direct mode theories and experiments are included in Tables 4-1 and 4-2.

C. Open Systems: Continuous and Semicontinuous Operation

In large-scale applications a continuous or semicontinuous system is preferable to a batch system. These have been called "open" systems as compared to the "closed" batch systems. The early recuperative mode systems[1090,1091] were semicontinuous systems with fresh feed introduction. The most extensive studies of continuous and semicontinuous systems have been with direct mode systems.[5,235,236,240,242,243,446,447,844,960] Continuous recuperative mode operation has been analyzed using the solute movement theory[1048] and was used commercially in the Sirotherm process (see Section III.E).

Many different continuous or semicontinuous systems could be used. One possible configuration is shown in Figure 4-15.[1048] This can represent either a single column, semicontinuous system, or a two-column continuous system. When two columns are used, this system is quite similar to the two-column PSA system shown in Figure 4-1. With open systems the reflux ratios at the top and bottom of the column become very important variables. The reflux ratios are defined as equivalent to internal reflux ratios in distillation.

$$R = \frac{\text{fluid returned to column}}{\text{fluid exiting column}} \tag{4-10}$$

When both R_T and R_B equal one, the system is at total reflux and becomes a batch system. As with distillation or PSA the higher the reflux ratio, the better the separation, but the lower the productivity.

The solute movement theory is easily applied to open systems.[1048] This can be done either for start-up or for the cyclic steady state. Start-up will be similar to batch operation (Figure 4-14B). At the cyclic steady state the cycle repeats itself every cycle. This is shown in Figure 4-15B for linear isotherms for the case where $u_{th} > u_s(T_h) > u_s(T_c)$, and in Figure 4-15C for nonlinear isotherms where $u_s(T_h, O) > u_{th} > u_s(T_c, c_F)$. In an open system with top feed the volumetric displacement during the cold (downward) portion of the cycle must

Table 4-1
APPLICATIONS OF LOCAL EQUILIBRIUM MODEL

Application	Operation	Ref.
Development of basic model for two-column system with linear equilibrium, and comparison with experiments.	PSA	506,900
More detailed linear equilibrium model. Carrier gas can adsorb.	PSA 2 Columns	224
Comparison of linear model to experiment. Discussion of situation where model fails.	PSA	1099
Continuation of model[224,900] including sorption effects. Linear isotherms. Required numerical solution.	PSA	599
Application of theory[224] to one-column system. Linear isotherms.	PSA	504
Theory for one-column system. Both linear and nonlinear isotherms. Comparison with results for air drying. Nonlinear isotherms required numerical solutions.	PSA	375
Use of theory[224] for multicomponent separation. Developed feasible regions to achieve separation.	PSA	758
Use of linear theory[224] for systems using two adsorbents to fractionate solutes.	PSA	598,1064
Use of linear theory to study transient performance.	PSA	600
Development of theory for rapid cycle, one-column system including pressure drop. Linear isotherms. Required numerical solutions. Comparison with data for N_2-CH_4 separations.	PSA	999
Study of purge step for both adiabatic and isothermal operation. Sample calculations for CO_2 and CH_4,N_2 or H_2. Isothermal assumption can seriously underpredict purge gas requirements.	PSA	913
Basic development of model for batch systems with linear isotherms. Graphical and analytical solution. Semiempirical data fitting. Direct mode.	PP	805
Generalization of solution.[805]	PP	58
Development of model and graphical solution for direct mode.	PP	962
Letter on effect of nonlinear isotherms.	PP	833
Effect of nonmixed reservoirs in direct model batch. Linear isotherms.	PP	979
Batch method for multicomponent separation direct mode linear isotherms.	PP	195
Comparison of sinusoidal and square waves. Very little difference when equal displacements are compared. Batch direct mode, linear isotherms.	PP	839
Donnan concentration effects, batch direct mode. Comparison with experiments.	PP	453
Model for coupled, nonlinear systems. Batch and open systems, direct mode.	PP	205
Application of linear model in design of direct mode system with shell-and-tube heat exchanger.	PP	940
Analysis of batch recuperative mode.	PP	961
Theory of direct mode open systems. Development of external equations. Linear isotherms.	PP	236,242,243,446
Analysis of open recuperative mode systems. Mixed and unmixed reservoirs. Comparison of energy use in direct and recuperative modes.	PP	1048
Analysis of open, direct mode systems nonmixed reservoirs with dead volume. Good agreement between theory and experiment.	PP	288
Application of model to direct mode, open system for separation of two noncoupled solutes. Theory overpredicts experimental separations.	PP	1100
Comparison of theory and experiment. Equilibrium parameters obtained from separate experiments. Theory overpredicts observed separations.	PP	9,788,879,895

Table 4-1 (continued)
APPLICATIONS OF LOCAL EQUILIBRIUM MODEL

Application	Operation	Ref.
Comparison of theory and experiment. Semiempirical fit since parameters obtained from PP experiments.	PP	5,235,242—244,805
Use of two adsorbents for binary and multicomponent separations. Direct and recuperative mode in open systems.	PP	1064
Development of theory and analytical equations for direct mode linear equilibrium.	CZA	464
Direct and traveling wave modes. Linear and nonlinear equilibria. Reasonable comparison of theory and experiment.	CZA	74
Traveling wave with linear equilibrium and Arrhenius temperature dependence. Optimized.	CZA	721
Traveling wave with Langmuir isotherm and Arrhenius temperature dependence. Reasonable agreement with experiments.	CZA	601
Multicomponent traveling wave system with focusing. Linear equilibrium.	CZA	1043
Direct and traveling wave modes, multicomponent separation. Linear and nonlinear equilibria for uncoupled solutes.	CZA	1044
Direct mode, concentrated systems with nonlinear equilibrium. Effect of layered beds and recycle.	CZA	388
Forced direct mode to simulate traveling wave of controlled velocity. Linear and nonlinear isotherms. Uncoupled, multicomponent. Reasonable comparison with experiment.	CZA	376
Use of linear theory for fractionation systems using two adsorbents.	CZA	1064
Comparison of theory with experiment.	CZA	903
Gas system with focusing. Large heat of adsorption couples thermal and solute wave. Reasonable agreement between theory and experiment. Development of feasibility conditions.	CZA	537—541
Theory for chromatothermography.	—	993

Table 4-2
MORE COMPLEX MODELS FOR CYCLIC SEPARATIONS

Model	Operation	Ref.
Used method of characteristics to obtain ODE which was solved by finite differences for two-column system. Assumed isothermal, linear equilibrium, ideal gas, negligible Δp, and no axial dispersion. Good agreement with experiment.[900]	PSA	736
Ignore dispersion and mass transfer. Assume linear equilibria and use D'Arcy's law for pressure drop. Solve one-column system by finite differences. Reasonable agreement with experiments.	PSA	969
Assume isothermal and ignore diffusion. Solve mass balance and rate equations by numerical integration. Good agreement with data on drying air in two-column system.	PSA	217
Nonisothermal and isothermal systems with linear equilibria. Use numerical integration. Adiabatic case shows considerably less separation then isothermal case.	PSA	258
Theory above[258] compared to experiments for air drying. To obtain agreement needed to use high values for mass transfer coefficients.	PSA	257
Linear, well-mixed cell model applied to one- and two-column systems. Sample calculations for recovery of He from the HE-CH_4 mixture.[249] Comparison between theory and experiment for one-column system[250] (Figure 4—5A).	PSA	249,250

Table 4-2 (continued)
MORE COMPLEX MODELS FOR CYCLIC SEPARATIONS

Model	Operation	Ref.
Nonlinear, well-mixed cell model. No details of model given, but results compared well with experimental data for drying air.	PSA	375
Equilibrium staged, countercurrent distribution type model. Few details. Agreed well with finite difference model.[b11] One-column system.	PSA	969
Equilibrium staged model for rapid cycle one-column system.	PSA	617
Overall mass balance model for two-column system. Assumes complete bed equilibrium at end of cycle. Agreement between theory and experiment poor, but improves with longer cycle times.	PSA	1067,1087
Overall mass balance model with MTZ of assumed length.	PSA	333
Mass balance model treating cyclic steady state as a steady countercurrent contactor. Used for N_2, CO, H_2, and air-drying examples.	PSA	952a
Pore diffusion model for mass transfer with thermal equilibrium (nonisothermal). Ideal gas law, loading ratio correlation equilibrium. Negligible Δp. Numerical integration. Applied to separation of concentrated H_2/CH_4 mixture,[1105,1106] $H_2/CH_4/H_2S$ separation,[220,1105] H_2/CO[1105] mixtures, and $H_2/CH_4/CO_2$[1105] separation.	PSA	220,1105,1106
Empirical approach for small-scale systems for drying compressed air. Comparison with experiments.	PSA	64
Numerical analysis using orthogonal collocation for isothermal system with one adsorbate and an inert carrier. Two-bed Skarstrom-type system.	PSA	819,965
Equilibrium and linear driving force model applied to H_2/CO mixtures.[221,1105] Nonisothermal but with thermal equilibrium. Hybrid Freundlich-Langmuir isotherm. Numerical integration. Application of equilibrium model to explore use of inert material to slow down thermal wave[1104] and to explore heat exchange between two columns.[1103]	PSA	221,1103,1105
Modeling of depressurization step. Numerical analysis. Applied to depressurization only, not complete cycle.	PSA	849
Detailed modeling of open system, thermal, recuperative mode PP for ion exchange resins. Used 2nd order Langmuir equation and included pore diffusion. Model solved by finite difference. Excellent agreement between theory and experiment. This remains the most detailed PP modeling.	PP	856
Solved PDE by finite difference. Include mass transfer inside particles and used an effective surface diffusivity. Thermal, direct mode basket PP.	Basket PP	703
Zero order reaction with Langmuir kinetics and first order reaction with linear kinetics. Utilizes Thomas-type solution method. Numerically calculates results.	PP with reaction	585
General integral analysis. No numerical calculations.	PP	516
Frequency response solutions of partial differential equations for thermal, direct mode. Effect of radial heat transfer. Calculation of optimal ultimate separation conditions. Good agreement with experiment.[844,845] Consideration of design problems.[844]	PP	377,378,836—838,840,844,845
Deans and Lapidus mixing cell model with 50 stages. Direct[1091,1092] and recuperative[1091] mode calculations. Calculation of maximum separation region.	PP	1091,1092
Mixing cell model for direct mode, thermal PP. Calculated number of cells from Van Deemter equation (see Equation	PP	465

Table 4-2 (continued)
MORE COMPLEX MODELS FOR CYCLIC SEPARATIONS

Model	Operation	Ref.
2—117), but to achieve fit to experimental data this had to be adjusted. Also developed a near equilibrium model and assumed each cell was at equilibrium. Matrix manipulation used for solution of both theories.		
Near equilibrium model for direct mode, thermal PP as chemical reactor	PP	52
STOP-GO algorithm. During GO step all fluid in cell is discretely transferred to next cell. During STOP step heat and mass transfer occur. Solve ODE during stop step numerically. Good agreement with data[1089,1092] for direct thermal mode.	PP	962
Apply STOP-GO algorithm to recuperative thermal mode. Good agreement with data.[856]	PP	961
Apply STOP-GO algorithm to gas PP. Modified to allow for pressure equalization and axial mixing. Model qualitatively agreed with data.	PP	552
Applied STOP-GO to thermal, direct mode PP in batch and open systems. Very good agreement for batch.	PP	960
Use STOP-GO algorithm to optimize several open, thermal, direct mode PP. Good agreement between theory and experiment.	PP	447
Use STOP-GO to compare continuous recuperative mode PP and conventional thermally regenerated adsorption.	PP	445
Analyzed open, pH PP with circulation using STOP-GO algorithm.	PP	239
Equilibrium staged analysis for direct mode PP. Models with continuous flow of fluids and with discrete transfer and equilibrium steps. Reasonable agreement with extraction PP.	PP	1038
Equilibrium stage analysis for direct mode system. Graphical solution.	PP	1030
Use of equilibrium staged analysis to stimulate direct mode multicomponent PP. Good agreement with experiments when number of stages was estimated from PP data.	PP	1100
Graphical analysis based on discrete transfer, equilibrium staged model.[1038] Developed analogy between PP and distillation.	PP	454,455
Use of graphical analysis[454,455] to simulate separation of Ag^+ and Ca^{2+}. Fit data to find HETP. Studied reflux effects.	PP	451
Use of graphical analysis[454,455] to simulate direct thermal mode PP.	PP	449,459,757,1030
Use of graphical analysis[454,455] for multicolumn, pH recuperative mode PP. Reasonable agreement between theory and experiment.	PP	245
Staged graphical analysis with one equilibrium stage for direct, thermal mode basket PP.	Basket PP	441
Equilibrium staged analysis applied to extraction PP. Used matrix formalism to study transient and cyclic steady states. Good agreement between theory and experiment.	Extraction PP	871
Equilibrium staged analysis applied to extraction PP with chemical reaction.	Extraction PP	440
Comparison of local equilibrium model, STOP-GO, integration of PDE and discrete transfer, discrete equilibrium model. Related parameters of different models.	PP	338
McCabe-Thiele type graphical analysis for distillative adsorption with equal[841] and unequal solids loadings.[604]	Adsorptive-distillation (3-phase PP)	604,841

Table 4-2 (continued)
MORE COMPLEX MODELS FOR CYCLIC SEPARATIONS

Model	Operation	Ref.
Pore diffusion model for Langmuir isotherm with Arrhenius temperature dependence. Numerical solution. Good agreement between theory and experiment for separation of CH_4-H_2 and CH_4-H_2-H_2S.	CZA	248,991,992
Perturbation analysis for sine wave temperature change in direct mode. Linear isotherms. Include diffusion of solute inside pores. Good agreement between theory and experiment. Extended to multizone systems.[597]	CZA	597,601,903
Equilibrium staged analysis with discrete transfer and equilibrium steps. Traveling wave mode. Good agreement between theory and experiment when pH wave velocity was determined from experiments.	CZA	194,331
Equilibrium staged analysis with discrete transfer and equilibrium for direct[1039] and traveling wave[1041] thermal operation in extraction. Good agreement with extraction experiments.	CZ extraction	1039,1041
Discrete transfer and equilibrium steps in equilibrium staged model applied to multicomponent CZA. Linear isotherms.	CZA	1043
Continuous flow equilibrium staged model for direct mode CZA. Good agreement between theory and experiment.	CZA	759
Mixing cell model applied to gas separation with focusing. Good fit with data, but a low estimate of gas heat capacity had to be used.	CZA	541
Mixing cell model. Prove that a cyclic steady state exists for cocurrent (CZA) flow. A cyclic steady state was found in numerical simulations for countercurrent flow (PP) but a general proof was not found.	PP, CZA	635

be greater than that in the hot portion. Thus with equal velocities the period of the cold portion of the cycle T_c must be greater than the hot portion T_h. The larger the shift in equilibrium with temperature the lower the reflux ratio can be and the more pure bottom product will be produced. In Figure 4-15B and C, a bottoms concentration of zero is predicted. In actual practice, some solute will be observed in the bottom product. In Figure 4-15C a case where $u_s(T_h, c_F) > u_{th} > u_s(T_c, c_F)$ is shown. The solute movement theory will predict infinite concentrations if linear isotherms are used. The use of nonlinear isotherms shows a large increase in concentration before the thermal wave during the hot cycle. This results in a shock wave as shown in the figure. Figure 4-15C is drawn for a nonoptimum case since solute waves can pass directly from the bottom to top reservoirs. With a much shorter cycle this would be prevented, a more concentrated top product could result, and R_B would be reduced, producing more product. Timing is very important in all the cyclic separation procedures.

Experimental results for the removal of dilute concentrations of phenol from water on Duolite ES 861 are shown in Figure 4-16[9] for a system similar to that shown Figure 4-15A. The local equilibrium model overpredicts separation after the first four cycles, but separation factors up to 10^3 were obtained experimentally. Larger separations would be obtained if the column were adiabatic. In large-diameter columns heat losses through the wall would be much less a problem.

In general, the product concentrations and energy use depend not only on the solute movement in the column but also on the details of the reservoir. If separation is not complete the separation can be increased by using nonmixed or segregated reservoirs which retain partial separation instead of mixing everything together.[979] Energy use can also be reduced with selective recycle.[1048] Thus, for the system shown in Figure 4-15A, energy use will be

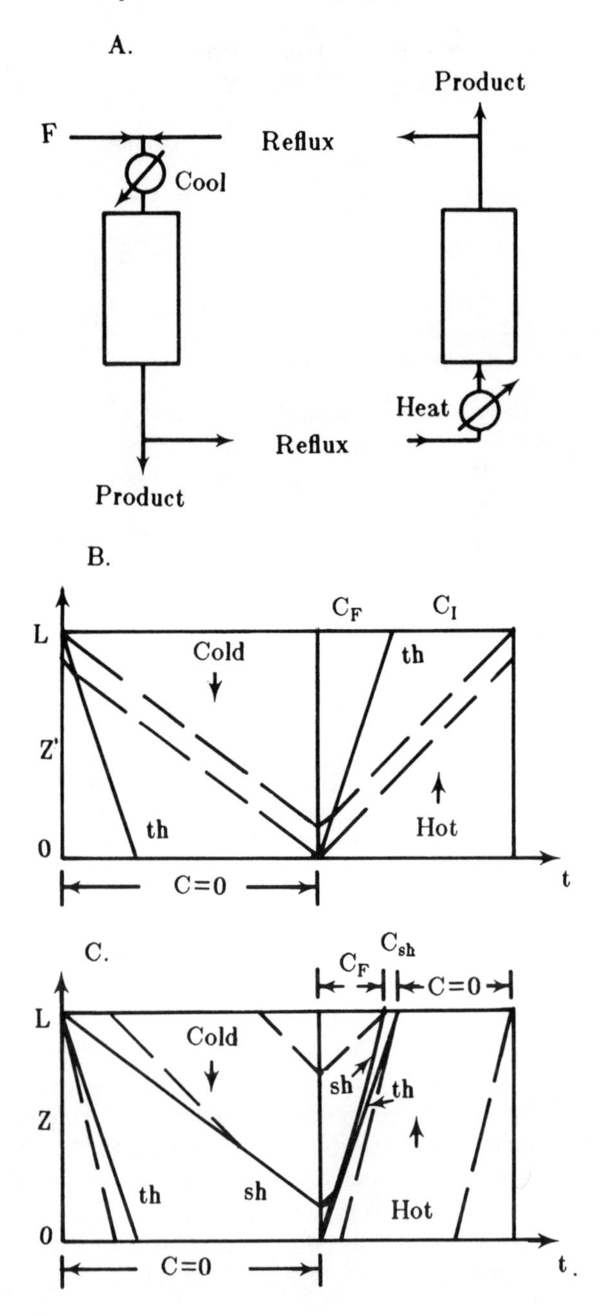

FIGURE 4-15. Semicontinuous recuperative mode system. (A) Cycle used; (B) solute movement diagram for linear isotherms; (C) solute movement diagram for nonlinear isotherms.

reduced if the products are always withdrawn after the thermal wave has broken through. This allows one to reflux all of the hot material for the next hot step and all of the cold material for the next cold step. Segregated reservoirs were used in the commercial Sirotherm process[535] for reflux of specific streams.

D. Modifications and Extensions of PP

A large number of modifications and extensions of parametric pumping (PP) have been

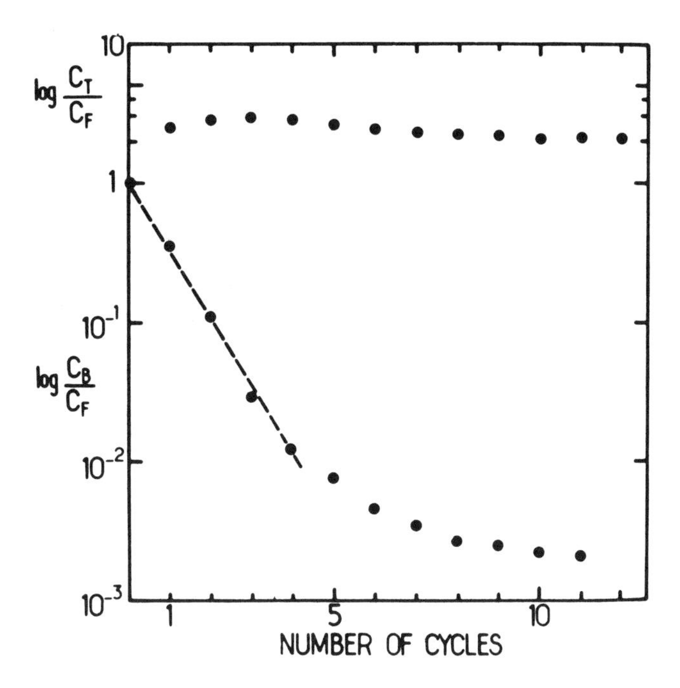

FIGURE 4-16. Recuperative thermal mode parametric pumping separation of phenol from water. Dotted line is local equilibrium theory. (From Almeida, F. et al., *Physicochemical Methods for Water and Wastewater Treatment,* Elsevier, Amsterdam, 1982, 169. With permission.)

developed since 1966. In this section we will briefly consider these. Theories are outlined in Section V.

Any thermodynamic variable which affects the equilibrium can be used in a parametric pump. These have included pH,[237,239,241,244,245,514,869,895] pH and electric field,[238,514] pH and ionic strength,[234] chemicals which complex with the solute,[225] and pressure (see Section II.C.2). The recuperative mode is used when chemicals are added. The added chemical can either compete with the solute for the adsorbent[869] or change the solute in free solution so that it is less adsorbed.[225,234,237-239,241,244,245,895] The latter approach has advantages since less chemical may be consumed and it may be possible to reuse the chemical. Chen and co-workers developed a number of different operating methods.[225,234,237-239,241,244,245,514] The simplest are one-column systems similar to Figure 4-16 but with a pH addition and control system replacing the heat exchanger.[244,869,895] The solute movement diagrams for these systems are similar to Figure 4-16C and D. Two-column center feed systems[237] should be useful for separating two proteins, but reverse separations were observed. This equipment is similar to the center feed systems used for separating mixtures of two ions[205,456] (except the ionic separations were direct mode operations). Multicolumn systems using both cation and anion exchangers were used for continuously separating proteins.[241,245] Fairly high separation factors were obtained with continuous operation. When a single resin is used, complicated cycles have been developed.[234,239] One such cycle[239] is shown in Figure 4-17. The circulation steps are used to make the concentrations and pH values equal. These protein separations by pH parametric pumping use pH values which bracket the isoelectric point of the desired protein. The protein will either strongly adsorb or not adsorb at all depending on the pH. The separation can be improved by applying an electrical field during the upflow steps.[514]

Adsorption or ion exchange is not necessary to develop a parametric pump. A thermal mode size exclusion parametric pump has been developed.[612] The apparent cause of the

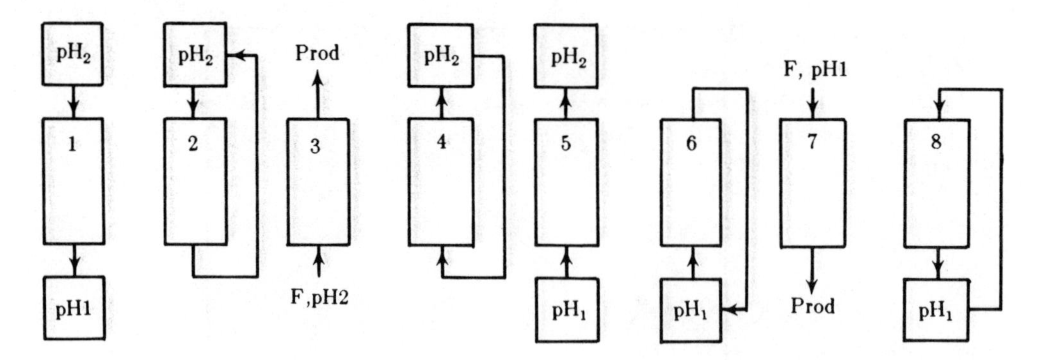

FIGURE 4-17. Cycle for pH parametric pumping protein separation.[239] Step 1, downflow; 2, circulate; 3, bottom feed; 4, circulate; 5, upflow; 6, circulate; 7, top feed; 8, circulate.

separation was changes in porosities of the size exclusion gel when the temperature was varied. If there is no adsorption, u_s is given by Equation 2-28 with $\Delta q/\Delta c = 0$. Changes in either α or ϵ will change the solute velocity.

Two-adsorbent systems with different selectivities can be used for completely separating a binary mixture or for fractionating two solutes.[1064] If only one adsorbent is available, fairly complex cycles can be used for fractionation by PP,[1100] but cycling zone adsorption appears to work better (see Section IV).

Electrochemical PP[774-777] has been used for removing sodium chloride from water and could potentially be used for desalting brackish water. The amount adsorbed on the carbon depends on the electrostatic charge and a parametric pump can be operated electrochemically. Since there are no Faradic reactions and no electron transfer, the energy usage could be very low. This technique is still under development.

Parametric pumping has been applied to the separation of gas systems.[505,551,552,788] The amount of separation achieved has been significantly less than that predicted by the local equilibrium theory.[788] Problems with gas systems are high dispersion, low heat transfer rates, and in closed systems raising the temperature to decrease adsorption also increases the pressure which increases adsorption. A combined PSA thermal PP process which eliminates the problems has been developed.[505] PSA and cycling zone adsorption (CZA) seem to work better with gas systems than thermal PP.

Other types of equipment have been used for PP. Instead of using pumps, an oscillating gravity flow system can be used.[844] Although simple in concept, in practice unequal flow in the two columns caused the separation to deteriorate. ''Tea bag''[959] or ''basket''[441,703] systems pack the sorbent in a bag or basket and then move this basket back and forth between the hot and cold reservoirs. This is a one-stage system and the amount of separation is limited. PP has also been extended to extraction systems[817,1038] and electrodialysis.[976-978]

Adsorptive-distillation can be considered a type of three-phase parametric pump.[604,841,842] Each stage undergoes the cycle:

- Adsorption on solid from liquid phase (equilibrium)
- Drain
- Vapor phase desorption using heat and vacuum
- Charge with liquid

This cycle is thus similar to drying of liquids (see Chapter 3 [Section IV.A]) except it is operated more like distillation. The vapor from desorption of stage n − 1 is condensed and mixed with the liquid from draining stage n + 1. This is then the charge to stage n for the next cycle. The flow of liquid and vapor in the opposite directions gives a countercurrent-type flow. At cyclic steady state the mass balances can be rearranged into forms which are essentially the same as the operating equations for distillation. The system was conveniently analyzed with a modified McCabe-Thiele diagram using a pseudo-equilibrium curve.[841,842] The pseudo-equilibrium depends on adsorption equilibrium and not vapor-liquid equilibrium as does distillation. Thus, adsorptive distillation may be favorable when relative volatilities are close to one, but selectivity in adsorption is large. Since only theoretical results have been reported, several important questions remain unanswered. These include equipment configurations which allow convenient construction and scale-up, and methods for energy reuse.

PP systems work best when there is a large change in the equilibrium when the thermodynamic variable is varied. For thermal cycling, equilibrium isotherms will often follow an Arrhenius relationship. Large values of ΔH give large changes in the isotherm. For many adsorbents the entropy change on reaction, $\Delta S \approx 0$. Then, if $|\Delta H|$ is large, the free energy change of the reaction, $|\Delta G|$, is also large and the adsorption is essentially irreversible. To have a large $|\Delta H|$ and a modest $|\Delta G|$, the entropy change on adsorption, ΔS, must be a large negative value. This is the case for some biochemical affinity systems.[230] For example, some monoclonal antibody systems for antigen purification have changes in the dissociation complex of two orders of magnitude when temperature is increased from 4 to 43°C. If denaturing does not occur, similar biochemical systems may present future applications for PP and CZA. Affinity chromatography has been tried in PP,[225,895] but the separations obtained were not outstanding. Careful choice of chemical systems for PP should result in much more impressive separations.

E. Commercial Use of Parametric Pumping: The Sirotherm Process

The one advertised commercial use of PP was the Sirotherm process[69,144,146,147,152,153,155,198,535,897,963,966] developed by CSIRO, ICI Australia, and Australian Technology Engineering and Processes (AUSTEP). The Sirotherm process uses a combination weak acid/weak base ion exchange resin which can be thermally regenerated.[574] The regeneration occurs because of the more than 30-fold increase in the ionization of water when the temperature is raised from 25 to 85°C. The major problems in development of the Sirotherm process have been in resin development. The earlier resins oxidized and lost their capacity fairly rapidly.

A variety of operating techniques[69,144,146,147,152,153,155,198,535,897,963,966] have been tried with Sirotherm. In small-scale plants the operation used counter-flow regeneration using the heated tails as regenerant. This is a form of recuperative mode PP although it was apparently developed independently of Wilhelm's work. Several small demonstration plants using this cycle have been operated in Australia for desalinating brackish water.[147] A temperature difference of about 60°C was used. Cycle period was about 1 hr and the yield of purified water was from 70 to 90%. The largest fixed-bed Sirotherm was a 600,000 ℓ/day unit at ICI Australia's Osborne plant which started up in 1976. This Sirotherm unit has apparently been shut down. It is convenient to combine Sirotherm with other ion exchange processes. For example, a presoftening system before the Sirotherm system is convenient since the presoftener increases the capacity of the Sirotherm process, and the presoftener can be regenerated with the concentrated Sirotherm effluent.

Co-flow operation, which is essentially a CZA-type procedure, has also been used.[198] In this application, co-flow regeneration was less efficient than counter-flow regeneration. The major problems with this plant[198] were in the pretreatment section. Because of the very low oxygen levels required by the early resins, vacuum degassing was required. This took 30% of capital and 40% of operating costs.[198] Newer resins are more tolerant of oxygen.[146,147,963]

In large-scale systems continuous operation appears to be cheaper than fixed-bed plants.[69,147,198,963] The major advantages of continuous plants are it is much easier to exclude

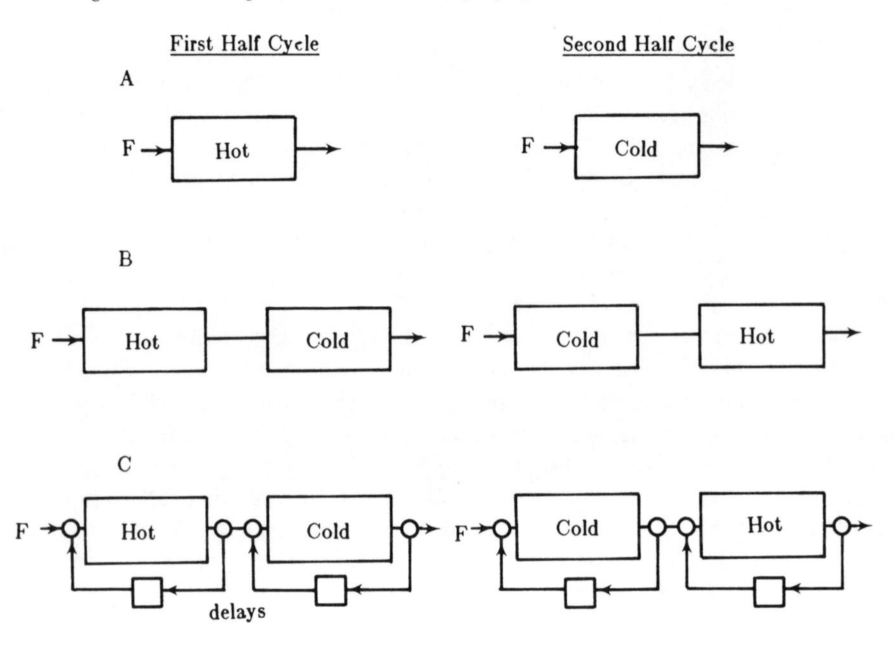

FIGURE 4-18. Direct thermal mode cycling zone adsorption. (A) Single zone; (B) two zone; (C) two zone with recycle and delays.

oxygen which translates into longer resin life, and the equipment will be smaller. These advantages are important since fixed-bed Sirotherm does not appear to be economic when compared to reverse osmosis or electrodialysis.[198,966] Ashai-type continuous systems with normal Sirotherm resins, pipe-flow reactors, and fluidized beds with magnetic resins have been used (see Chapter 6). The future of this process appears to focus on continuous plants, and thus PP apparently has lost its one commercial application.

IV. CYCLING ZONE ADSORPTION (CZA) AND CHROMATOTHERMOGRAPHY

Cycling zone adsorption (CZA) and chromatothermography were developed independently and had different purposes. CZA was developed as a large-scale system for bulk separation while chromatothermography was developed as a type of analytical chromatograph. I have juxtaposed these methods because they are very similar. Both processes use unidirectional flow and both methods can use focusing to increase the separation.

A. Direct Mode Cycling Zone Adsorption

The basic ideas of CZA were first presented by Pigford and co-workers,[74,806] and several reviews have been written.[450,1051,1058] In the direct thermal mode the columns are externally heated or cooled, and the temperature is cycled between the hot and cold temperatures. A single zone device is shown in Figure 4-18A. Note that feed enters the column continuously and reflux is not used. Since the change of the isotherm with temperature is usually modest, the separation obtained in a single zone will be modest. This separation can be multiplied by adding additional zones as shown in Figure 4-18B. When the adsorbent is hot, solute is weakly adsorbed and a concentrated solution exits the column. If this solution is sent to a cold column which already contains solute, the concentration will be increased further when the column is heated since solute must desorb. Thus a peak of material which has been concentrated twice will exit the hot column in Figure 4-18B if the fluid and adsorbent have undergone two temperature changes from cold to hot. Thus correct timing is essential.

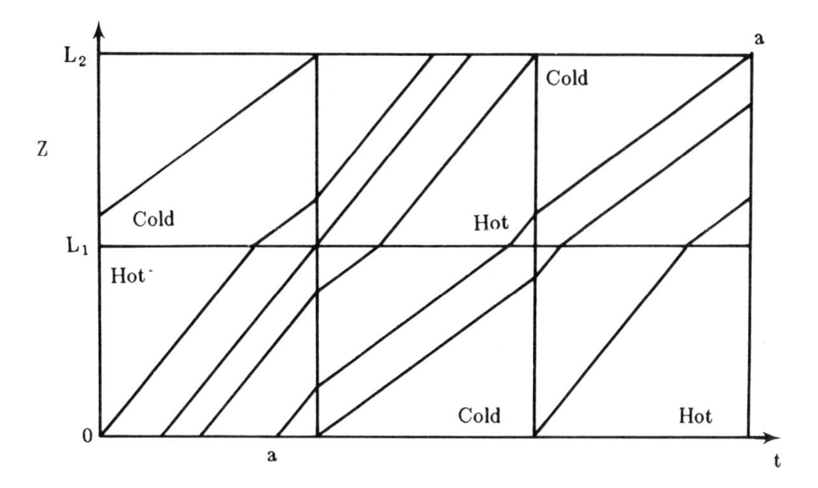

FIGURE 4-19. Solute movement theory for two-zone, direct mode CZA with linear isotherms.

Additional separation can be obtained by controlled recycle of portions of the product stream[388,850] as shown in Figure 4-18C. The concentrated peaks leaving the column when the column is hot need to be delayed during recycle so that the peaks enter the column while it is cold. Then, when the column is heated for the next cycle, the concentrated peaks will be further concentrated. Delay can be obtained with a column packed with an inert solid or with a coil of small-diameter tubing. Further separation can be obtained by packing the bed with alternating layers of packing with different temperature dependence.[388] By using two adsorbents with different selectivities, a modified CZA system with recycle can be used to fractionate mixtures of solutes.[1064]

The solute movement model is easily applied to CZA.[74,806,1058] The solute movement diagram which results is shown in Figure 4-19 for a system with linear isotherms. The basics of the calculation are essentially the same as for PP except there is no flow reversal. When fluid enters zone 2, which is at a different temperature than zone 1, the solute velocity changes because k_i changes; however, the fluid concentration does not change since the adsorbent in zone 2 has not changed temperature. By counting the number of times the adsorbent changes temperature, the concentration of any characteristic can be determined. Thus, characteristic a-a undergoes two changes of adsorbent temperature from hot to cold and, according to Equation 2-46, its concentration will be doubly diluted. The diagram for the single-zone system shown in Figure 4-18A will look like the zone 1 portion of Figure 4-19. Extension to nonlinear isotherms is straightforward.

Closed form solutions can be obtained for systems with linear isotherms. When the system is properly timed, the maximum separation factor is[464]

$$\frac{c_{conc}}{c_{dilute}} = 2\left[\frac{1 + k(T_c)}{1 + k(T_h)}\right]^m - 1 \qquad (4\text{-}11)$$

where m is the number of zones. This is the same exponential increase that is observed with batch PP but the number of zones has replaced the number of cycles. If dispersion or nonlinear isotherms are included, the batch direct mode PP can ultimately obtain more separation than CZA. The reason is that because of flow reversal, PP keeps the product end of the column free of solute, and the mass transfer zone (MTZ) stays inside the column. In CZA the MTZ moves through the column and all parts of the column see high and low solute concentrations. If there is significant tailing or dispersion complete cleanup is difficult. However, as we will see in the next two sections, CZA has advantages in some cases.

A number of experimental separations have illustrated direct mode CZA. These include the separation of acetic acid from water on activated carbon with one[74,806] and two zones,[74] separation of methane from helium on carbon,[806] separation to toluene and *n*-heptane on silica gel,[850] separation of oxygen from air with 5 Å molecular sieve,[1010] separation of methane from nitrogen on silica gel,[142] and the removal of sodium chloride from water using mixed beds.[420,633,903] Direct mode CZA has been extended to extraction where diethylamine was removed from water using toluene as solvent.[1039] Direct mode CZA can also be operated by changing electrical potential of the adsorbent.[347,644] In all cases the outlet product exits as a peak concentration followed by a valley of low concentration. If timing is incorrect, shoulders with a concentration equal to the feed concentration will appear.

B. Traveling Wave CZA and Focusing

Thermal traveling wave CZA is analogous to recuperative mode PP; the feed is heated or cooled and a thermal wave passes through the column. The equipment for this is quite simple and is illustrated in Figure 4-20A. Additional separation can be obtained by adding additional zones and heat exchangers in series or with recycle, but timing is crucial. A lag may have to be built into the connection between zones or in the recycle line. For a liquid system where $u_{th} > u_s$ the concentrated portion of the wave should always undergo temperature changes from cold to hot while the dilute part of the wave should go from hot to cold.

Analysis by the solute movement theory is straightforward. For liquids where u_{th} is usually greater than u_s, Equation 2-45 predicts there is more concentration in a single-zone traveling wave system than in a single-zone direct mode system (Equation 2-46). This is collaborated by experimental results for thermal traveling wave CZA with liquid systems.[74,601,806] This is also true in thermal traveling wave cycling zone extraction.[1041] Similar results are obtained when a thermodynamic variable other than temperature is used if the wave velocity is greater than the solute velocity. This was observed for separation of dipeptides from water using waves of dichloroacetic acid.[760]

As the thermal wave velocity* approaches the solute wave velocity the solute movement theory (Equation 2-45) predicts better separation. The solute movement diagram for a linear system with $u_s(T_H) > u_{th} > u_s(T_c)$ is shown in Figure 4-20B. Once a solute wave intersects the thermal wave it is "trapped" there. If the solute moves faster than the thermal wave, the solute will enter a cold region and must slow down. If the solute moves slower than the thermal wave, the solute will be heated up and must speed up. An infinite concentration is predicted at the thermal wave and a zero concentration everywhere else. This result is physically impossible. Equation 2-45 predicts negative concentrations under these circumstances. The reason why the local equilibrium theory predicts physically impossible results is the linear isotherm is not adequate at high concentrations. When a nonlinear favorable isotherm such as a Langmuir isotherm is used, physically reasonable results are obtained. This is shown in Figure 4-20C.[1044] The concentration at the thermal wave builds up to a high enough level so that the solute wave velocity in the cold region becomes greater than the thermal wave velocity. This leads to a region of very high concentration, c_I which exits *ahead* of the thermal wave. This is shown in Figure 4-20D.[1044] The wave velocities in Figure 4-20C are in the order,

$$u_s(T_H,c_F) > u_s(T_c,c_I) > u_{sh}(T_c,c_I) > u_{th} > u_s(T_c,c_F) > u_{sh}(T_c,c_F) > u_s(T_c,0) \quad (4\text{-}12)$$

A long period of zero concentration will follow the diffuse wave. Thus with the appropriate thermal wave velocity a single-zone system can achieve an excellent separation. This con-

* Other thermodynamic variables can be used instead of temperature.

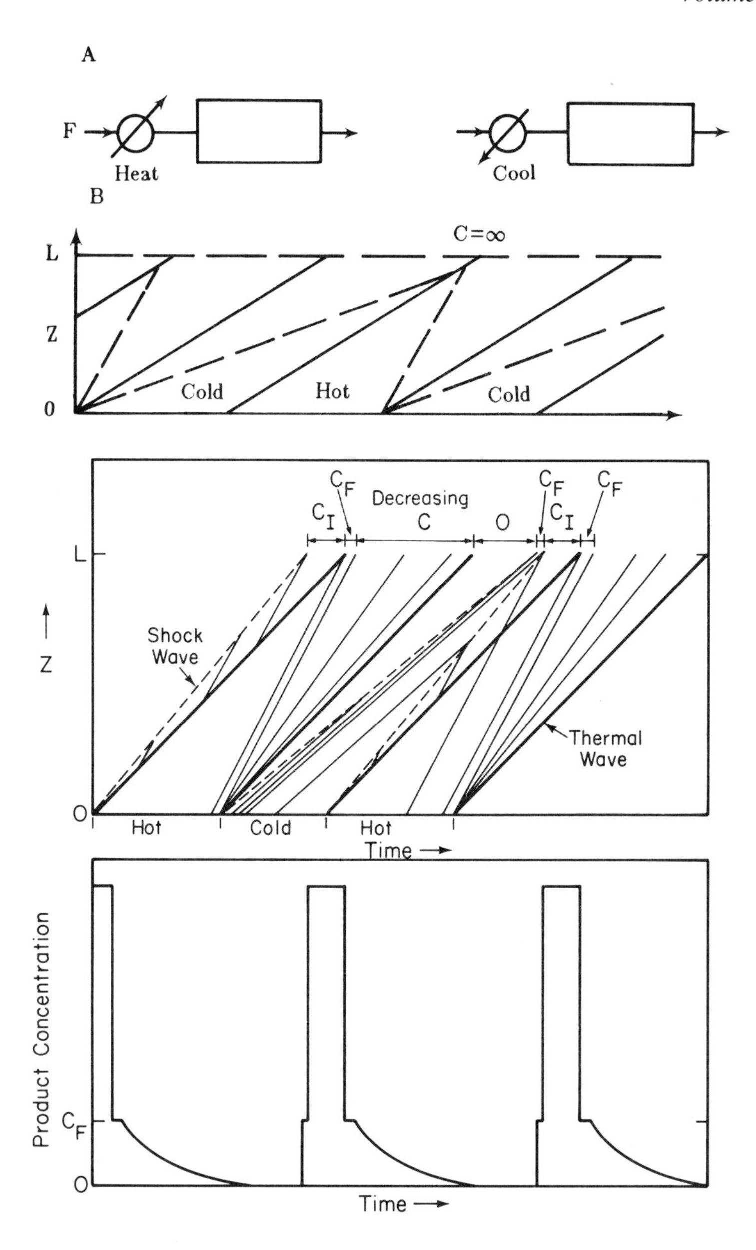

FIGURE 4-20. Single-zone traveling wave cycling zone adsorption. (A) Equipment; (B) solute movement theory for linear isotherms with $u_s(T_H > u_{th} > u_s(T_c)$; (C) nonlinear favorable isotherms; (D) product concentrations predicted in part C. (From Wankat, P. C., *Chem. Eng. Sci.*, 32, 1283, 1977. With permission.)

centrating effect has been called focusing, trapping, or the Guillotine effect, and it is a powerful tool.

In liquid systems the natural thermal wave velocity is usually too high to obtain focusing, but other thermodynamic waves will illustrate focusing. The separation of fructose or glucose from water using pH waves is an examples of this.[194,331] The removal of fructose from water is shown in Figure 4-21.[759] These results are at the cyclic steady state. Note that there is a long period of almost complete removal of the fructose and a concentrated peak which exits ahead of the high pH wave. The theoretical results were obtained with an equilibrium staged theory with four stages.[759] Focusing was also obtained in cycling zone extracting of dyes from water using Na_2CO_3 concentration as the cyclic variable.[1063]

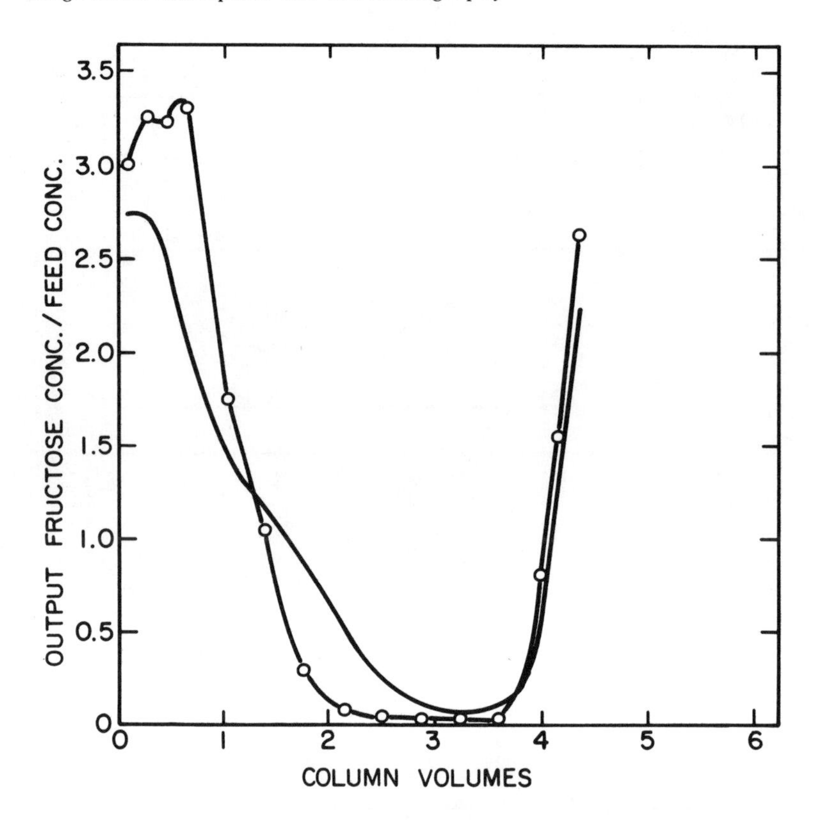

FIGURE 4-21. Separation of fructose from water by traveling wave CZA using pH waves. o——o, Data ——. equilibrium staged theory. (From Nelson, W. C. et al., *Ind. Eng. Chem. Fundam.*, 17, 32, 1978. With permission.)

Focusing can occur in gas systems but the analysis is more difficult since the concentration and thermal waves are usually coupled by the large heat of adsorption (see Chapter 2 [Section IV.D.2]). Focusing was observed in adsorption studies.[113,784,986] In thermal traveling wave CZA in gas systems focusing has been more extensively studied than in liquid systems.[537-541,987,992] If a temperature wave of the right amplitude is used for a gas within the correct concentration range, focusing will occur. This is illustrated in Figure 4-22[541] for the removal of *n*-pentane from isopentane on 5 Å molecular sieve. Note that the cyclic steady state is rapidly attained. The product is roughly half pure iso-pentane and half a pulse concentrated in *n*-pentane. The pulse exits before the peak in the thermal wave. Because of energy losses, the thermal wave is considerably attenuated. Separation will be easier in a large-diameter column where the heat losses are proportionally less.

The conditions for occurrence of focusing in gas systems have been extensively studied.[539-541,991] Essentially, the hot and cold temperatures must bracket the "reversal temperature" T_r and the concentration of strongly adsorbed compound must not be too high. The reversal temperature is defined as the temperature at which the initial slope of the isotherm equals the ratio of the heat capacities of packing to mobile phase.[541]

$$\left(\frac{dq}{dp}\right)_{\substack{\bar{p}=0 \\ T=T_r}} = \frac{C_{ps}}{C_{pg}}\left(\frac{M_1}{M_0}\right)\frac{1}{p} \tag{4-13}$$

where $\bar{p}$ = partial pressure, p = total pressure, M_1 and M_0 are the molecular weights of the adsorbed and nonadsorbed species, C_{ps} and C_{pg} are the mass specific heats of the solid and the nonadsorbed gas. If we define f(p) as

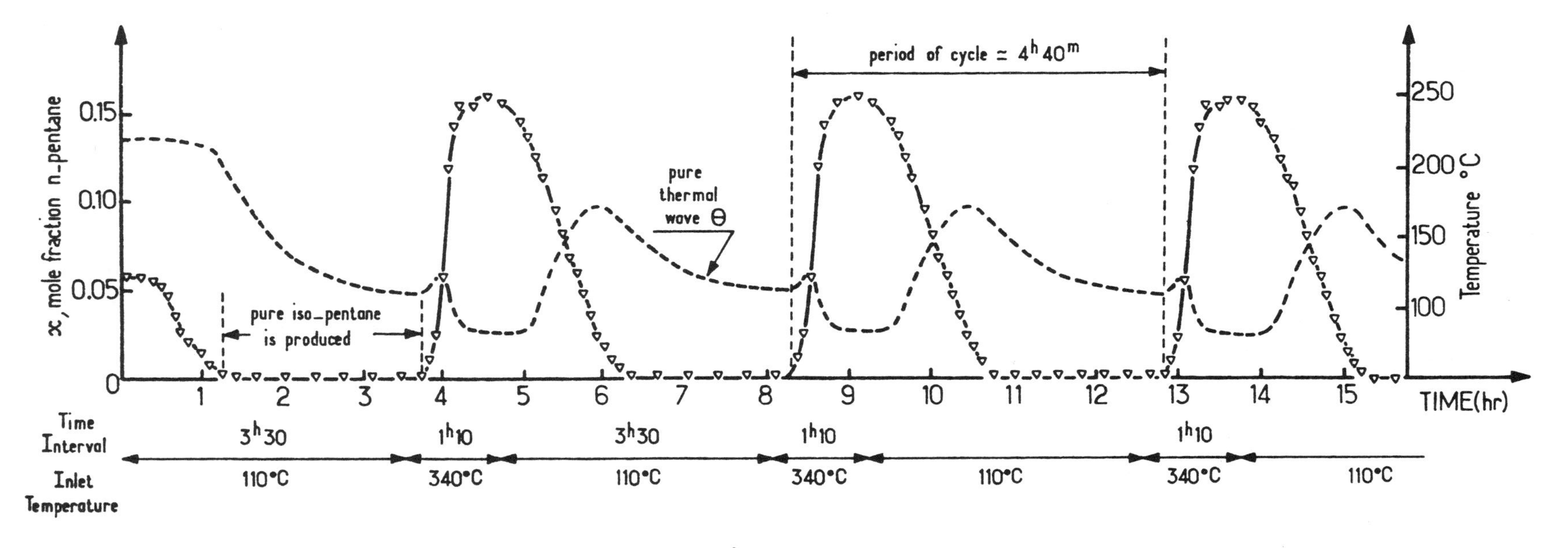

FIGURE 4-22. Traveling wave CZA of *n*-pentane and iso-pentane on 5 Å molecular sieve. (From Jacob, P. and Tondeur, D., *Chem. Eng. J.*, 31, 23, 1985. With permission.)

$$f(p) = \frac{M_I}{M_0} \frac{C_{ps}}{C_{pg}} \frac{\bar{p}}{p} \tag{4-14}$$

this function can be plotted on a graph of q vs. p along with the isotherms. If the cold isotherm lies entirely above the curve and the hot isotherm lies entirely below f(p) up to the partial pressure of the feed, then focusing can occur. This condition sets limits on hot and cold temperatures and the feed concentration. The recovery of pure component can be increased by using recycle.[541]

In Section II.A we saw that PSA systems had an optimum operation temperature. Gas CZA systems will have an optimum operating pressure. As pressure increases the equilibrium loading increases and with constant mass flow rate the gas velocity decreases which means a longer contact time.[992] These phenomena favor increased adsorption. However, pore diffusion will be slower at higher pressures, particularly for tight pores. Adsorbents with large pores will have high optimum pressures while those with small pores will have lower optimum pressures.

The gas separations with focusing have considerable promise for large-scale application. Large systems have been designed, but apparently none have been built yet.

C. Multicomponent CZA

Focusing can be used to develop multicomponent CZA systems. Essentially, the idea is to focus each solute at a different wave boundary. This was first illustrated theoretically[1043] using the solute movement theory and has since been shown experimentally[331,376,1036,1063] and with more detailed theories.[331,759] The basic idea is illustrated in Figure 4-23 for the local equilibrium theory with linear isotherms. (Obviously, nonlinear isotherms are physically more realistic as shown in Figure 4-20.) A series of steps in temperature is input as in Figure 4-23A. The temperature levels are selected so that

$$u_A(T_2) > u_A(T_1) > u_{th} > u_A(T_c)$$

$$u_B(T_2) > u_{th} > u_B(T_1) > u_B(T_c) \tag{4-15}$$

Then solute A will be focused at the $T_c - T_1$ boundary and solute B at the $T_1 - T_2$ boundary. If a continuous temperature gradient were used, each solute would concentrate at the temperature where the wave velocity of the solute equaled the thermal wave velocity.

There are difficulties in putting this idea into practice. In many liquid systems the natural thermal wave velocity is too high and the criterion for focusing cannot be met. If waves other than temperature are used, focusing can occur. This was first illustrated by fractionating dyes using cycling zone extraction with Na_2CO_3 concentration as the cyclic variable.[1063] In a packed column partial fractionation of fructose and glucose was obtained using pH as the cyclic variable.[331] Considerable dispersion of the pH wave and interaction between the sugars decreased the separation somewhat. The method is applicable to gases also. For the system hydrogen, methane, and hydrogen sulfide on activated carbon, a region of almost pure hydrogen, a region quite concentrated in methane with no H_2S, and a region with all three gases was obtained.[1036] This would be useful for feeds containing small amounts of H_2S.

Excellent fractionation of three solutes was obtained by overriding the natural thermal wave velocity.[376] This was done by using a column with a jacket divided into sections as shown in Figure 4-24A.[376] The temperatures in each jacket section are changed as shown in Figure 4-24B. This direct mode system simulates a traveling wave with any desired thermal wave velocity and is similar to chromatothermography (see next section). One of the resulting separations for the separation of acenaphthylene, anthracene, and pyrene is shown in Figure 4-24C.[376] Note that the feed enters the column continuously, and the outlet

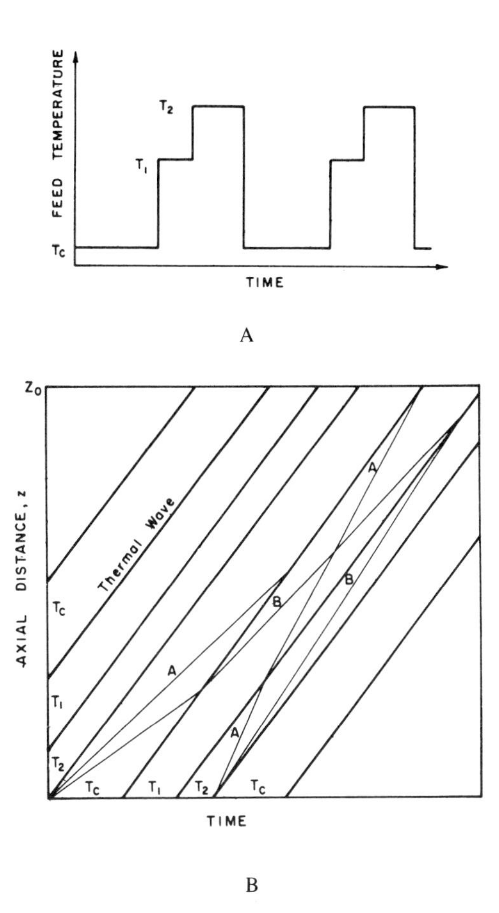

FIGURE 4-23. Multicomponent CZA. (A) Feed temperature; (B) solute movement diagram for linear isotherms. (From Wankat, P. C., *Ind. Eng. Chem. Fundam.*, 14, 96, 1975. With permission.)

products exit as peaks which are considerably concentrated. These results are shown at the cyclic steady state which was usually reached after the fourth cycle of operation. A direct mode system like this will require more energy than the traveling wave mode. In large-scale systems slow radial heat transfer might force one to build the column as the tubes of a shell-and-tube heat exchanger. Currently, this direct mode system does represent the best CZA fractionation.

D. Chromatothermography

Chromatothermography was developed as a gas chromatography method for continuously fractionating several solutes.[746,771,773,993,1107,1119,1120,1123] Chromatothermography uses a furnace which moves along the column. This furnace develops both radial and axial temperature gradients. The axial gradient can move at any desired velocity. Each solute will focus at the temperature where its velocity is equal to the velocity of the furnace. Thus, if a dilute feed continuously enters the column the solutes will be fractionated and separated. The solute movement theory with nonlinear isotherms has been applied to chromatothermography.[993] Small-scale gas chromatothermography systems were commercially available in the U.S.S.R. for a period.[746] Current interest in the method appears to be quite low, although specialty applications as an analytical method have been reported.[771,1107]

The forced traveling wave CZA system[376] is obviously quite similar to chromatother-

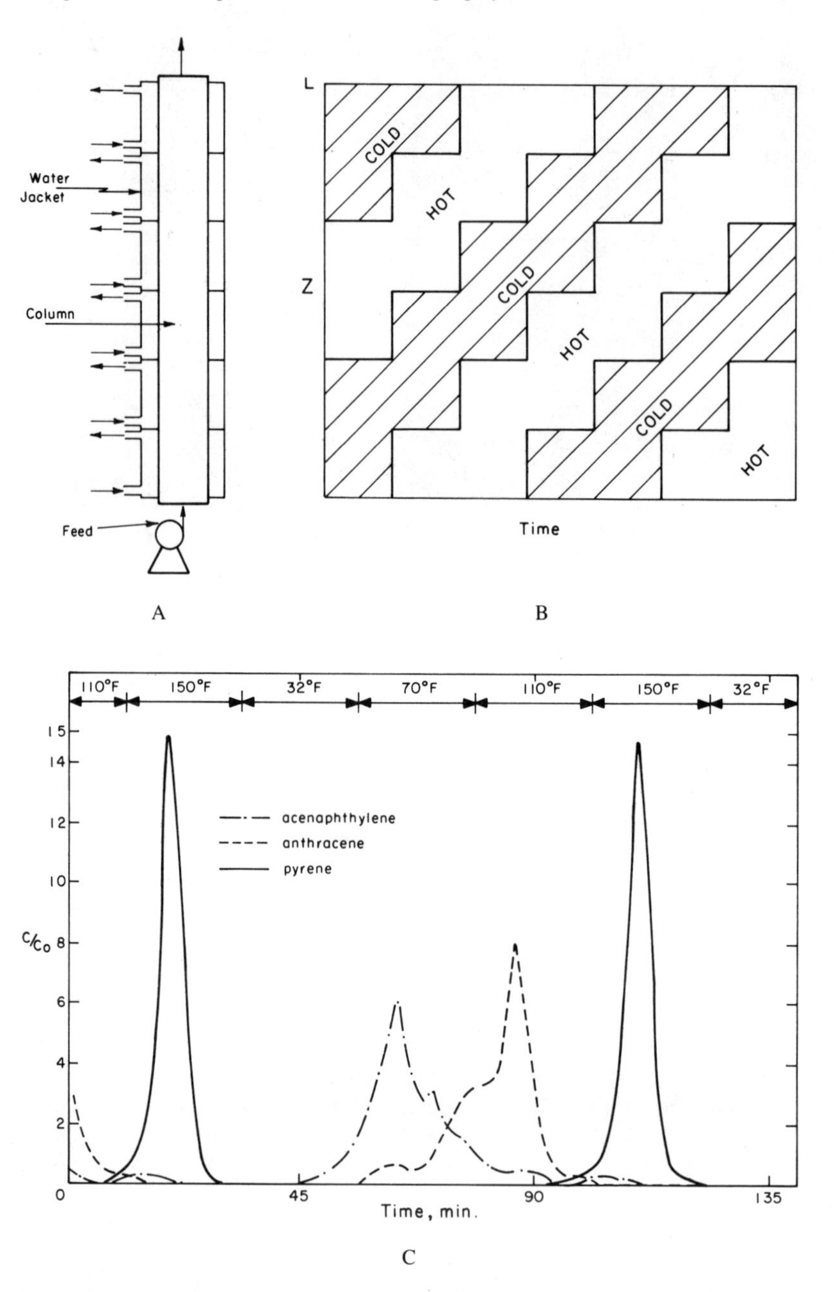

FIGURE 4-24. Forced traveling wave, multicomponent CZA. (A) Column system; (B) theoretical temperature profile; (C) separation of 3 solute system in 18-section column. Solvent is 2-propanol and adsorbent is polyvinyl pyrrolidone resin. Temperatures shown are jacket temperatures of the last section in the column. (From Foo, S. C. et al., *Ind. Eng. Chem. Fundam.*, 19, 86, 1980. With permission.)

mography. The CZA system would be easier to scale-up. Traveling wave CZA systems would be even easier to scale-up and would have lower energy requirements, but the thermal wave velocity may be too high or too low. There is more interest in the CZA-type systems for industrial gas separations than in trying to scale-up chromatothermography.

Chromatographic distillation[1121,1122] has some similarities with chromatothermography in that a temperature gradient is often employed. However, chromatographic distillation usually uses inert packings and thus is really a distillation operated like a chromatography. The separation obtained depends on the volatility of the compound. Analytical applications have been reported.

V. THEORIES FOR CYCLIC SEPARATIONS

The most commonly used theory for PSA, PP, and CZA has been the local equilibrium or solute movement theory. This theory is simple enough to allow simple analytical or graphical solutions. Thus this theory is very useful for visualizing the separation and as a conceptual tool in developing new cycles. The applications of this theory are outlined in Table 4-1.

The major drawbacks of the local equilibrium theory are it does not include dispersion or rate limitations. Thus it is inadequate for design or when the separation is based on kinetics instead of equilibrium. A variety of more detailed models have been applied to cyclic separations and are outlined in Table 4-2. These theories can be classified as either continuous contact or staged-type models. The continuous contact models treat the column as a continuum and solve the partial differential equations which result. Numerical integrations may first simplify the equations to ordinary differential equations or may numerically integrate the complete set of equations. Frequency response methods are an alternate method of solving the equations after making some limiting assumptions. They have been used for extensive studies of the optimum operating conditions of direct, thermal parametric pumps.

Staged and cell theories divide the column into a number of discrete stages or cells. Equilibrium staged models assume that each stage is at equilibrium; thus, they are not appropriate when the separation is based on kinetics instead of equilibrium. The equilibrium staged models are certainly appropriate for staged systems such as extraction, but are less useful for column systems where an HETP or number of stages must be measured. These models have been used to develop a useful graphical analogy between distillation and PP.[448,449,451,454,455] Mixing cell models also divide the column into sections, but they do not assume equilibrium. One modification, the STOP-GO algorithm,[962] has been used for extensive simulations and comparisons with experimental PP results. The STOP-GO method transfers all of the fluid from one cell to the next during the GO period. During the STOP the resulting ordinary differential equations are solved for mass transfer. In the limit of long times this model reduces to an equilibrium staged model with discrete transfer steps.[1038]

Usually, all of the detailed models will give approximately the same results if the parameters are adjusted accordingly. Several models have been compared and the relationships between parameters developed for PP.[338]

VI. THE FUTURE FOR CYCLIC SEPARATIONS

Obviously, pressure swing and vacuum swing adsorption are now and will remain popular separation methods for gases. I expect that the current trends towards systems with higher productivity or higher recoveries will continue. These two trends represent lower capital and lower operating costs, but they are not always compatible. The goal is to achieve both simultaneously. I also expect that PSA, VSA, or pressure-swing chromatography will be used for fractionation of gases.

Parametric pumping has not fulfilled its initial, perhaps oversold, promise. Except for a few special applications such as Sirotherm, I doubt that parametric pumping of liquid systems will have much large-scale application. For gases, systems combining aspects of PSA and PP[581] may be used on a large scale.

Cycling zone adsorption with focusing may well be used for large-scale separation of gases. Application to liquid separations is likely to be limited to specialized circumstances where the chemicals used for the traveling wave already occur in the manufacturing process.

APPENDIX A

NOMENCLATURE

Symbol	Definition
a,b	Langmuir equilibrium constants (see Equation 2-14)
a_{io}, B_{io}, β_i, a_{ij}, b_{ij}	Constants in combined Langmuir-Freundlich Equation 2-16)
a_p	Surface area of particles/unit volume
a_I, a_{II}	Interfacial area/volume between phases V and W, and W and L
A_1, A_2, A_{1i}, A_{2i}	Equilibrium constants (see Equation 2-12)
A_c	Cross-sectional area of column
A_w	Area of wall for heat transfer
A,B	Van Deemter equation constants (see Equations 2-117 to 2-119)
b_1, b_2, b_3	Constants in linear equilibrium (see Equations 6-10 and 6-28)
c	Concentration (mol/ℓ)
c	Total equivalents per liter of ions in liquid (see Equations 2-20, 2-21, and 2-23)
c_a, c_b	Concentrations after and before shock
c_{conc}, c_{dilute}	Concentrations in CZA (see Equation 4-11)
c_h	High concentration (see Figure 2-6)
c_i^*	Concentration inside pores
c_i, c_{init}	Initial concentration
c_{imax}	Maximum concentration in pulse
c_l	Low concentration (see Figure 2-6)
c_{RA}, c_{Ri}	Concentration of ion A on resin (see Equations 2-20 and 2-21)
c_S, c_{SM}, c_M	Constants in expansion of Van Deemter Equations 2-121 to 2-124
C, C_M, C_{SM}, C_S	Van Deemter equation constants (see Equations 2-117 to 2-119)
C_{PB}	Bulk heat capacity of solid (+ pore fluid) (see Equation 2-77)
C_{P_f}, C_{P_s}, C_{P_w}	Heat capacities of fluid, solid, and wall, respectively
C_{Rtotal}	Total resin capacity (see Equations 2-20,2-21, and 2-23)
d_f	Film thickness
d_p	Particle diameter
D_M	Molecular diffusivity
D_T	Total effective axial diffusivity (see Equation 2-112)
D_T	Thermal diffusivity
D_θ	Diffusivity in θ direction
$D_{\theta T}$	Thermal diffusivity in θ direction
$\mathrm{erf}(u)$	Error function (Equation 2-102)
E_D	Eddy diffusivity
F_i	Pulse of moles of feed (see Equations 2-84 and 2-86)
$f_{(p)}$	Function in Equation 4-14
g	Acceleration due to gravity
G	Mass velocity of gas
ΔG	Free energy change of reaction
h	Height in column
h_p	Heat transfer coefficient for particle (see Equations 2-69 and 8-5)
h_W	Heat transfer coefficient for walls (see Equation 2-68)
H, HETP	Height equivalent to theoretical place (see Equation 2-19)
ΔH	Heat of adsorption
$\Delta H'_{ads}$	Heat of adsorption as function of q'
HTU_{VW}, HTU_{LW}	Height of a transfer unit (see Equation 6-23 and 6-24)

i Subscript for solute
j Subscript for stage number
k Constant in LRC (Equation 2-11)
k_i Linear equilibrium constant (Equation 2-29)
k_{inert} Linear equilibrium constant for inerts (Equation 4-3)
k_i' $k_i V_s/V_m$ relative retention (Equation 2-127)
$\bar{k}_i'$ Average relative retention of two solutes
k_m Mass transfer resistance in mobile film (Equation 2-120)
k_s Mass transfer resistance in solid (Equation 2-120)
k_1, k_2 Rate constants (Equation 2-110)
k_A First order rate constant (Equation 3-11)
k_{A_i},k_{A_p},k_{B_d}, k_{B_p} Proportionality constants (Equation 5-5)
k_T Mass transfer coefficient (Equation 2-67)
K Constant in BET (Equation 2-8)
K Permeability (Equations 3-1 to 3-3)
K_A, K_i Langmuir equilibrium constant (Equations 2-4, 2-7, and 2-13)
K_{AB}, K_{DB} Equilibrium constants for ion exchange (Equations 2-20, 2-22, and 2-23)
K_d Fraction of pores available to molecule (Equation 2-3)
K_o Constant in Arrhenius relationship (Equation 2-5)
K_{LW}, K_{VW} Mass transfer coefficients (Equations 6-20 and 6-21)
K_1, K_2, K_3 Linear equilibrium constants (Equations 6-10 and 6-28)
l_{port} Distance between switching ports in SMB
L Column length
L Solid flow rate (kilogram of dry adsorbent per hour [Section II.A in chapter 6])
L_{MTZ} Mass transfer zone length
LUB Length of unused bed
m Mass of adsorbent (Equations 2-17, 2-19, 2-132, and 3-11)
m Constant in LRC (Equation 2-11)
m Number of zones in CZA (Equation 4-11)
M_s Mass stationary phase/plate (Equation 2-82)
M_1, M_0 Molecular weights of adsorbed and nonabsorbed species (Equations 4-13 and 4-14)
n Equilibrium constant (Equation 2-12)
n Number of components (Equations 2-7 and 2-62)
n Freundlich equilibrium constant (Equation 2-15)
n Number of cycles
n_{feed}, n_{purge} Moles of feed and purge gas, PSA (Equation 4-5)
n_o Original total number of moles (Equation 2-17)
n_1^s, n_2^s Number of moles adsorbed (Equation 2-17)
$(n_i^s)_m$ Number of moles for monolayer coverage (Equation 3-18)
N Number of equilibrium stages in plate models
N Loading on zeolite (Equation 2-11)
N_o Maximum loading on zeolite (Equation 2-11)
NTU_{VW}, NTU_{LW} Number of transfer units (Equations 6-23 and 6-24)
p Pressure
p_{after}, p_{before} Pressure after or before pressure change in PSA
$\bar{p}$ Partial pressure of solute
$\bar{p}_v$ Partial pressure of solute during vacuum regeneration

p°	Vapor pressure of pure component
p_H, p_L	High and low pressures
Pe_z	$Lv/(D_M + E_D)$, Peclet number
PSA	Pressure swing adsorption
q, q_i	Amount adsorbed (moles per kilogram)
q_a, q_b	Amount adsorbed after and before shock
$\overline{q}$	Total adsorption (Equations 2-50 to 2-52)
q_i°	Amount adsorbed from pure gas (Equation 2-9)
q_i'	Amount adsorbed and in stagnant fluid for single porosity model (Equations 2-73 to 2-81)
q_{max}	Maximum monolayer coverage (Langmuir Equation 2-6)
q_{mono}	Monolayer coverage (Equation 2-8)
q_{sat}	Bed capacity in equilibrium with feed
Q	Volumetric flow rate of gas (Equation 3-11)
r	Radial coordinate
r_o	Inner radius
R	Gas constant
R	Resolution (Equation 2-125)
R, R_T, R_B	Reflux ratios in parametric pump, R_T is at top and R_B is at bottom of column (Equation 4-10)
Re	Reynolds number, $vd_p\rho_f/\mu$
R_T	Cross-over ratio (Equation, 2-65)
Sc	Schmidt number, $\mu/D_M\rho$
Sh_m	Sherwood number for mass transfer, k_Ld_p/D_M
S_i	Slope of solute i in two-dimensional system (Equations 8-6 and 8-7)
S_{th}	Slope of thermal wave in 2D system (Equation 8-9)
t_{band}	Period band exits in displacement development (Equation 2-64)
t_{cycle}	Period of cycle (Equation 5-3)
t_F	Period feed is introduced (Equations 2-64 and 5-3)
t_{MTZ}	Time mass transfer zone takes to exit column
t_{port}	Time between switching ports in SMB
t_{Ri}	Retention time (Equation 2-94)
T	Temperature
T*	Temperature of stagnant fluid
T_{amb}	Ambient temperature
T_c,T_h	Cold and hot temperatures
T_f,T_s,T_w,T_{ref}	Temperatures of fluid, solid, wall, and reference
$u_{A_{cc}}$, $u_{B_{cc}}$	Solute velocity in countercurrent system (Equation 6-2)
u_{large}	Velocity of large molecules (Equation 5-7)
u_{port}	Average velocity of ports in SMB (Equation 6-34) and in moving feed chromatography
u_{small}	Velocity of small molecules (Equation 5-8)
u_s	Solute wave velocity
u_{sh}	Shock wave velocity
u_{Steam}	Velocity of steam wave
u_{th}	Thermal wave velocity
u_{wave}	Wave velocity when fluid velocity changes (Equations 2-55, 2-56, 2-57, and 2-59)
v	Interstitial fluid velocity
v_a, v_b	Velocity after and before wave
v_{in}	Inlet interstitial fluid velocity (Equations 2-53 and 2-54)

v_o	Velocity at inner radius of chromatofuge
$v_{solids, avg}$	Average solids velocity with intermittent flow (Equation 6-32)
v_{solid}	Solids velocity
v_{super}	Superficial fluid velocity
V	Volume carrier gas (Equation 2-82)
V	Upward flowing fluid flow rate (moles per hour [Section II. A, Chapter 6])
V	$\epsilon' v A_c t$, elution volume in Equations 2-101 to 2-108
$\overline{V}$	Volume solution to saturate column (Equations 2-101 to 2-108)
V_{col}	Volume empty column (Equation 2-1)
V_e	Elution volume (Equation 2-1)
V_{feed}	Volume of feed gas
V_{feed}, V_{purge}	Volumes of feed and purge gas at column conditions, PSA
V_i	Internal void volume (Equation 2-1)
V_o	External void volume (Equation 2-1)
V_M	Volume mobile phase/plate (Equation 2-82)
V_{purge}	Volume of purge gas
VSA	Vacuum swing adsorption
W	Weight column wall/length column
W	Flow rate fluid in pores (Chapter 6, Section II. A)
w_i	Dimensionless peak width (Equation 2-89)
w_p	Width of constant pattern (Equation 5-4)
w_d	Width of diffuse wave (Equation 5-4)
X	Deviation from peak maximum (Equation 2-96)
X(L, V)	Breakthrough solution (Equations 2-103 to 2-105)
Δx_1	Decrease in mole fraction in liquid (Equations 2-17 and 2-19)
x'	Weighted average mole fraction (Equations 6-9 and 6-26)
x_E, y_E	Mole fraction in equilibrium with bulk concentration, z
y_i, x_i	Mole fractions
y_{in}	Inlet mole fraction
y^*_{n+1}	Hypothetical concentration in Kremser equation (Equation 6-13)
y_{STM}	Mole fraction water in steam (Equations 3-9 and 3-10)
y_{w_b}	Mole fraction water in the bed before regeneration (Equations 3-9 and 3-10)
z	Mole fraction in phase W (fluid in pores)
z	Axial distance
z_{after}, z_{before}	Position of solute wave after and before pressure change in PSA

Greek

α	Interparticle porosity (Equation 2-1)
α_{12}	Separation factor (Equation 2-10)
α_{21}	$k'_2/k'_1 = k_2/k_1$ selectivity (Equation 2-128)
β	Term in Kremser equation (Equation 6-12)
β_A	Ratio inert to adsorbate held in column (Equation 4-3)
ϵ	Intraparticle porosity (Equation 2-1)
ϵ'	Single porosity (Equation 2-75)
γ	Labyrinth factor (Equation 2-115)
γ	Volumetric purge-to-feed ratio $= V_{purge}/V_{feed}$ in PSA
λ	Constant in expansion of D_T (Equation 2-116)
λ_w	Latent heat of water
ν_i	Dimensionless cumulative flow of mobile phase (Equation 2-85)

ρ_B	Bulk density including fluid in pores for single porosity equations (Equations 2-73 to 2-81)
$\bar{\rho}_f$	Molar density of fluid (Equations 2-50, 2-51, 2-52, and 2-53)
ρ_f	Fluid density
ρ_s	Structural solid density
σ,σ_t,σ_l	Standard deviation (Equations 2-90, 2-97, and 2-98)
τ	Deviation from center of constant pattern (Equation 2-133)
μ	Viscosity
θ	Angular coordinate
ω	Angular velocity

Other

$\mathcal{F}$	Term in Colburn equation (Equation 6-30)

REFERENCES

1. **Absolom, D. R.,** Affinity chromatography, *Sep. Purific. Methods,* 10, 239, 1981.
2. Ace Glass, Inc., *Liquid Chromatography,* Vineland, N.J., 1983.
3. **Adams, F. M.,** *p*-Xylene recovery by the Parex process, *Eur. Chem. News,* Oct. 13, 1972.
4. **American Water Works Assoc.,** Adsorption techniques practical applications, *Am. Water Works Assoc. J.,* 71,(11), 1979.
5. **Ahmed, Z. M.,** Separation of glucose-fructose-water systems via thermal parametric pumping, *(Paper F2-2), Reprints AIChE-GVC Joint Meeting,* Vol. IV, DECHEMA, Munich, 1974.
6. **Alexander, F. and McKay, G.,** Kinetics of the removal of basic dye from effluent using silica. I. Batch experiments. II. Fluidized bed experiments, *Chem. Eng. (London),* 319, 243, 1977.
7. **Alexis, R. W.,** Upgrading hydrogen via heatless adsorption, *Chem. Eng. Prog.,* 63(5), 69, 1967; *Chem. Eng. Prog. Symp. Ser.,* 63(74), 50, 1967.
8. **Alhert, R. C. and Gorgol, J. F.,** Competitive adsorption by components of landfill leachate, *Environ. Prog.,* 2, 54, 1983.
9. **Almeida, F., Costa, C., Rodrigues, A. E., and Grevillot, G.,** Removal of phenol from wastewater by recuperative mode parametric pumping, in *Physicochemical Methods for Water and Wastewater Treatment,* Pawlowski, L., Ed., Elsevier, Amsterdam, 1982, 169.
10. **Alsop, R. M., Barker, P. E., and Vlachogiannis, G. J.,** Efficient production of clinical dextran from dextran hydrolysate, *Chem. Eng. (London),* 399, 24, 1984.
11. **Altshuller, D.,** Design equations and transient behavior of the counter-current moving bed chromatographic reactor, *Chem. Eng. Commun.,* 19, 363, 1983.
12. Aluminum Company of America, *Activated Alumina Technical Bulletins,* F37-14370, GB2A, 1969.
13. American Norit Co., Inc. Introduction to Activated Carbon, Purification of Gases and Liquids using Norit Granular Activated Carbon, 420 Agmac Ave., Jacksonville, Fla., (no date).
14. Amicon Corp. Amicon Products for Separation Technology, (no date); Columns for Chromatgraphy in Industrial Purification and Laboratory Use, 1985, Danvers Mass.,
15. Amicon Corp., United States Ultrafiltration and Chromatography Price List, Danvers, Mass., Jan 1, 1984.
16. **Amphlett, C. B.,** *Inorganic Ion Exchangers,* Elsevier, Amsterdam, 1964.
17. **Anderson, R. A.,** Molecular sieve adsorbent applications: state of the art, *Molecular Sieves-II,* Katzer, J. R., Ed., ACS Symp. Ser., #40, American Chemical Society, Washington, D.C., 1977, 637.
18. **Anderson, R. E.,** Ion-exchange separations, in *Handbook of Separation Techniques for Chemical Engineers,* Schweitzer, P. A., Ed., McGraw-Hill, New York 1979, Sect. 1.12.
19. **Anderson, N.J., Dixon, D. R. and Swinton, E. A.,** Continuous ion exchange using magnetic shell resins. II. Dealkalisation — pilot study, *J. Chem. Technol. Biotechnol.,* 29, 332, 1979.
20. **Andrew, S. P. S.,** Some problems and opportunities in separation process technology in chemicals and manufacturing, *J. Sep. Process. Technol.,* 2(3), 13, 1981.
21. **Andrews, G. F. and Tien, C.,** The interaction of bacterial growth, adsorption, and filtration in carbon columns treating liquid wastes, *AIChE Symp. Ser.,* 71(152), 164, 1974.
22. **Andrews, G. F. and Tien, C.,** Bacteria film growth on adsorbent surfaces, *AIChE J.,* 27, 396, 1981.
23. **Andrews, G. F. and Tien, C.,** An analysis of bacterial growth in fluidized-bed adsorption columns, *AIChE J.,* 28, 182, 1982.
24. **Andrews, G. F. and Trapasso, R.,** A novel adsorbing bioreactor for wastewater treatment, *Environ. Prog.,* 3, 57, 1984.
25. **Angelino, H., Gibert, H., and Gardy, H.,** Study of a fluidized bed adsorber, *Inst. Chem. Eng. Symp. Ser.,* 30, 67, 1968.
26. **Anon.,** Ion exchange system, *Chem. Eng.,* 91(17), 59, 1984.
27. **Anon.,** Solvent recovery/activated carbon system saves $1.3 million, *Chem. Process.,* Jan. 1983.
28. **Anon.,** Activated-carbon fiber aids in solvent recovery, *Chem. Eng.,* 88(17), 63, 1981.
29. **Anon.,** Nonwovens with high activated carbon content, *Chem. Eng. News.,* 31, April 23, 1984.
30. **Anon.,** Adsorption unit is a cost-saver for petrochemical feedstocks, *Chem. Eng.,* 87(17), 49, 1980.
31. **Anon.,** Supercritical fluids: still seeking acceptance, *Chem. Eng.,* 92(3), 14, 1985.
32. **Anon.,** Waterfilters, *Consumer Reports,* 48(2), 68, 1983.
33. **Anon.,** Simple gas adsorber features self-regeneration, *Chem. Eng.,* 88(24), 71, 1981.
34. **Anon.,** Normal paraffin extraction, *Hydrocarbon Process.,* 53(9), 204, 1974.
35. **Anon.,** Molecular sieves offer low-cost oxygen source, *Chem. Eng.,* 77(21), 54, 1970.
36. **Anon.,** Heatless dryer solves freeze-up problems, *Actual Specif. Eng.,* 28(4), 125, 1972.
37. **Anon.,** Sieving air to yield nitrogen., *Chem. Eng.,* 9(22), 41, 1982.
38. **Anon.,** Hydrogen recovery by pressure swing adsorption, *Chem. Eng. (London),* 345, 395, 1979.
39. **Anon.,** Whatman Prep 25. Advertisement in *Process. Biochem.,* 19, March/April: 1985.
40. **Anon.,** Instrumentation 83, *Chem. Eng. News,* 60, March 21, 1983.

41. **Anon.,** Instrumentation 84, *Chem. Eng. News,* 72, March 19, 1984.
42. **Anon.,** Instrumentation 85, *Chem. Eng. News,* 58, March 11, 1985.
43. **Anon.,** Production Scale GC developed, *Anal. Chem.,* 52, 481A, 1980.
44. **Anon.,** Otherwise impossible separations may now be handled by gas chromatography, *Chem. Process.,* Nov. 1980.
45. **Anon.,** Gas chromatography tackles production jobs, *Chem. Eng.,* 88, 70, 1981.
46. **Anon.,** "Asahi Chemical Industries", *J. Magnet. Trend,* 1(3), 26, 1984.
47. **Anon.,** Beaded carbon ups solvent recovery, *Chem. Eng.,* 39, Aug. 29, 1977.
48. **Anon.,** Fluidized recovery system nabs carbon disulfide, *Chem. Eng.,* 70(8), 92, 1963.
49. **Anon.,** DuPont is expanding waste treatment service, *Chem. Eng. News,* 7, June 11, 1984.
50. **Anon.,** Linear mono-olefins (Pacol-Olex), *Hydrocarb. Process.,* 52(11), 145, 1973.
51. **Anon.,** Gas liquid recovery men watch solid adsorption for future, *Petrol. Process.,* 11(6), 15, 1956.
52. **Apostolopoulos, G. P.,** The parametric pump as a chemical reactor, *Ind. Eng. Chem. Fundam.,* 14, 11, 1975.
53. **Applebaum, S. B.,** *Demineralization by Ion Exchange,* Academic Press, New York, 1968
54. **Applegate, G.,** The thermal wheel or how to recover exhaust heat, *Chem. Eng. (London),* 284, 222, 1974.
55. **Arbuckle, W. B.,** PAC for priority pollutant spill control, *AIChE Symp Ser.,* 76(197), 61, 1980.
56. **Arden, T. V.,** *Water Purification by Ion Exchange,* Butterworths, London, 1968.
57. **Arehart, T. A., Bresee, J. C., Hancher, C. W., and Jury, S. H.,** Countercurrent ion exchange, *Chem. Eng. Prog.,* 52, 353, 1956.
58. **Aris, R.,** Equilibrium theory of the parametric pump, *Ind. Eng. Chem. Fundam.,* 8, 603, 1969.
59. **Aris, R. and Amundson, N. R.,** *Mathematical Methods in Chemical Engineering,* Vol. 2, *First Order partial Differential Equations with Applications,* Prentice Hall, Englewood Cliffs, N.J., 1973.
60. **Arnaldos, J., Casal, J., and Puigjaner, L.,** Le lit fluidise stabilise magnetiquement: caracteristiques et possibilities, *Powder Technol.,* 36, 33, 1983.
61. **Arnold, F. H., Blanch, H. W., and Wilke, C. R.,** Analysis of affinity separations. I. Predicting the performance of affinity adsorbers, *Chem. Eng. J.,* 30, B9, 1985.
62. **Asher, W. J., Campbell, M. L., Epperly, W. R., and Robertson, J. L.,** Desorb *n*-paraffins with ammonia, *Hydrocarb. Process.,* 48(1), 134, 1969.
63. **Aubert, J. H. and Tirrell, M.,** Flow rate dependence of elution volumes in size exclusion chromatography: a review, *J. Liq. Chromatogr.,* 6(Suppl. 2), 219, 1983.
64. **Ausikaitis, J. P.,** Pneumatic system air drying by pressure swing adsorption, in *Molecular Sieves-* II, Katzer, J. R., ACS Symp. Ser #40, American Chemicals Society, Washington, D.C., 1977, 681.
65. **Avery, D. A. and Tracey, D. H.,** The application of fluidised beds of activated carbon to solvent recovery from air or gas streams, *Inst. Chem. Eng. Symp. Ser.,* 30, 28, 1968.
66. **Avery, W. F. and Lee. M. N. Y.,** ISOSIV process goes commercial, *Oil Gas J.,* 60, 121, 1962.
67. AMF Molcular Separations Div., Zeta Prep Cartridge DEAE, QAE and SP Chromatrgraph, 400 Research Parkway, Meriden, Conn., 1/15/84.
68. AMF Molecular Separations Division, Zetaprep Chromatography, *Exchanges,* Vol. 1, 400 Research Parkway, Meriden, Conn., 06450, 1984.
69. AUSTEP, Clean Clear Water. New Processes in Water Treatment, Melbourne, Australia (no date).
70. **Bailey, J. E., Jr.,** Purification of D&C Red number 33 by preparative high performance liquid chromatography, *J. Chromatogr.,* 314, 379, 1984.
71. **Bailly, M. and Tondeur, D.,** Continuous countercurrent thermal ion exchange fractionation, *Inst. Chem. Eng. Symp. Ser.,* 54, 111, 1978.
72. **Bailly, M. and Tondeur, D.,** Two-way chromatography. Flow reversal in nonlinear preparative liquid chromatography, *Chem. Eng. Sci.,* 36, 455, 1981.
73. **Bailly, M. and Tondeur, D.,** Recycle optimization in non-linear productive chromatography. I, *Chem. Eng. Sci.,* 37, 1199, 1982.
74. **Baker, B. and Pigford, R. L.,** Cycling zone adsorption: quantitative theory and experimental results, *Ind. Eng. Chem. Fundam.,* 10, 283, 1971.
75. **Bandermann, F. and Harder, K.-B.,** Production of H_2 via thermal decomposition of H_2S and separation of H_2 and H_2S by pressure swing adsorption, *Int. J. Hydrogen Energy,* 7, 471, 1982.
76. **Barford, R. A., McGraw, R., and Rothbart, H. L.,** Large sample volumes in preparative chromatography, *J. Chromatogr.,* 166, 365, 1978.
77. **Barker, P. E.,** Production scale separations by continuous circular chromatography, *Br. Chem. Eng.,* 11, 202, 1966.
78. **Barker, P. E.,** Continuous chromatographic refining, in *Preparative Gas Chromatography,* Zlatkis, A. and Pretorius, V., Eds., Wiley-Interscience, New York, 1971, chap. 10.
79. **Barker, P. E.,** Development in continuous chromatographic refining, in *Developments in Chromatography,* Vol. 1, Knox, J. H., Ed., Applied Science Publishers, Barking, Essex, England, 1978, 41.

80. **Barker, P. E. and Al-Madfai, S.,** Continuous chromatographic refining using a new compact chromatographic machine, *J. Chromatogr. Sci.,* 7, 425, 1969.
81. **Barker, P. E., Barker, S. A., Hatt, B. W., and Somers, P. J.,** Separation by continuous chromatography, *Chem. Process. Eng.,* 52(1), 64, 1971.
82. **Barker, P. E., Bell, D. M., and Deeble, R. E.,** Mass and heat transfer effects in continuous production scale gas-liquid chromatographic equipment, *Chromatographia,* 12, 277, 1979.
83. **Barker, P. E., Bell, D. M., and Deeble, R. E.,** The mathematical modeling of a production scale sequential continuous chromatographic refiner unit, *Chromatographia,* 13, 334, 1980.
84. **Barker, P. E and Chuah, C. H.,** A sequential chromatographic process for the separation of glucose/fructose mixtures, *Chem. Eng. (London),* 371, 389, 1981.
85. **Barker, P. E. and Critcher, D.,** The separation of volatile liquid mixtures by continuous gas-liquid chromatography, *Chem. Eng. Sci.,* 13, 82, 1960.
86. **Barker, P. E. and Deeble, R. E.,** Production scale organic mixture separation using a new sequential chromatographic machine, *Anal. Chem.,* 45, 1121, 1973; reprinted in *Advances in Chromatography 1973,* Zlatkis, A., Ed., Chromatography Symposium, University of Houston, Tex., 1973.
87. **Barker, P. E. and Deeble, R. E.,** Sequential chromatographic equipment for the separation of a wide range of organic mixtures, *Chromatographia,* 8(2), 67, 1975.
88. **Barker, P. E., Ellison, F. J., and Hatt, B. W.,** Continuous chromatography of macromolecular solutes, in *Chromatography of Synthetic and Biological Polymers,* Vol. 1, Epton, R., Ed., Ellis Horwood, Chichester, 1978, 218.
89. **Barker, P. E., Ellison, F. J., and Hatt, B. W.,** A new process for the continuous fractionation of dextran, *Ind. Eng. Chem. Process Des. Develop.,* 17, 302, 1978.
90. **Barker, P. E., England, K., and Vlachogiannis, G. J.,** Mathematical model for the fractionation of dextran on a semi-continuous countercurrent simulated moving bed chromatograph, *Chem. Eng. Res. Design. Trans. Inst. Chem. Eng. (London),* 61, 241, 1983.
91. **Barker, P. E., Gould, J. C. and Irlam, G. A.,** Production scale chromatography for the continuous separation of fructose from carbohydrate mixtures, Inst. Chem. Engrs. Jubilee Conference Proceedings, London, April 1982, 19.
92. **Barker, P. E., Hatt, B. W., and Williams, A. N.,** Theoretical aspects of a preparative continuous chromatograph, *Chromatographia,* 10, 377, 1977.
93. **Barker, P. E., Howari, M. I., and Irlam, G. A.,** Further developments in the modelling of a sequential chromatographic refiner unit, *Chromatographia,* 14,192, 1981.
94. **Barker, P. E., Howari, M. I., and Irlam, G. A.,** The separation of fatty acid esters by continuous gas-liquid chromatographic refining, *J. Sep. Process. Technol.,* 2(2), 33, 1981.
95. **Barker, P. E. and Huntington, D. H.,** *Dechema Monogr.,* 62, 153, 1966.
96. **Barker, P. E. and Huntington, D. H.,** The preparative scale separation of multicomponent mixtures by continuous gas liquid chromatography, *J. Gas. Chromatogr.,* 4, 59, 1966.
97. **Barker, P. E. and Huntington, D. H.,** A circular chromatography machine for the preparative separation of liquid or gaseous mixtures, in *Gas Chromatography 1966,* Littlewood, A. B., Ed., Institute of Petroleum, London, 1967, 135.
98. **Barker, P. E., Irlam, G. A. and Abusabah, E. K. E.,** Continuous chromatographic separation of glucose-fructose mixtures using anion-exchange resins, *Chromatographia,* 18, 567, 1984.
99. **Barker, P. E. and Liodakis, S. E.,** Separation of fatty acid esters and fatty acid derivatives using preparative scale sequential gas-liquid chromatographic equipment, *Chromatographia,* 11, 703, 1978.
100. **Barker, P. E., Liodakis, S. E., and Howari, M. I.,** Separation of organic mixtures by sequential gas-liquid chromatography, *Can. J. Chem. Eng.,* 57, 42, 1979.
101. **Barker, P. E. and Lloyd, D. I.,** The separation of azeotropic and close-boiling mixtures by continuous gas-liquid chromatography, *Symp. Less Common Means of Separation,* Institute of Chemical Engineers, London, 1964, 68.
102. **Barker, P. E. and Thawait, S.,** Separation of fructose from carbohydrate mixtures by semi-continuous chromatography, *Chem. Ind.,* p. 817, Nov. 7, 1983.
103. **Barker, P. E., Vlachogiannis, G. and Alsop, R. M.,** Dextran fractionation on a semi-continuous countercurrent simulated moving bed chromatograph, *Chromatographia,* 17, 149, 1983.
104. **Barnebey, H. L. and Davis, W. L.,** Costs of solvent recovery systems, *Chem. Eng., 65(26),* 51, 1958.
105. **Barrer, R. M.,** *Zeolites and Clay Minerals as Sorbents and Molecular Sieves,* Academic Press, New York, 1978.
106. **Barrese, J. and Bacchetti, J. A.,** Compressed air dryer uses no external energy, *Chem. Process.,* 62, March 1983.
107. **Barry, H. M.,** Fixed-bed adsorption, Chem. Eng., 67(3), 105, 1960.
108. **Bartels, C. R., Kleiman, G., Korzun, J. N., and Irish, D. B.,** A novel ion-exchange method for the isolation of streptomycin, *Chem. Eng. Prog.,* 54(8), 49, 1958.

109. **Basmadjian, D.,** On the possibility of omitting the cooling step in thermal gas adsorption cycles, *Can. J. Chem. Eng.*, 53, 234, 1975.
110. **Basmadjian, D.,** Rapid procedures for the prediction of fixed-bed adsorber behavior. I. Isotherm sorption of single gases with arbitrary isotherms and transport modes: principles and recommended methods, *Ind. Eng. Chem. Process Des. Develop.*, 19, 129, 1980.
111. **Basmadjian, D.,** Rapid procedures for the prediction of fixed-bed adsorber behavior. II. Adiabatic sorption of single gases with arbitrary isotherms and transport modes, *Ind. Eng. Chem. Process. Des. Develop.*, 19, 137, 1980.
112. **Basmadjian, D.,** The adsorption drying of gases and liquids, in *Advances in Drying*, Vol. 3, Mujundar, A. S., Ed., Hemisphere, Washington, D. C., 1983, chap. 8.
113. **Basmadjian, D., Dan Ha, K., and Pan, C. Y.,** Nonisothermal desorption by gas purge of single solutes in fixed adsorbers. I. Equilibrium theory, *Ind. Eng. Chem. Process Des. Develop.*, 14, 328, 1975.
114. **Basmadjian, D. and Karayannopoulos, C.,** Rapid procedures for the prediction of fixed-bed adsorber behavior. III. Isothermal sorption of two solutes from gases and liquids, *Ind. Eng. Chem. Process Des. Develop.*, 24, 140, 1985.
115. **Begovich, J. M., Byers, C. H., and Sisson, W. G.,** A high capacity pressurized continuous chromatograph, *Sep. Sci. Technol.*, 18, 1167, 1983.
116. **Begovich, J. M. and Sisson, W. G.,** Rotating annular chromatograph for continuous metal separations and recovery, *Resour. Conserv.*, 9, 219, 1982.
117. **Begovich, J. M. and Sisson, W. G.,** A rotating annular chromatograph for continuous separations, *AIChE J.*, 30, 705, 1984.
118. **Begovich, J. M. and Sisson, W. G.,** Continuous ion exchange separation of zirconium and hafnium using an annular chromatograph, *Hydrometallurgy*, 10, 11, 1983.
119. **Belfort, G., Altshuler, G. L., Thallam, K. K., Feerick, C. P., Jr., and Woodfield, K. L.,** Selective adsorption of organic homologues onto activated carbon from dilute aqueous solutions: solvophobic interaction approach. IV. Effect of simple structural modifications with aliphatics, *AIChE J.*, 30, 197, 1984.
120. **Belter, P. A.,** Ion exchange and adsorption in pharmaceutical manufacturing, *AIChE Symp. Ser.*, 80(233), 110, 1984.
121. **Belter, P. A., Cunningham, F. L., and Chen, J. W.,** Development of a recovery process for Novobiocin, *Biotechnol. Bioeng.*, 15, 533, 1973.
122. **Bennett, B. A., Cloete, F. L. D., and Streat, M.,** A systematic analysis of the performance of a new continuous ion-exchange technique, in *Ion Exchange in the Process Industries*, Society of Chemical Industry, London, 1970, 133.
123. **Benson, R. E. and Courouleau, P. H.,** European practice for solvent recovery in printing industry, *Chem. Eng. Prog.*, 44, 459, 1948.
124. **Berg, C.,** Hypersorption process for separation of light gases, *Trans. AIChE*, 42, 665, 1946.
125. **Berg, C.,** Hypersorption design. Modern advancements, *Chem Eng. Prog.*, 47, 585, 1951.
126. **Berg, C.,** Design of process systems utilizing moving beds, *Chem. Eng. Prog.*, 51, 326, 1955.
127. **Berg, C.,** Design and control techniques in moving solids bed systems, in *Fluidization*, Othmer, D. F., Ed., Reinhold, New York, 1956, chap. 7.
128. **Berger, C. V.,** Match Isomer with Parex, *Hydrocarb. Process.*, 52(9), 122, 1973.
129. **Berglof, J. H. and Cooney, J. M.,** Scale-up of laboratory systems for large scale chromatography, Biotech 83, London, May 4—6, 1983.
130. **Berlin, N. H.,** U.S.P. 3,280,536, Oct. 25, 1966.

130a. **Bernardin, F. E., Jr.,** Experimental design and testing of adsorption and adsorbates, in *Adsorption Technology*, Slejko, F. L., Ed., Marcel Dekker, New York, 1985, 37.

131. **Bertsch, W.,** Methods in high resolution gas chromatography: two-dimensional techniques, *J. High Resol. Chromatogr. Chromatogr. Commun.*, 1, 85, 187, 289, 1978.
132. **Bhattacharya, K. N.,** Centrifugal chromatography — a rapid microanalytical technique for chemists, *Chem. Era*, 18(2), 47, 1980.
133. **Bidlingmeyer, B. A., Ed.,** *Solving Problems with Preparative Liquid Chromatography*, Elsevier, Amsterdam, in press.
134. **Bidlingmeyer, B. A. and Warren, F. V., Jr.,** Column efficiency measurements, *Anal. Chem.*, 56, 1583A, 1984.
135. **Bieser, H. J. and de Rosset, A. J.,** Continuous countercurrent separation of saccharides with inorganic adsorbent, *Die Starke*, 29(11), 392, 1977.
136. **Bieser, H. J., Winter, G. R., and Barbu, G.,** Parex boosts *p*-Xylene yield without added isomerization, *Oil Gas. J.*, 73(32), 74, 1975.
137. **Bilello, L. J. and DeJohn, P. B.,** Total process design and economics, in *Activated Carbon Adsorption for Wastewater Treatment*, Perrich, J. R., Ed., CRC Press, Boca Raton, Fla., 1981, chap. 7.
138. **Bird, R. B., Stewart, W. L., and Lightfoot, E. N.,** *Transport Phenomena*, Wiley, New York, 1960, 196.

139. **Birss, R. R. and Parker, M. R.,** High intensity magnetic separation, in *Progress in Filtration and Separation,* Vol. II, Wakeman, R. J., Ed., Elsevier, Amsterdam, 1981, 171.
140. **Block, H. S.,** A new route to linear alkylbenzes, paper presented at Symposium on *n*-paraffins, Institution of Chemical Engineers, Manchester, England, Nov. 16, 1966.
141. **Bloom, R., Jr., Joseph, R. T., Friedman, L. D., and Hopkins, C. B.,** New technique cuts carbon regeneratin costs, *Environ. Sci. Technol.,* 3, 214, 1969.
142. **Blum, D. E.,** Cycling Zone Adsorption Separation of Gas Mixtures, Ph.D. thesis, University of California-Berkeley, Calif., 1971.
143. **Bohle, W. and van Swaay, W. P. M.,** The influence of gas adsorption on mass transfer and gas mixing in a fluidized bed, in *Fluidization,* Proceedings of the Second Engineering Foundation Conference, Davidson, J. F. and Keairns, D. L., Eds., Cambridge Univ. Press, England, 1978, 167.
144. **Bolto, B. A.,** Sirotherm desalination, ion exchange with a twist, *Chemtech,* 5, 303, 1975.
145. **Bolto, B. A.,** Magnetic micro-ion-exchange resins, in *Ion Exchange for Pollution Control.,* Vol. II, Calmon C. and Gold, H., Eds., CRC Press, Boca Raton, Fla., 1979, chap. 23.
146. **Bolto, B. A.,** Novel water treatment processes which involve polymers, in *Polymeric Separation Media,* Cooper, A. R., Ed., Plenum Press, New York, 1982, 211.
147. **Bolto, B. A.,** Oxidation-resistant resin for the Sirotherm desalination process, *CSIRO, Division of Chemical Technology Research Review, 1982,* CSIRO, Clayton, Victoria, Australia, 1982, 1.
148. **Bolto, B. A., Dixon, D. R., Eldridge, R. J., Kolarik, L. O., Priestley, A. J., Raper, W. G., Rowney, J. E., Swinton, E. A., and Weiss, D. E.,** Continuous ion exchange using magnetic micro resins, in *The Theory and Practice of Ion Exchange,* Slater, M. J., Ed., Society of Chemical Industry, London, 1976, 27.7.
149. **Bolto, B. A., Dixon, D. R., Eldridge, R. J., Swinton, E. A., Weiss, D. E., Willis, D., Battaerd, H. A. J., and Young, P. H.,** The use of magnetic polymers in water treatment, *J. Polym. Sci. Polym. Symp.,* 49, 211, 1975.
150. **Bolto, B. A., Dixon, D. R., Priestley, A. J., and Swinton, E. A.,** Ion exchange in a moving bed of magnetised resin, *Prog. Water Technol.,* 9, 833, 1977.
151. **Bolto, B. A., Dixon, D. R., Swinton, E. A., and Weiss, D. E.,** Continuous ion-exchange using magnetic shell resins. I. Dealkalisation — laboratory scale, *J. Chem. Technol. Biotechnol.,* 29, 325, 1979.
152. **Bolto, B. A., Eppinger, K. H., Ho, P. S. K., Jackson, M. B., Pilkington, M. H., and Siudak, R. V.,** An ion exchange process with thermal regeneration. XII. Desalting of sewage effluents, *Desalination,* 25, 45, 1978.
153. **Bolto, B. A., Eppinger, K., MacPherson, A. S., Siudak, R., Weiss, D. E., and Willis, D.,** An ion-exchange process with thermal regeneration. IX. A new type of rapidly reacting ion-exchange resin, *Desalination,* 13, 269, 1973.
154. **Bolto, B. A., Swinton, E. A., Nadebaum, P. R., and Murtagh, R. W.,** Desalination by continuous ion exchange based on thermally regenerated magnetic microresins, *Water Sci. Technol.,* 14, 523, 1982.
155. **Bolto, B. A. and Weiss, D. E.,** The thermal regeneration of ion-exchange resins in *Progress in Ion Exchange and Solvent Extraction,* Marinsky, J. A. and Marcus, Y., Eds., Vol. 7, Marcel Dekker, New York, 1976, 221.
156. **Bombaugh, K. J. and Almquist, P. W.,** The role of thruput in high-performance liquid chromatography, *Chromatographia,* 8, 109, 1975.
157. **Bond, R. G. and Straub, C. P., Eds.,** *Handbook of Environmental Control.,* Vol. 1, *Air Pollution,* CRC Press, Boca Raton, Fla., 1972, 453.
158. **Bonmati, R., Chapelet-Letourneax, G., and Guiochon, G.,** Gas chromatography: a new industrial process of separation. Application to essential oils, *Sep. Sci. Technol.,* 19, 113, 1984.
159. **Bonmati, R. G., Chapelet-Letournex, G., and Margulis,. J. R.,** Gas chromatography — analysis to production, *Chem. Eng.,* 87, 70, 1980.
160. **Bonmati, R. and Guiochon, G.,** Gas chromatography as an industrial process operation — application to essential oils, *Perfum. Flavor.,* 3(5), 17, 1978.
161. **Bonmati, R., Marqulis, J. R., and Chapelet-Letournex, G.,** Gas chromatography. A new industrial separation process, Elf Technologies, New York, no date.
162. **Boschetti, E., Girot, P., and Pepin, O.,** Trisacryl polymers — new supports for liquid chromatography of biological macromolecules, *Sci. Tools,* 30(2), 27, 1983.
163. **Bouchard, J.,** Development of the Degremont-Asahi continuous ion-exchange process, in *Ion Exchange in the Process Industries,* Society of Chemical Industry, London, 1970, 90.
164. **Bourges, J., Madic, C., and Koehly, G.,** Transplutonium elements production programs. Extraction chromatographic process for plutonium irradiated targets, in *Actinide Separations,* Navratil, J. D. and Schulz, W. W., Eds., ACS Symp. Ser. #117, Americal Chemical Society, Washington, D.C., 1979, 33.
165. **Bowen, J. H.,** Sorption processes, in *Chemical Engineering,. Vol. 3, 2nd ed., Richardson, J. F. and Peacock, D. G., Eds.,* Pergamon Press, Oxford, 1979, chap. 7.

166. **Brauch, V. and Schlunder, E. U.,** The scale-up of activated carbon columns for water purification, based on results from batch tests. II. Theoretical and experimental determination of breakthrough curves in activated carbon columns, *Chem. Eng. Sci.,* 30, 539, 1975.
167. **Bray, L. A. and Fullan, H. T.,** Recovery and purification of Cesium-137 from Purex wastes using synthetic zeolites, in *Molecular Sieve Zeolites-I,* Flanigen, E. M. and Sand, L. M., Eds., ACS Adv. Chem. Ser., #101, American Chemical Society, Washington, D.C.,, 1971, 450.
168. **Breck, D. W.,** *Zeolite Molecular Sieves,* Wiley, New York, 1974.
169. **Brian, P. L. T.,** *Staged Cascades in Chemical Processing,* Prentice-Hall, Englewood Cliffs, N.J., 1972.
170. **Bridger, J. S. and Dixon, D. R.,** Continuous ion exchange using magnetic shell resins. IV. Pilot plant studies using unclarified feed, *J. Appl. Chem. Biotechnol.,* 28, 17, 1978.
171. **Bristow, P. A.,** Computer-controlled preparative liquid chromatograph, *J. Chromatogr.,* 122, 277, 1976.
172. **Broughton, D. B.,** Molex: case history of a process, *Chem. Eng. Prog.,* 64(8), 60, 1968.
173. **Broughton, D. B.,** A method for simulating continuous countercurrent contact between granular solids and fluids, 156th Nat'l. Meet. ACS, Atlantic City, N.J., Sept. 1968.
174. **Broughton, D. B.,** Continuous adsorptive processing — a new separation technique, 34th Annu. Meet. Soc. Chem. Eng., Tokyo, April, 1969.
175. **Broughton, D. B.,** Applications of adsorption to hydrocarbon separations, 65th Annu. Meet. AIChE, Houston, Tex., Nov. 1970.
176. **Broughton, D. B.,** Separation d'hydrocarbures par procede d'adsorption en continu, *Chim. Ind. Genie Chim.,* 104(19), 2443, 1971.
177. **Broughton, D. B.,** Adsorptive separations(liquids), in *Kirk-Othmer Encyclopedia of Chemical Technology,* Vol. 1, 3rd ed., Wiley-Interscience, New York, 1978, 563.
178. **Broughton, D. B.,** Continuous Desalination Process, U.S. Patent 4,447,329, May 8, 1984.
179. **Broughton, D. B.,** Production-scale adsorptive separations of liquid mixtures by simulated moving-bed technology, *Sep. Sci. Technol.,* 19, 723, 1984—85.
180. **Broughton, D. B. and Berg, R. C.,** Olefins by dehydrogenation-extraction, *Hydrocarb. Process.,* 48(6), 115, 1969.
181. **Broughton, D. B. and Carson, D. B.,** The Molex process, *Petrol. Refiner,* 38(4), 130, 1959.
182. **Broughton, D. B. and Gembicki, S. A.,** Adsorptive separation by simulated moving bed technology: the Sorbex process, in *Fundamentals of Adsorption,* Myers, A. L., and Belfort, G., Eds., Engineering Foundation, New York, 1984, 115.
183. **Broughton, D. B. and Gembicki, S. A.,** Production-scale adsorptive separations of liquid mixtures, *AIChE Symp. Ser.,* 80(233), 62, 1984.
184. **Broughton, D. B. and Gerhold, C. G.,** Continuous sorption process employing fixed beds of sorbent and moving inlets and outlets, U.S. Patent 2,985,589, May 23, 1961.
185. **Broughton, D. B. and Lickus, A. G.,** Production of high-boiling normal paraffins from petroleum, *Proc. Am. Petrol. Inst.,* 41, 111, 237, 1961.
186. **Broughton, D. B., Neuzil, R. W., Pharis, J. M., and Brearley, C. S.,** The Parex process for recovering paraxylene, *Chem. Eng. Prog.,* 66(9), 70, 1970.
187. **Brown, G. N., Tarrence, S. I., Repik, A. J., Stryker, J. L., and Bull, F. J.,** SO_2 recovery via activated carbon, *Chem. Eng. Prog.,* 68(8), 55, 1972.
188. **Brunauer, S., Deming, L. S., Deming, W. E., and Teller, E.,** On a theory of the van der Waals adsorption of gases, *J. Am. Chem. Soc.,* 62, 1723, 1940.
189. **Brunauer, S., Emmett, P. H., and Teller, E.,** Adsorption of gases in multimolecular layers, *J. Am. Chem. Soc.,* 60, 309, 1938.
190. **Buck, D. M. and Wilkinson, E. L.,** VSA and PSA systems as nitrogen supply alternatives, AIChE Meeting, San Francisco, Calif., Nov. 1984.
191. **Burgoyne, R. F. and Tarven, T. L.,** Preparative scale purification of ACTH, Pittsburgh Conference and Expositon on Analytical Chemistry and Applied Spectroscopy, Atlantic City, N.J., 1984, Abstr. 512.
192. **Burke, D. J.,** Adsorption separation technologies via ion exchange resins, paper presented at AIChE meeting, May 12, 1983.
193. **Burney, G. A.,** Separations of macro-quantities of actinide elements at Savannah River by high-pressure cation exchange, *Sep. Sci. Technol.,* 15, 763, 1980.
194. **Busbice, M. E. and Wankat, P. C.,** pH cycling zone separation of sugars, *J. Chromatogr.,* 114, 369, 1975.
195. **Butts, T. J., Gupta, R., and Sweed, N. H.,** Parametric pumping separations of multicomponent mixtures: an equilibrium theory, *Chem. Eng. Sci.,* 27, 855, 1972.
196. **Butts, T. J., Sweed, N. H., and Camero, A. A.,** Batch fractionation of ionic mixtures by parametric pumping, *Ind. Eng. Chem. Fundam.,* 12, 467, 1973.
197. **Bywater, R. and Marsden, N. V. B.,** Gel chromatography, in *Chromatography. Fundamentals and Applications of Chromatographic and Electrophoretic Methods, Part A: Fundamentals and Techniques,* Heftmann, E., Ed., Elsevier, Amsterdam, 1983, chap. 8.

198. **Cable, P. J., Murtagh, R. W., and Pilkington, N. R.,** An Australian desalination process becomes established, *Chem. Eng. (London),* #324, 624, 1977.
199. **Calderbank, P. H. and Mackrakis, N. S.,** Some fundamentals of the hypersorption process, *Trans. Inst. Chem. Eng.,* 34, 320, 1956.
200. California Research Corp., *Cyclic adsorption process, British Patent 33,142, 1948.*
201. **Calmon, C. and Gold, H., Eds.,** *Ion exchange for Pollution Control,* CRC Press, Boca Raton, Fla., 1979.
202. **Calmon, C. and Simon, G. P.,** The ion exchangers, in *Ion Exchange for Pollution Control,* Vol. 1, Calmon, C. and Gold, H., Eds., CRC Press, Boca Raton, Fla., 1972, 23.
203. **Calmon, C.,** Specific ion exchangers, in *Ion Exchange for Pollution Control,* Vol. II, Calmon, C. and Gold, H., Eds., CRC Press, Boca Raton, Fla., 1979, 151.
204. **Calton, G. J.,** Immunosorbent chromatography for recovery of protein products, Paper MBT 92, ACS Meeting, Washington, D.C., Sept. 2, 1983.
205. **Camero, A. A. and Sweed, N. H.,** Separation of nonlinearly sorbing solutes by parametric pumping, *AIChE J.,* 22, 369, 1976.
206. **Campbell, J. M.,** *Gas Conditioning and Processing,* 2nd ed., J. M. Campbell, Norman, Okla., 1970, chap. 19.
207. **Canon, R. M., Begovich, J. M., and Sisson, W. G.,** Pressurized continuous chromatography, *Sep. Sci. Technol.,* 15, 655, 1980.
208. **Canon, R. M. and Sisson, W. G.,** Operation of an improved continuous annular chromatograph, *J. Liq. Chromatogr.,* 1, 427, 1978.
209. **Cantwell, A. M., Calderone, R., and Sienko, M.,** Process scale-up of a β-lactam antibiotic purification by high performance liquid chromatography, *J. Chromatogr.,* 316, 133, 1984.
210. **Capes, P.,** Filling the solvent recovery energy gap, *Process Eng.,* 64(1), 33, 1983.
211. Cargocaire Engineering Corp., Honeycombe Dehumidifier Selection Guidelines, Honeycombe Industrial Dehumidifiers. Model HC-150, (no date), Amesbury, Mass. 01913.
212. Cargocaire Engineering Corp., Econocaire Total Enthalpy Heat Exchangers, Bull. EC-685 (2/74), Amesbury, Mass. 01913.
213. **Carnes, B. A.,** Laboratory simulation and characterization for water-pollution control, *Chem. Eng.,* 79(28), 97, 1972.
214. **Carpenter, B. M. and Le Van, M. D.,** A model for steam regeneration of adsorption beds. Paper 90d presented at AIChE Meeting, Los Angeles, Nov. 18, 1982.
215. **Carra, S., Morbidelli, M., Storti, G., and Paludetto, R.,** Experimental analysis and modeling of adsorption separation of chlorotoluene isomer mixtures, in *Fundamentals of Adsorption,* Myers, A. L. and Belfort, G., Eds., Engineering Foundation, New York, 1984, 143.
216. **Carrubba, R. V., Urbanic, J. E., Wagner, N. J., and Zanitsch, R. H.,** Perspectives of activated carbon — past, present and future, *AIChE Symp. Ser.,* 80(233), 76, 1984.
217. **Carter, J. W. and Wyszynski, M. L.,** The pressure-swing adsorption drying of compressed air, *Chem. Eng. Sci.,* 38, 1093, 1983.
218. **Cassidy, R. T.,,** Polybed pressure-swing adsorption hydrogen processing, in *Adsorption and Ion Exchange with Synthetic Zeolites,* Flank, W. H., Ed., ACS Symp. Ser. 135, American Chemical Society, Washington, D.C., 1980, 247.
219. **Cassidy, R. T. and Holmes, E. S.,** Twenty-five years of progress in "adiabatic" adsorption processes, *AIChE Symp. Ser.,* 80(223), 68, 1984.
220. **Cen, P.-L., Chen, W.-N., and Yang, R. T.,** Ternary gas mixture spearation by pressure swing adsorption: a combined hydrogen-methane separation and acid gas removal process, *Ind. Eng. Chem. Process Des. Devel.,* 24, 201, 1985.
221. **Cen. P.-L. and Yang, R. T.,** Bulk gas separation by pressure swing adsorption: equilibrium and linear-driving-force models for separating a binary mixture, *Chem. Eng. Sci.,* in press.
222. **Chadyk, W. A. and Snoeylnk, V. L.,** Bioregeneration of activated carbon saturated with phenol., *Environ. Sci. Technol.,* 18, 1, 1984.
223. **Chakravorti, R. K. and Weber, T. W.,** A comprehensive study of the adsorption of phenol in a packed bed of activated carbon, *AIChE Symp. Ser.,* 71(151), 392, 1975.
224. **Chan, Y. N. I., Holl, F.B., and Wong, Y. W.,** Equilibrium theory of a pressure-swing adsorption process, *Chem. Eng. Sci.,* 36, 243, 1981.
225. **Chao, J. F., Huang, J. J., and Huang, C. R.,** Continuous multiaffinity separation of proteins; cyclic processes, *AIChE Symp. Ser.,* 78(219), 39, 1982.
226. **Chapelet-Letournex, G. and Perrut, M.,** Preparative supercritical fluid chromatography (PSFC), Pittsburgh Conference and Exposition on Analytical Chemistry and Applied Spectroscopy, Atlantic City, N.J., 1984, Abstr. 513.
227. **Charest, F. and Chornet, E.,** Wet oxidation of active carbon, *Can. J. Chem. Eng.,* 54, 190, 1976.

228. **Charm, S. E. and Matteo, C. C.,** Scale-up of protein isolation, in *Methods in Enzymology*, Vol. 22, *Enzyme Purification and Related Techniques*, Jakoby, W. B., Ed., Academic Press, New York, 1971, chap. 37.
229. **Charm, S. E. and Wong, B. L.,** Application of immunoadsorbents for isolation of placental alkaline phosphatase, carboxypeptidase G-1, and serum hepatitis antigen, *J. Macromol. Sci. Chem.*, A10, 53, 1976.
230. **Chase, H. A.,** Affinity separations utilizing immobilized monoclonal antibodies — a new tool for the biochemical engineer, *Chem. Eng. Sci.*, 39, 1099, 1984.
231. **Chase, H. A.,** Prediction of the performance of preparative affinity chromatography, *J. Chromatogr.*, 297, 179, 1984.
232. Chemical Separations Corp. Ion Exchange Systems for the Treatment of Industrial Municipal Water and Wastes. Demineralization, Dealkalization, Softening with Chem-Seps Down-Flow Design Continuous Ion Exchange Systems, Oak Ridge Turnpike, Oak Ridge, Tenn., no date.
233. **Chen, H. T.,** Parametric pumping, in *Handbook of Separation Techniques for Chemical Engineers*, Schweitzer, P. A., Ed., McGraw-Hill, New York, 1979, 1—467.
234. **Chen, H. T., Ahmed, Z. M., and Follen, V.,** Parametric pumping with pH and ionic strength: enzyme purification, *Ind. Eng. Chem. Fundam.*, 20, 171, 1981.
235. **Chen, H. T. and D'Emidio, V. J.,** Separation of isomers via thermal parametric pumping, *AIChE J.*, 21, 813, 1975.
236. **Chen, H. T. and Hill, F. B.,** Characteristics of batch, semicontinuous and continuous equilibrium parametric pumps, *Sep. Sci.*, 6, 411, 1971.
237. **Chen, H. T., Hsieh, T. K., Lee, H. C., and Hill, F. B.,** Separation of proteins via semicontinuous pH parametric pumping, *AIChE J.*, 23, 695, 1977.
238. **Chen, H. T., Kerobo, C. O., Hollein, H. C., and Huang, C. R.,** Research on parametric pumping, *Chem. Eng. Educ.*, 15(4), 166, 1981.
239. **Chen, H. T., Yang, W. T., Wu, C. M., Kerobo, C. O.,, and Jajalla, V.,** Semicontinuous pH parametric pumping: process characteristics and protein separations, *Sep. Sci. Technol.*, 16, 43, 1981.
240. **Chen, H. T. and Manganaro, J. A.,** Optimal performance of equilibrium parametric pumps, *AIChE J.*, 20, 1020, 1974.
241. **Chen, H. T., Pancharoen, U., Yang, W. T., Kerobo, C. O., and Parisi, R. J.,** Separation of proteins via pH parametric pumping, *Sep. Sci. Technol.*, 15, 1377, 1980.
242. **Chen, H. T., Rak, J. L., Stokes, J. D., and Hill, F. B.,** Separation via continuous parametric pumping, *AIChE J.*, 18, 356, 1972.
243. **Chen, H. T., Reiss, E. H., Stokes, J. D., and Hill, F. B.,** Separations via semicontinuous parametric pumping, *AIChE J.*, 19, 589, 1973.
244. **Chen, H. T., Wong, Y. W., and Wu, S.,** Continuous fractionation of protein mixtures by pH parametric pumping, *AIChE J.*, 25, 320, 1979.
245. **Chen, H. T., Yang, W. T., Pancharoen, U., and Parisi, R.,** Separation of proteins via multicolumn pH parametric pumping, *AIChE J.*, 26, 839, 1980.
246. **Chen, H. T., Lin, W. W., Stokes, J. D., and Fabisiak, W. R.,** Separation of multicomponent mixtures via thermal parametric pumping, *AIChE J.*, 20, 306, 1974.
247. **Chen, J. W., Cunningham, F. L., and Buege, J. A.,** Computer simulation of plant-scale multicolumn adsorption processes under periodic countercurrent operation, *Ind. Eng. Chem. Process Des. Devel.*, 11, 430, 1972.
248. **Chen. W. N. and Yang, R. T.,** Adsorption of gas mixtures and modeling cyclic processes for bulk multicomponent gas separation, in *Recent Developments in Separation Science*, Vol. IX, Li, N. N. and Calo, J. M., Eds., CRC Press, Boca Raton, Fla., 1984.
249. **Cheng, H. C. and Hill, F. B.,** Recovery and purification of light gases by pressure swing adsorption, in *Industrial Gas Separations*, Whyte, T. E., Jr., Yon, C. M., and Wagener, E. H., Eds., ACS Symposium Ser. #223, American Chemical Society, Washington, D.C., 1983, 195.
250. **Cheng, H. C. and Hill, F. B.,** Separation of helium-methane mixtures by pressure swing adsorption, *AIChE J.*, 31, 95, 1985.
251. **Cheremisinoff, P. N. and Ellerbusch, F., Eds.,** *Carbon Adsorption Handbook*, Ann Arbor, Science, Ann Arbor, Mich., 1978.
252. **Chi, C. W.,** A design method for thermal regeneration of hydrated 4A molecular sieve bed, *AIChE Symp. Ser.*, 74(179), 42, 1978.
253. **Chi, C. W. and Cummings, W. P.,** Adsorption separations (gases), in *Kirk-Othmer Encyclopedia of Chemical Technology*, Vol. 1, 3rd ed., Wiley-Interscience, New York, 1978, 544.
254. **Chi, C. W. and Lee, H.,** Separation of n-tetradecane from i-octane n-tetradecane mixture by 5A molecular sieves, *Chem. Eng. Prog. Symp. Ser.*, 65(96), 65, 1969.
255. **Chihara, K., Matsui, I., and Smith, J. M.,** Regeneration of powdered activated carbon. II. Steam carbon reaction kinetics, *AIChE J.*, 27, 220, 1981.

256. **Chihara, K., Smith, J. M., and Suzuki, M.,** Regeneration of powdered activated carbon. I. Thermal decomposition kinetics, *AIChE J.*, 27, 213, 1981.
257. **Chihara, K. and Suzuki, M.,** Air drying by pressure swing adsorption, *J. Chem. Eng. Jpn.*, 16, 293, 1983.
258. **Chihara, K. and Suzuki, M.,** Simulation of nonisothermal pressure swing adsorption, *J. Chem. Eng. Jpn.*, 16, 53, 1983.
259. **Chihara, K., Suzuki, M., and Smith, J. M.,** Cyclic regeneration of activated carbon in fluidized beds, *AIChE J.*, 28, 129, 1982.
260. **Ching, C. B.,** A theoretical model for the simulation of the operation of the semicontinuous chromatographic refiner for separating glucose and fructose, *J. Chem. Eng. Jpn.*, 16(1), 49, 1983.

260a. **Ching, C. B. and Ruthven, D. M.,** Separation of glucose and fructose by simulated counter-current adsorption, *AIChE Symp. Ser.*, 81(242), 1, 1985.

261. **Cines, M. R., Haskell, D. M., and Houser, C. G.,** Molecular sieves for removing H_2S from natural gas, *Chem. Eng. Prog.*, 72(8), 89, 1976.
262. **Claesson, S.,** Theory of frontal analysis and displacement development, *Discuss. Faraday Soc.*, 7, 34, 1949.
263. **Clayton, R. C.,** Systematic analysis of continuous and semi-continuous ion-exchange techniques and the development of a continuous system, in *Ion Exchange in the Process Industries,* Society of Chemical Industry, London, 1970, 98.
264. **Clemence, L. J., Eldridge, R. J., and Lydiate, J.,** Preparation of magnetic ion exchangers by cerium-initiated graft polymerization on crosslinked poly (vinyl alcohol), *React. Polym.*, 2, 197, 1984.
265. **Cloete, F. L. D.,** Comparative engineering and process features of operating continuous ion exchange plants in Southern Africa, in *Ion Exchange Technology,* Naden, D. and Streat, M., Eds., Ellis Horwood, Chichester, England, 1984, 661.
266. **Cloete, F. L. D., Frost, C. R., and Streat, M.,** Fractional separations by continuous ion exchange, in *Ion Exchange in the Process Industries,* Society of Chemical Industry, London, 1970, 182.
267. **Cloete, F. L. D. and Streat, M.,** A new continuous solid-fluid contacting technique, *Nature (London),* 200 (4912), 1199, 1963.
268. **Cochrane, G. S.,** Molecular sieves for gas drying, *Chem. Eng.*, 66(17), 129, 1959.
269. **Cofield, E. P., Jr.,** Solvent extraction of oil seed, *Chem. Eng.*, 58(1), 127, 1951.
270. **Cohen, J. M. and English, J. N.,** Activated carbon regeneration, *AIChE Symp. Ser.*, 70(144), 326, 1974.
271. **Cohen, Y. and Metzner, A. B.,** Wall effects in laminar flow of fluids through packed beds, *AIChE J.*, 27, 705, 1981.
272. **Colin, H., Krstulovic, A. M., Excoffler, J.-L., and Guiochon, G.,** *A Guide to the HPLC Literature,* Vols. 1 to 3, Wiley, New York, 1984.
273. **Collins, J. J.,** The LUB/equilibrium section concept for fixed-bed adsorption, *Chem. Eng. Prog. Symp. Ser.*, 63(74), 31, 1967.
274. **Collins, J. J., Fornoff, L. L., Manchanda, K. D., Miller, W. C., and Lovell, D. C.,** The Pura Siv S Process for SO_2 removal and recovery, 66th Annual AIChE Meeting, Philadelphia, Nov. 15, 1973.
275. **Conder, J. R.,** Production-scale gas chromatography, in *New Developments in Chromatography,* Purnell, J. H., Ed., Wiley-Interscience, New York, 1973, 137.
276. **Conder, J. R.,** Performance optimisation for production gas chromatography, *Chromatographia,* 8, 60, 1975.
277. **Conder, J. R.,** Design procedures for perparative and production gas chromatography, *J. Chromatogr.*, 256, 381, 1983.
278. **Conder, J. R., Fruitwala, N. A., and Shingari, M. K.,** Design of a production gas chromatograph for separating heat-sensitive materials, *Chromatographia,* 16, 44, 1982.
279. **Conder, J. R. and Shingari, M. K.,** Throughput and band overlap in production and preparative chromatography, *J. Chromatogr. Sci.*, 11, 525, 1973.
280. **Conrard, P., Caude, M., and Rosset, R.,** Separation of close species by displacement development on ion exchangers. I. A theoretical equation of steady-state permutation front. Experimental verification in the case of the separation of boron iostopes, *Sep. Sci.*, 7, 465, 1972.
281. **Cooney, D. O.,** Numerical investigation of adiabatic fixed-bed adsorption, *Ind. Eng. Chem. Process Des. Devel.*, 13, 368, 1974.
282. **Cooney, D. O.,** *Activated Charcoal: Antidotal and Other Medical Uses,* Marcel Dekker, New York, 1980.
283. **Cooney, D. O. and Lightfoot, E. N.,** Existence of asymptotic solutions to fixed-bed separations and exchange equations, *Ind. Eng. Chem. Fundam.*, 4, 233, 1965.
284. **Cooney, D. O., Nagerl, A., and Hines, A. L.,** Solvent regeneration of activated carbon, *Water Res.*, 17, 403, 1983.
285. **Coq, B., Cretier, G., Gonnet, C., and Rocca, J. L.,** How to approach preparative liquid chromatography, *Chromatographia,* 12, 139, 1979.

286. **Coq, B., Cretier, G., and Rocca, J. L.,** Preparative liquid chromatography: sample volume overload, *J. Chromatogr.*, 186, 457, 1979.
287. **Corr, F., Dropp, F., and Rudelstorfer, E.,** Pressure-swing adsorption produces low-cost high-purity hydrogen, *Hydrocarb. Process.*, 58(3), 119, 1979.
288. **Costa, C. A. V., Rodriques, A. E., Grevillot, G., and Tondeur, D.,** Purification of phenolic wastewater by parametric pumping: nonmixed dead volume equilibrium model, *AIChE J.*, 28, 73, 1982.
289. **Cox, M.,** A fluidized adsorbent air-drying plant, *Trans. Inst. Chem. Eng.*, 36, 29, 1958.
290. **Cuatrecasas, P. and Anfinsen, C.,** Affinity chromatography, in *Methods in Enzymology,* Vol. 22, Jakoby, W. B., Ed., Academic Press, New York, 1971, 345.
291. **Culp, R. L., Wesner, G. M., and Culp, G. L.,** *Handbook of Advanced Wastewater Treatment,* 2nd ed., Van Nostrand-Reinhold, New York, 1978, chap. 5.
292. **Cumberland, R. F. and Broadbent, F.,** Applications of the basket centrifuge, *Chem. Process Eng.*, 50(8), 55, 1969.
293. **Curling, J. M.,** Industrial scale gel filtration and ion exchange chromatography with special reference to plasma protein fractionation, in *Chromatography of Synthetic and Biological Polymers,* Vol. 2, Epton, R., Ed., Ellis Horwood, Chichester, 1978, chap. 6.
294. **Curling, J. M., Ed.,** *Separation of Plasma Proteins,* Joint Meeting of the 19th Congress of the International Society of Haematology and the 17th Congress of the International Society of Blood Transfusion, Budapest, August 1 to 7, 1982, Pharmacia Fine Chemicals AB, Uppsala, Sweden, 1983.
295. **Curling, J. M. and Cooney, J. M.,** Operation of large scale gel filtration and ion-exchange systems, *J. Parent. Sci. Technol.*, 36(2), 59, 1982.
296. **Curling, J. M., Gerglof, J. H., Eriksson, S., and Cooney, J. M.,** Large Scale Production of Human Albumin by an All-Solution Chromatographic Process, Joint meeting of the 18th Congress of the International Society of Heamatology and the 16th Congress of the International Society of Blood Transfusion, Montreal, Quebec, Aug. 16 to 22, 1980.
297. **Danielson, J. A., Ed.,** *Air Pollution Engineering Manual,* EPA Publ. #AP-40, Research Triangle Park, New York, 1973, 189.
298. **Darbyshire, J.,** Large scale enzyme extraction and recovery, in *Topics in Enzyme and Fermentation Biotechnology,* Vol. 5, Wiseman, A. Ed., Ellis Horwood, Chichester, 1981, chap. 3.
299. **Darji, J. D. and Michalson, A. W.,** Ion-exchange and adsorption equiment, in *Perry's Chemical Engineer's Handbook,* 6th ed., Perry, R. H. and Green, D., Eds., McGraw-Hill, New York, 1984, 19-40.
300. **Davis, J. C.,** Oxygen separated from air using molecular sieve, *Chem. Eng.*, 79(23), 88, 1972.
301. **Davis, K. G. and Manchanda, K. D.,** Unit operations for drying fluids, *Chem. Eng.*, 81(19), 102, 1974.
302. Davy Bamag Ltd., Solvent Recovery Waste Air Purification, Removal and Recovery of Volatile Solvents from Exhaust Gases by Carbon Adsorption, Hydrocarbon Recovery. Improving the Economy of Chemical Production Processes, 213 Oxford St., London W1R 2DE, England, no date.
303. **Deans, D. R.,** An improved technique for back-flushing gas chromatographic columns, *J. Chromatogr.*, 18, 477, 1965.
304. **Deans, D. R.,** Use of heart-cutting on gas chromatography: a review, *J. Chromatogr.*, 203, 19, 1981.
305. Dedert/Lurgi, Solvent Recovery & Exhaust Air Purification: Supersorbon Process, 20000 Governors Dr., Olympia Fields, Ill., no date.
306. **Delaney, R. A. M.,** Industrial gel filtration of proteins, in *Applied Protein Chemistry,* Grant. R. A., Ed., Applied Science, London, 1980, 233.
307. **Delaney, R. A. M., Donnelly, J. K., and Kerney, R. D.,** Industrial applications of gel filtration, *Process Biochem.*, 8(3), 13, 1973.
308. **Dell'Osso, L., Jr., Ruder, J. M., and Winnick, J.,** Mixed gas adsorption and vacuum desorption of carbon dioxide on molecular sieve. Bed performance and data analysis, *Ind. Eng. Chem. Process Des. Devel.*, 8, 477, 1969.
309. **Dell'Osso L., Jr. and Winnick J.,** Mixed-gas adsorption and vacuum desorption of carbon dioxide on molecular sieve. Bed design for use in a humid atmosphere, *Ind. Eng. Chem. Process Des. Devel.*, 8, 469, 1969.
310. Deltech Engineering Corp., Deltech Air Dryer, Bulletin 300C, 1967; Reactivation Requirements, Bulletin 308C, no date; Deltech Heatless Air Dryer, Bulletin 312B, no date; Heatless Air Dryer Dimensions Bulletin 327B, no date; Deltech Filter-Dryer Combination Assures Quality Film Processing, no date, New Castle, Del. 19720.
311. **Determann, H.,** *Gel Chromatography,* 2nd ed., Springer-Verlag, Berlin, 1969.
312. **deFilippi, R. P., Krukonis, V. J., Robey, R. J., and Modell, M.,** Supercritical fluid regeneration of activated carbon for adsorption of pesticides, U.S. EPA Report No. EPA-600/2-80-054, EPA, Office of Research and Development, Research Triangle Park, N.C., 1980.
313. **DeLasa, H. and Gau, G.,** Solutions approximatives pour l'adsorption dans un solide poreux, *Chem. Eng. Sci.*, 28, 1885, 1973.

314. **DeMarco, J., Miller, R., Davis, D., and Cole, C.,** Experiences in operating a full-scale granular activated carbon system with on-site regeneration, in *Treatment of Water by Granular Activated Carbon,* McGuire, M. J. and Suffet, I. H., Eds., American Chemical Society, Washington, D.C., 1983, chap. 23.
315. **deRosset, A. J., Neuzil, R. W., Korous, D. J., and Rosback, D. H.,** Separation of C_8 aromatics by adsorption, 161st Natl. Meeting ACS, Los Angeles, Calif., March 1971.
316. **de Rosset, A. J. and Kaufman, J. G.,** Molecular sieves, a new separation and preparative tool, Society of Cosmetic Chemists, Assoc. Official Analytical Chemists, Washington, D.C., Oct. 1974.
317. **de Rosset, A. J. Neuzil, R. W., and Korous, D. J.,** Liquid column chromatography as a predictive tool for continuous countercurrent adsorptive separations, *Ind. Eng. Chem. Proc. Des. Devel.,* 15, 261, 1976.
318. **de Rosset, A. J., Neuzil, R. W., Tajbl, D. G., and Braband, J. M.,** Separation of ethylbenzene from mixed xylenes by continuous adsorptive processing, *Sep. Sci. Technol.,* 15, 637, 1980.
319. **deRosset, A. J., Neuzil, R. W., and Broughton, D. B.,** Industrial Applications of preparative chromatography, in *Percolation Processes, Theory and Applications,* Rodrigues, A. E. and Tondeur, D., Eds., Sijthoff and Noordhoff, Alphen aan den Rijn, the Netherlands, 1981, 249.
320. **De Vault, D.,** The theory of chromatography, *J. Am. Chem. Soc.,* 65, 532, 1943.
321. **Deitz, V. R.,** *Bibliography of Solid Adsorbents, 1900—1942* National Bureau of Standards, Washington, D.C., 1944.
322. **Dietz, V. R.,** *Bibliography of Solid Adsorbents, 1943—1953,* National Bureau of Standards, Washington, D.C., 1956.
323. **Dinelli, D. Polezzo, S., and Taramasso, M.,** Rotating unit for preparative-scale gas chromatography, *J. Chromatogr.,* 7, 477, 1962.
324. **Dixon, D. R. and Hawthorne, D. B.,** Continuous ion exchange using magentic shell resins. III. Treatment of effluents containing heavy metal ions, *J. Appl. Chem. Biotechnol.,* 28, 10, 1978.
325. **Dixon, D. R. and Lydiate, J.,** Selective magnetic adsorbents, *J. Macromol. Sci. Chem.,* A14, 153, 1980.
326. **DiCesare, J. L., Dong, M. W., and Ettre, L. S.,** Very-high-speed liquid column chromatography. The system and selected applications, *Chromatographia,* 14, 257, 1981.
327. **DeGiano, F. A.,** Toward a better understanding of the practice of adsorption, *AIChE Symp. Ser.,* 76(197), 86, 1980.
328. **Dodds, R., Hudson, P. I., Kershenbaum, L., and Streat, M.,** The operation and modelling of a periodic countercurrent, solid-liquid reactor, *Chem. Eng. Soc.,* 28, 1233, 1973.
329. **Dolphin, R. J., and Willmot, F. W.,** Band-broadening effects in a column-switching system for HPLC, *J. Chromatog., Sci.,* 14, 584, 1976.
330. **Domine, D. and Hay, L.,** Process for separating mixtures of gases by isothermal adsorption: possibilities and applications in *Molecular Sieves,* Barrer, R. M., Ed., Society Chemical Industry, London, 1968, 204.
331. **Dore, J. C. and Wankat, P. C.,** Multicomponent cycling zone adsorption, *Chem. Eng. Sci.,* 31, 921, 1976.
332. **Dorfner, K.,** *Ion Exchangers, Principles and Applications,* Ann Arbor Science, Ann Arbor, Mich., 1972.
333. **Doshi, K. J., Katira, C. H., and Stewart, H. A.,** Optimization of a pressure swing cycle, *AIChE Symp. Ser.,* 67(117), 90, 1971.
334. **Dow Chemical Co.,** *Dowex: Ion Exchange,* Midland, Mich., 1964.
335. **Driscoll, J. N. and Krill, I. S.,** Improved GC separations with chemically bonded supports, *Am. Lab.,* 15(5), 42, 1983.
336. **Duarte, P. E. and McCoy, B. J.,** Stationary spatial temperature gradients in gas chromatography, *Sep. Sci. Technol.,* 17, 879, 1982.
337. **Dunnill, R. and Lilly, M. D.** in *Biotechnol. Bioeng. Symp.,* Gaden E., Ed., (3), 47, 1972.
338. **Dunsap, G., Bugarel, R., and Gros, J. B.,** Modelisation d'un system de pompage parametrique. Fonctionment en discontinu. I. Relations entre les differents modeles publies, *Chem. Eng. J.,* 20, 107, 1980.

338a. **Dwyer, J. L.,** Scaling-up bio-product separation with high performance liquid chromatography, *Bio/Technology,* November 1984.

339. **Eagle, S. and Rudy, C. E., Jr.,** Separation and desulfurization of cracked naphtha. Application of cyclic adsorption process. *Ind. Eng. Chem.,* 42, 1294, 1950.
340. **Eagle, S. and Scott, J. W.,** Refining by adsorption, *Petrol. Process.,* 4, 881, 1949.
341. **Earls. D. E. and Long, G. N.,** Multiple bed rapid pressure swing adsorption for oxygen, U.S. Patent 4,194,891, March 25, 1980.
342. Eco-Tec Ltd., Eco-Tec Ion Exchange Systems, 1983, and Recoflo - A breakthrough in water deionization systems, 1984, 925 Brock Rd. South, Pickering, Toronto, Ontario, Canada L1W 2X9.
343. **Edwards, V. H. and Helft, J. M.,** Gel chromatography: improved resolution through compressed beds, *J. Chromatogr.,* 47, 490, 1970.
344. **Eisele, J. A., Colombo, A. F., and McClelland, G. E.,** Recovery of gold and silver from ores by hydrometallurgical processing, *Sep. Sci. Technol.,* 18, 1081, 1983.
345. **Eisenacher, K. and Neumann, U.,** A process for the further treatment of waste water with powdered activated carbon, *German Chem. Eng.,* 7, 45, 1984.

346. **Eisenbeiss, E. and Hauke, G.,** New large dimension HPLC/LC column design for laboratory and pilot scale applications, Pittsburg Conference and Exposition on Analytical Chemistry and Applied Spectroscopy, Atlantic City, N.J., 1984. Abstr. 515.
347. **Eisinger, R. S. and Alkire, R. C.,** Separation by electrosorption of organic compounds in a flow through porous electrode. II. Experimental validation of model, *J. Electrochem. Soc.,* 130, 93, 1983.
348. **Ek, L.,** *Process Biochem.,* 3(9), 25, 1968.
349. **Eketorp, R.,** Affinity chromatography in industrial blood plasma fractionation, in *Affinity Chromatography and Related Techniques,* Gribnau, T. C. J., Visser, J., and Nivard, R. J. F., Eds., Elsevier, Amsterdam, 1982, 263.
350. **El-Rifai, M. A., Saleh, M. A., and Youssef, H. A.,** Steam regeneration of a solvent adsorber, *Chem. Eng. (London),* 269, 36, 1973.
351. Elf Aquitane Development, Series 300 LC. Design, Operation and Performances of a Industrial Scale High Performance Liquid Chromatography (HPLC) Systems — Preliminary Technical Notes; Elf Series 1000 HPLC. Automatic Lab-Scale Preparative High Performance Liquid Chromatography with Axial Compression Technology, 9 W. 57th St., New York, 10019, no date.
352. Elf-Aquitane Development, Announcing a New Development in Production-Scale Gas Chromatography That Really Deserves to be Called a Breakthrough, 9 W. 57th St., New York, 10019, no date.
353. **Emneus, N. I. A.,** A procedure for gel filtration of viscous solutions, *J. Chromatogr.,* 32, 243, 1968.
354. **Engelhardt, H.,** *High Performance Liquid Chromatography,* Springer-Verlag, New York, 1979, chap. 2.
355. **Erdman, J. F., Kunke, J., and Roberts, L. R.,** Achieving best available control technology, *Chem. Eng. Prog.,* 78(6), 56, 1982.
356. **Erdos, E. and Szekely, G.,** Continuous chromatography on a packing of plastic fibres, *J. Chromatogr.,* 241, 103, 1982.
357. **Ermenc, E. D.,** Wisconsin process system for recovery of dilute oxides of nitrogen, *Chem. Eng. Prog.,* 52(11), 488, 1956.
358. **Ermenc, E. D.,** Designing a Fluidized Adsorber, *Chem. Eng.,* 68(11), 87, 1961.
359. **Ersson, B.,** Large-scale prepartion of a lectin from sunn hemp seeds *(Crotaleria juncea), Biotechnol. Bioeng.,* 22, 79, 1980.
360. **Eschrich, H. and Ochesenfield, W.,** Application of extraction chromatography to nuclear fuel reprocessing, *Sep. Sci. Technol.,* 15, 697, 1980.
361. **Etherington, L. D., Fritz, R. J., Nicholson, E.W., and Scheeline, H. W.,** Fluid char adsorption process, *Chem. Eng. Prog.,* 52, 274, 1956.
362. **Ettre, L. S.,** American instrument companies and the early development of gas chromatography, *J. Chromatogr. Sci.,* 15, 90, 1977.
363. **Ettre, L. S.,** Preparative liquid chromatography: history and trends, supplemental remarks, *Chromatographia,* 12, 302, 1979.
364. **Everett, D. H.,** The Sir Eric Rideal lecture. Adsorption from solution, in *Adsorption From Solution,* Ottewill, R. H., Rochester, C. H., and Smith, A. L., Eds., Academic Press, New York, 1983, 1.
365. **Everett, D. H.,** Thermodynamics of adsorption from solution, in *Fundamentals of Adsorption,* Myers, A. L. and Belfort, G., Eds., Engineering Foundation, New York, 1984, 1.
366. **Farooqui, A. A. and Harrocks, L. A.,** Heparin-Sepharose affinity chromatography, in *Advances in Chromatography,* Vol. 23, Giddings, J. C., Grushka, E., Cazes, J., and Brown, P. R., Eds., Marcel Dekker, New York, 1984, 127.
367. **Feeney, E. C.,** Removal of organic materials from wastewaters with polymeric adsorbents, in *Ion Exchange for Pollution Control,* Vol. II, Calmon C. and Gold, H., Eds, CRC Press, Boca Raton, Fla., 1979, chap. 4.
368. **Feissinger, F., Mallevialle, J., and Benedek, A.,** Interaction of adsorption and bioactivity in full-scale activated carbon filters: the Mont Valerien experiment, in *Treatment of Water by Granular Activated Carbon,* McGuire, M. J. and Suffet, I. H., Eds., American Chemical Society, Washington, D.C., 1983, chap. 14.
369. **Fickel, R. G.** Continuous adsorption — a chemical engineering tool, *AIChE Symp. Ser.,* 69(135), 65, 1973.
370. **Filan, J., Labaw, C., Tramper, A., and Dougherty, J.,** The pilot scale purification of cefonicid on the Waters process liquid chromatograph. Paper MBT 91 presented at ACS Meeting, Washington, D.C., Sept. 2, 1983.
371. **Filipi, T. J.,** Production liquid chromatography (PLC), a tool for product purification, AIChE Meeting, Philadelphia, Pa., 1984.
372. **Findlay, A. and Creighton, H. J. M.,** The influence of colloids and fine suspensions on the solubility of gases in water. I. Solubility of carbon dioxide and nitrous oxide, *J. Chem. Soc,. (London),* 97, 536, 1910.
373. **Finley, J. W., Krochta, J. M., and Heftmann, E.,** Rapid preparative separation of amino acids with chromatofuge, *J. Chromatogr.,* 157, 435, 1978.
374. **Fitch, G. R., Probert, M. E., and Tiley, P. F.,** Preliminary studies of moving-bed chromatography, *J. Chem. Soc. (London),* p.4875, 1962.

375. **Flores-Fernandez, G. and Kenney, C. N.,** Modeling of the pressure-swing air separation process, *Chem. Eng. Sci.,* 38, 827, 1983.
376. **Foo, S. C., Bergsman, K. H. and Wankat, P. C.,** Multicomponent fractionation by direct, thermal mode cycling zone adsorption, *Ind. Eng. Chem. Fundam.,* 19, 86, 1980.
377. **Foo, S. C. and Rice, R. G.,** On the prediction of ultimate separation in parametric pumps, *AIChE J.,* 21, 1149, 1975.
378. **Foo, S. C. and Rice, R. G.,** Steady state predictions for nonequilibrium parametric pumps, *AIChE J.,* 23, 120, 1977.
379. **Ford, M. A.,** The simulation and process design of NIMCIX contactors for the recovery of uranium, in *Ion Exchange Technology,* Naden, D. and Streat, M., Eds., Ellis Horwood, Chichester, 1984, 668.
380. **Fornwalt, H. J. and Hutchins, R. A.,** Purifying liquids with activated carbon, *Chem. Eng.,* 73(10), 155, 1966.
381. **Fox, C. R.,** Removing toxic organics from waste water, *Chem. Eng. Prog.,* 75 (8), 70, 1979.
382. **Fox, C. R.,** Plant uses prove phenol recovery with resins, *Hydrocarb. Process.,* 57(11), 269, 1978.
382a. **Fox, C. R.,** Industrial wastewater control and recovery of organic chemicals by adsorption, in *Adsorption Technology,* Slejko, F. L., Ed., Marcel Dekker, New York, 1985, 167.
382b. **Fox, C. R. and Kennedy, D. C.,** Conceptual design of adsorption systems, in *Adsorption Technology,* Slejko, F. L., Ed., Marcel Dekker, New York, 1985, 91.
383. **Fox, J. B.,** Continuous chromatography apparatus. II. Operation, *J. Chromatogr.,* 43, 55, 1969.
384. **Fox, J. B., Calhoun, R. C., and Eglinton, W. J.,** Continuous chromatography apparatus. I. Construction, *J. Chromatogr.,* 43, 48, 1969.
385. **Frenz, J. and Horvath, C.,** Movement of components on reversed-phase chromatography. III. Regeneration policies in liquid chromatography, *J. Chromatogr.,* 282, 249, 1983.
386. **Freund, M., Benedek, P., and Szepesy, L.,** Chemical engineering design of a unit for continuous gas chromatography (hypersorption), in *Vapour Phase Chromatography,* Desty, D. H., Ed., Academic Press, New York, 1957, 359.
387. **Freundlich, H.,** *Colloid and Capillary Chemistry,* Methuen, New York, 1926.
388. **Frey, D. D.,** A model of adsorbent behavior applied to the use of layered beds in cycling zone adsorption, *Sep. Sci. Technol.,* 17, 1485, 1982—83.
389. **Friday, D. K. and LeVan, M. D.,** Solute condensation in adsorption beds during thermal regeneration, *AIChE J.,* 28, 86, 1982.
390. **Friday, D. K. and LeVan, M. D.,** Thermal regeneration of adsorption beds; equilibrium theory for solute condensation, *AIChE J.,* 30, 679, 1984.
391. **Friday, D. K. and LeVan, M. D.,** Hot purge gas regeneration of adsorption beds with solute condensation: experimental studies, *AIChE J.,* 31, 1322, 1985.
392. **Fritz, W. and Schluender, E. V.,** Simultaneous adsorption equilibria of organic solutes in dilute aqueous solution on activated carbon, *Chem. Eng. Sci.,* 29, 1279, 1974.
393. **Frost, A. C.,** Slurry bed adsorption with molecular sieves, *Chem. Eng. Prog.,* 70(5), 70, 1974.
394. **Frost, F. and Glasser, D.,** An apparatus for fractional ion-exchange separation, in *Ion Exchange in the Process Industries,* Society of Chemical Industry, London, 1970, 189.
395. **Fulker, R. D.,** Adsorption, in *Processes for Air Pollution Control,* Nonhebel, G., Ed., CRC Press, Boca Raton, Fla., 1972, chap. 9.
396. **Fulton, S. P.,** Use of Amicon products in production and purification of interferons, Amicon Application Notes Publ. 552, Amicon Corp., Scientific Systems Div., Danvers, Mass., 01923, 1981.
397. **Fulton, S. P. and Carlson, E. R.,** Use of synthetic chemical ligands for the affinity chromatography of proteins, in *Polymeric Separation Media,* Cooper, A. R., Ed., Plenum Press, New York, 1982, 93.
397a. **Gallo, P., Olsson, O., and Siden, A.,** Chromato-focusing of biological fluids, *LC,* 3, 534, 1985.
398. **Ganho, R., Gibert, H., and Angelino, H.,** Cinetique de l'adsorption du phenol en couche fluidisee de charbon actif, *Chem. Eng. Sci.,* 30, 1231, 1975.
399. **Garg, D. R., Ausikaitis, J. P., and Yon, C. M.,** Adsorption heat recovery processes for breaking the ethanol-water azeotrope. Paper 90f presented at IAChE Annual Meeting, Los Angeles, Nov. 19, 1982.
400. **Gareil, P., Durieux, C., and Rosset, R.,** Optimization of production rate and recovered amount in linear and non-linear preparative elution liquid chromatography, *Sep. Sci. Technol.,* 18, 441, 1983.
401. **Gareil, P. and Rosset, R.,** La chromatographie en phase liquide preparative par developpement par elution. I. Notions fondamentals. Processes lineaire et non lineaire, *Analusis,* 10, 397, 1982.
402. **Gareil, P. and Rosset, R.,** La chromatographie in phase liquide preparative par developpement par elution. II. Aspects pratiques, strategie at appareillage, *Analusis,* 10, 445, 1982.
403. **Garg, D. R. and Ausikaitis, J. P.,** Molecular sieve dehydration cycle for high water content streams, *Chem. Eng. Prog.,* 79, (4), 60, 1983.
404. **Garg, D. R. and Yon, C. M.,** The adsorptive heat recovery drying system, *Chem. Eng. Prog.,* 82(2), 54, 1986.

405. **Geeraert, E. and Verzele, M.,** Preparative liquid chromatography: history and trends, *Chromatographia,* 11, 640, 1978.
406. **Geeraert, E. and Verzele, M.,** Peak doubling in preparative liquid chromatography, *Chromatographia,* 12, 50, 1979.
407. **Gelbin, D., Bunke, G., Wolff, H. J., and Neinass, J.,** Adsorption separation efficiency in the cyclic steady state. A continuous countercurrent analogy and an experimental investigation. *Chem. Eng. Sci.,* 38, 1993, 1983.
408. **Gembicki, S.,** Private communication, 1984.
409. **General Cable Apparatus Div. Westminister Co.,** Heatless Air Dryers for Ultra Dry Air, Bull. AP-976A, no date; Carbon Dioxide Adsorbers, Bull. AP-983A, no date; HE 200 Series Heatless Dryers for Hydrocarbon Adsorption, Bull. AP 984-3, no date; Series, HF 3000 Heatless Dryers, Installation, Operation and Maintenance Manual. Bull. M-73, no date; Drying Compressed Air and Other Gases with Pure-Gas Heatless Dryers, Bull. AP 963E, 1976, P. O. Box 666, 5600 West 88th Ave., Westminister, Colo. 80030.
410. **George, D. R. and Rosenbaum, J. B.,** New developments and applications of ion-exchange techniques for the mineral industry, in *Ion Exchange in the Process Industries,* Society of Chemical Industry, London, 1970, 155.
411. **George, D. R., Ross, J. R., and Prater, J. D.,** Byproduct uranium recovered with new ion-exchange techniques, *Min. Eng.,* 20, 1, 1968.
412. **Gerber, R. and Birss, R. R.,** *High Gradient Magnetic Separation,* Research Studies Press, Chichester, 1983.
413. **Gere, D. R.,** Supercritical fluid chromatography, *Science,* 222, 253, 1983.
414. **Gerhold, C. G.,** High efficiency continuous separation process, U.S. Patent 4,402,832, Sept. 6, 1983.
415. **Gidaspow, D. and Onischak, M.,** Regenerative sorption of nitric oxide: a concept for environmental control and kinetics for ferrous sorbents, *Can. J. Chem. Eng.,* 51, 337, 1973.
416. **Giddings, J. C.,** *Dynamics of Chromatography, Part I, Principles and Theory,* Marcel Dekker, New York, 1965.
417. **Giddings, J. C.,** Theoretical basis for a continuous, large-capacity gas chromatographic apparatus, *Anal. Chem.,* 34, 37, 1962.
418. **Gilbert, B. R., Fox, J. M., Bohannan, W. R., and White, R. A.,** Cost reduction ideas for an LNG Project, *Chem. Eng. Prog.,* 81 (4), 17, 1985.
419. **Gilwood, M. E.,** Saving capital and chemical with countercurrent ion exchange, *Chem. Eng.,* 74 (26), 83, 1967.
420. **Ginde, V. R. and Chu, C.,** An apparatus for desalination with ion exchange resins, *Desalination,* 10, 309, 1972.
421. **Gitchel, W. B., Meidl, J. A., and Burant, W., Jr.,** Powdered activated carbon by wet air oxidation, *AIChE Symp. Ser.,* 71(151), 414, 1975.
422. **Glajch, J. L. and Kirkland, J. J.,** Optimization of selectivity in liquid chromatography, *Anal. Chem.,* 55, 319A, 1983.
423. **Glasser, D.,** A new design for a continuous gas chromatography, in *Gas Chromatography 1966,* Littlewood, A. B., Ed., Institute of Petroleum, London, 1967, 119.
424. **Gledhill, C. and Slater, M. J.,** The design of deep fluidized bed continuous ion exchange equipment, in *The Theory and Practice of Ion Exchange,* Slater, M. J., Ed., Society of Chemical Industry, London, 1976, 46.1.
425. **Glueckauf, E.,** Contributions to the theory of chromatography, *Proc. R. Soc. London,* A186, 35, 1946.
426. **Glueckauf, E.,** Theory of chromatography. VII. The general theory of two solutes following non-linear isotherms, *Discuss. Faraday Soc.,* 7, 12, 1949.
427. **Glueckauf, E. and Coates, J. I.,** Theory of chromatography. IV. The influence of incomplete equilibrium on the front boundary of chromatograms and on the effectiveness of separation, *J. Chem. Soc.,* p. 1315, 1947.
428. **Godbille, E. and Devaux, P.,** Description and performance of an 8 cm id. column for preparative scale high pressure liquid-solid chromatography, *J. Chromatogr. Sci.,* 12, 564, 1974.
429. **Godbille, E. and Devaux, P.,** Use of an 18 mm id column for analytical and semipreparative-scale high pressure liquid chromatography, *J. Chromatogr.,* 122, 317, 1976.
430. **Gohr, E. J.,** Background, history and future of fluidization, in *Fluidization,* Othmer, D. E., Ed., Reinhold, New York, 1956, Chap. 4.
431. **Gold, H. and Calmon, C.,** Ion exchange: present status, needs and trends, *AIChE Symp. Ser.,* 76(192), 60, 1980.
432. **Gold, H. and Sonin, A. A.,** Design considerations for a truly continuous moving-bed ion exchange process, *AIChE Symp. Ser.,* 71(152), 48, 1975.
433. **Gold, H., Todisco, A., Sonin, A., and Probstein, R F.,** The Avco continuous moving bed ion exchange process and large scale desalting, *Desalination,* 17(1), 97, 1975.

434. **Golovoy, A. and Braslaw, J.,** Adsorption of automotive paint solvents on activated carbon. II. Adsorption kinetics of single vapors, *Environ. Prog.*, 1(2), 89, 1982.
435. **Gomez-Vaillard, R. and Kershenbaum, L. S.,** The performance of continuous cyclic ion-exchange reactors. II. Reactions with intraparticle diffusion controlled kinetics, *Chem. Eng. Sci.*, 36, 319, 1981.
436. **Gomez-Vaillard, R., Kershenbaum, L. S., and Streat, M.,** The performance of continuous, cyclic ion-exchange reactions. I. Reactions with film diffusion controlled kinetics, *Chem. Eng. Sci.*, 36, 307, 1981.
437. **Gondo, S., Itai, M., and Kushunoki, K.,** Computational and experimental studies on a moving ion exchange bed, *Ind. Eng. Chem. Fundam.*, 10, 140, 1971.
438. **Goodboy, K. P. and Fleming, H. L.,** Trends in adsorption with aluminas, *Chem. Eng. Progr.*, 80, (11), 63, 1984.
439. **Gordon, B. M., Rix, C. E., and Borgerding, M. E.,** Comparison of state-of-the-art column switching techniques in high resolution gas chromatography, *J. Chromatogr. Sci.*, 23, 1, 1985.
440. **Goto, S. and Matsubara, M.,** Extraction parametric pumping with reversible reaction, *Ind. Eng. Chem. Fundam.*, 16, 193, 1977.
441. **Goto, S., Sato, N., and Teshima, H.,** Periodic operation for desalting water with thermally regenerable ion-exchange resin, *Sep. Sci. Technol.*, 14, 209, 1979.
442. **Grant, J. T., Joyce, R. S., and Urbanic, J. E.,** The effect of relative humidity on the adsorption of water-immiscible organic vapors on activated carbon, in *Fundamentals of Adsorption*, Myers, A. L. and Belfort, G., Eds., Engineering Foundation, New York, 1984, 219.
443. **Graves, D. J. and Wu, Y.-T.,** On predicting the results of affinity procedures, in *Methods in Enzymology*, Vol. 34, *Affinity Techniques, Enzyme Purification*, Jakoby, W. B. and Wilchek, M., Eds., Academic Press, New York, 1974, 140.
444. **Graves, D. J. and Wu, Y.-T.,** The rational design of affinity chromatography separation processes, in *Advances in Biochemical Engineering*, Vol. 12, Ghose, T, K., Fiechter, A., and Blakebrough, N., Eds., Springer-Verlag, New York, 1979, 219.
445. **Gregory, R. A.,** Comparison of parametric pumping with conventional adsorption, *AIChE J.*, 20, 294, 1974.
446. **Gregory, R. A. and Sweed, N. H.,** Parametric pumping: behavior of open systems. I. Analytical solutions, *Chem. Eng. J.*, 1, 207, 1970.
447. **Gregory, R. A. and Sweed, N. H.,** Parametric pumping; behavior of open systems. II. Experiment and computation, *Chem. Eng. J.*, 4, 139, 1972.
448. **Grevillot, G.,** Equivalent staged parametric pumping. III. Open systems at steady state, McCabe-Thiele diagrams, *AIChE J.*, 26, 120, 1980.
449. **Grevillot, G.,** Principles of parametric pumping, in *Handbook for Heat and Mass Transfer Operations*, Cheremisinoff, N. P., Ed., Gulf Publishing, West Orange, N.J., chap. 36, in press.
450. **Grevillot, G., Bailly, M., and Tondeur, D.,** Thermofractionation: a new class of separation processes which combine the concepts of distillation, chromatography and the heat pump, *Int. Chem. Eng.*, 22, 440, 1982. Translated from *Entropie*, 16, #93, 3, 1980.
451. **Grevillot, G. and Dodds, J.,** Separation by thermal parametric pumping in adsorbing systems: a new approach, Inst. Chem. Eng. Symp. Series #54, *Alternatives to Distillation*, Rugby, England, 1978, 153.
452. **Grevillot, G., Dodds, J. A., and Marques, S.,** Separation of silver-copper mixtures by ion-exchange parametric pumping. I. Total reflux separations, *J. Chromatogr.*, 201, 329, 1980.
453. **Grevillot, G., Marques, S., and Tondeur, D.,** Donnan partition parametric pumping. Concentration and dilution effects on ion exchange parametric pumping, *React. Polym.*, 2, 71, 1984.
454. **Grevillot, G. and Tondeur, D.,** Equilibrium staged parametric pumping. I. Single transfer step per half cycle and total reflux — the analogy with distillation, *AIChE J.*, 22, 1055, 1976.
455. **Grevillot, G. and Tondeur, D.,** Equilibrium staged parametric pumping. II. Multiple transfer steps per half cycle and reservoir staging, *AIChE J.*, 23, 840, 1977.
456. **Grevillot, G. and Tondeur, D.,** Silver-copper separation by continuous ion exchange parametric pumping, in *Ion Exchange Technology*, Naden, D. and Streat, M., Eds., Ellis Horwood, Chichester, 1984, 653.
457. **Gribnau, T. C. J., Visser, J., and Nivard, R. J. F., Eds.,** *Affinity Chromatography and Related Techniques*, Elsevier, Amsterdam, 1982.
458. **Grimmett, E. S. and Brown, B. P.,** A pulsed column for countercurrent liquid-solids flow, *Ind. Eng. Chem.*, 54(11), 24, 1962.
459. **Gros, J. B., Lafaille, J. P., and Bugarel, R.,** Separation of binary mixtures by temperature effects of adsorption, *Inst. Chem. Eng. Symp. Ser.*, #54, p.97, 1978.
460. **Grubner, O. and Kucera, E.,** Countercurrent gas-liquid chromatography, *Collect. Czech. Chem. Commun.*, 29, 722, 1964.
461. **Guiochon, G., Beaver, L. A., Gonnord, M. F., Siouffi, A. M., and Zakaria, M.,** Theoretical investigation of the potentialities of the use of a multidimensional column in chromatography, *J. Chromatogr.*, 255, 415, 1983.

462. **Guiochon, G., Gonnord, M. F., Zakaria, M., Beaver, L. A., and Siouffi, A. M.,** Chromatography with a Two-Dimensional Column, *Chromatographia,* 17, 121, 1983.
463. **Guiochon, G., Jacob, L., and Valentin, P.,** Transformation des signaux finis dans une colonne chromatographique. VI. Optimisation numerique de la separation de deux corps, *Chromatographia,* 4, 6, 1971.
464. **Gupta, R. and Sweed, N. H.,** Equilibrium theory of cycling zone adsorption, *Ind. Eng. Chem. Fundam.,* 10, 280, 1971.
465. **Gupta, R. and Sweed, N. H.,** Modeling of nonequilibrium effects in parametric pumping, *Ind. Eng. Chem. Fundam.,* 12, 335, 1973.
466. **Haarhoff, P. C., Van Berge, P. C., and Pretorius, V.,** Role of the sample inlet volume in preparative chromatography, *Trans. Faraday Soc.,* 57, 1838, 1961.
467. **Hales, G. E.,** Drying reactive fluids with molecular sieves, *Chem. Eng. Prog.,* 67, (11), 49, 1971.
468. **Hall, K. B.,** Homestake uses carbon-in-pulp to recover gold from slimes, *World Mining,* p. 44, November 1974.
469. **Handelman, M. and Rogge, R. H.,** Ion-exchange refining of dextrose solutions, *Chem. Eng. Prog.,* 44, 583, 1948.
470. **Hanford, R., Maycock, W. d'A., and Vallet, L.,** Large scale purification of human plasma enzymes for clinical use, in *Chromatrography of Synthetic and Biological Polymers,* Vol. 2, Epton, R., Ed., Ellis Horwood, Chichester, 1978, chap. 9.
471. **Harper, J. I., Olsen, J. L., and Shuman, F. R.,** The Arosorb process, *Chem. Eng. Prog.,* 48, 276, 1952.
472. **Harwell, J. H., Liapis, A. I., Litchfield, R., and Hanson, D. T.,** A non-equilibrium model for fixed-bed multi-component adiabatic adsorption, *Chem. Eng. Sci.,* 35, 2287, 1980.
473. **Hashimoto, K., Miura, K., and Watanabe, T.,** Kinetics of thermal regeneration reaction of activated carbons used in waste water treatment, *AIChE J.,* 28, 737, 1982.
474. **Hassler, J. W.,** *Purification with Activated Carbon,* 3rd ed., Chemical Publishing, New York, 1974.
475. **Hatch, W. H., Balinski, G. J., and Ackermann, S.,** Magnetically Stabilized Bed Temperature, Partial Pressure Swing Hydrogen Recovery Process, U.S. Patent 4,319,893, March 16, 1982.
476. **Heaton, W. B.,** U.S. Patent 3,077, 103, 1963.
477. **Heck, J. L.,** First U.S. Polybed PSA unit proves its reliability, *Oil Gas J.,* 78(6), 122, 1980.
478. **Heckendorf, A., Ashore, E., Tracy, C., and Rausch, C. W.,** Process scale chromatography, the new frontier in HPLC. Paper MBT 90 presented at ACS Meeting, Washington, D.C., Sept. 2, 1983.
479. **Hedge, J. A.,** Separation of 2,7 dimethylnaphthalene from 2,6 dimethylnaphthalene with molecular sieves, in *Molecular Sieve Zeolites-II,* Flanigen, E. M., and Sand, L. M., Eds., ACS Advances in Chem. Ser., #102, American Chemical Society, Washington, D.C., 1971, 238.
480. **Heftmann, E.,** Liquid chromatography — past and future, *J. Chromatogr. Sci.,* 11, 295, 1973.
481. **Heftmann, E., Krochta, J. M., Farkas, D. F., and Schwimmer, S.,** The chromatofuge, an apparatus for preparative rapid radial column chromatography, *J. Chromatogr.,* 66, 365, 1972.
482. **Heikkila, H.,** Separating sugars and amino acids with chromatography, *Chem. Eng.,* 90(2), 50, 1983.
483. **Helbig, W. A.,** Adsorption from solution by activated carbon, in *Colloid Chemistry. Theoretical and Applied,* Vol. VI, Alexander, J., Ed., Reinhold, New York, 1946, 814.
484. **Helfferich, F.,** *Ion Exchange,* McGraw-Hill, New York, 1962.
485. **Helfferich, F. G.,** Conceptual view of column behavior in multicomponent adsorption or ion-exchange systems, *AIChE Symp. Ser.,* 80 (233), 1, 1984.
486. **Helfferich, F. and Klein, G.,** *Multicomponent Chromatography,* Marcel Dekker, New York, 1970.
487. **Hengstebeck, R. J.,** *Petroleum Processing. Principles and Applications,* McGraw-Hill, New York, 1959, chap. 4.
488. **Henley, E. J. and Seader, J. D.,** *Equilibrium-Stage Separation Operations in Chemical Engineering,* Wiley, New York, 1981.
489. **Hernandez, L. A. and Harriott, P.,** Regeneration of powdered active carbon in fluidized bed, *Environ. Sci. Technol.,* 10, 454, 1976.
490. **Herve, D.,** Ion exchange in the sugar industry -2, *Process Biochem.,* (5), 31, 1974.
491. **Hibshman, H. J.,** Separation of iso- and normal paraffins by adsorption, *Ind. Eng. Chem.,* 42, 1310, 1950.
492. **Hiester, N. K., Fields, E. F., Phillips, R. C., and Radding, S. B.,** Continuous countercurrent ion exchange with trace components, *Chem. Eng. Prog.,* 50, 139, 1954.
493. **Hiester, N. K. and Michalson, A. W.,** Ion-exchange and adsorption equipment, in *Chemical Engineer's Handbook,* 5th ed., Perry, R. H. and Chilton, C. H., Eds., McGraw-Hill New York, 1973, 19—33.
494. **Hiester, N. K. and Phillips, R. C.,** Ion exchange, *Chem. Eng.,* 61(10), 161, 1954.
495. **Hiester, N. K., Phillips, R. C., Fields, E. F., Cohen, R. K., and Redding, S. B.,** Ion exchange of trace components in a countercurrent equilibrium stage contactor, *Ind. Eng. Chem.,* 45, 2402, 1953.
496. **Higgins, I. R.,** Counter-Current Liquid-Solid Mass Transfer Methods and Apparatus, U.S. Patent 2,815,332, Dec. 1957.

497. **Higgins, I. R.,** Continuous ion-exchange equipment, *Ind. Eng. Chem.*, 52, 635, 1961.
498. **Higgins, I. R.,** Use ion exchange when processing brine, *Chem. Eng. Prog.*, 60(11), 60, 1964.
499. **Higgins, I. R. and Chopra, R. C.,** Chem-Seps continuous ion-exchange contactor and its applications to demineralization processes, in *Ion Exchange in the Process Industries,* Society of Chemical Industry, London, 1970, 121.
500. **Higgins, I. R. and Chopra, R. C.,** Municipal waste effluent treatment, in *Ion Exchange for Pollution Control,* Vol. II, Calmon, C. and Gold, H., Eds., CRC Press, Boca Raton, Fla., 1979, chap. 7.
501. **Higgins, I. R. and Roberts, J. T.,** A countercurrent solid-liquid contactor for continuous ion exchange, *Chem. Eng. Prog. Symp. Ser.*, 50 (#14), 87, 1954.
502. **Hildebrand, F. B.,** *Advanced Calculus for Engineers,* Prentice-Hall, Englewood Cliffs, N.J., 1948, 444.
503. **Hill, E. A. and Hirtenstein, M. D.,** Affinity chromatography: its application to industrial scale processes, in *Advances Biotechnological Processes,* Vol. 1, Mizrahi, A. and Van Wezel, A. L., Eds., Alan R. Liss, New York, 1983, 31.
504. **Hill, F. B.,** Recovery of a weakly adsorbed impurity by pressure swing adsorption, *Chem. Eng. Commun.*, 7, 37, 1980.
505. **Hill, F. B., Wong, Y. W., and Chan, Y. N. I.,** A temperature swing process for hydrogen isotope separation, *AIChE J.*, 28, 1, 1982.
506. **Hill, F. B., Wong, Y. W., and Chan, Y. N.,** Tritium removal using vanadium hydride, in *Proc. 15th DOE Nucl. Air Cleaning Conf.*, Vol. 1, First, M. W., Ed., Department of Energy, Washington, D.C., 1978, 167.
507. **Himmelstein, K. J., Fox, R. D., and Winter, T. H.,** In-place regeneration of activated carbon, *Chem. Eng. Prog.*, 69 (11), 65, 1973.
507a. **Himsley, A.,** Performance of Himsley continuous ion exchange system, Society of Chemical Industry Symposium — Hydrometallurgy, 81, Manchester, June 30 — July 3, 1981, paper E3.
508. **Himsley, A. and Bennett, J. A.,** A new continuous packed-bed ion exchanged system applied to treatment of mine water, in *Ion Exchange Technology,* Naden D. and Streat, M., Eds., Ellis Horwood, Chichester, 1984, 144.
509. **Himsley, A. and Farkas, E. J.,** Operating and design details of a truly continuous ion exchange system, in *The Theory and Practice of Ion Exchange,* Streat, M., Ed., Society of Chemical Industry, London, 1976, 45.7.
510. **Hines, A. L. and Maddox, R. N.,** *Mass Transfer Fundamentals and Applications,* Prentice-Hall, Englewood Cliffs, N.J., 1985, chap. 14.
511. **Hiroda, A. L. and Shioda, K.,** A Method for the Elimination of Oligosacchrides, Japanese Patent Office Patent Journal. Kokai Patent No. SH055, U1980e-48400, April 7, 1980.
512. Hitachi, Ltd., Pressure Swing-type adsorption columns, Japan Kokai Tokkyo Koho, 80 61 915, May 10, 1980.
513. **Hoffman, P. G., Zapf, M., O'Keefe, J., and Chapelet, G.,** Application of preparative gas chromatography in the flavor industry. Paper 51a presented at AIChE Meeting, Philadelphia, Pa., Aug. 1984.
513a. **Holland, C. D. and Liapis, A. I.,** *Computer Methods for Solving Dynamic Separation Problems,* McGraw-Hill, New York, 1983.
514. **Hollein, H. C., Ma, H.-C., Huang, C.-R., and Chen, H. T.,** Parametric pumping with pH and electric field: protein separations, *Ind. Eng. Chem. Fundam.*, 21, 205, 1982.
515. **Hopf, P. P.,** Radial chromatography in industry, *Ind. Eng. Chem.*, 39, 938, 1947.
516. **Horn, F. J. M. and Lin, C. H.,** On parametric pumping in linear columns under conditions of equilibrium and non-dispersive flow, *Ber. Bunsenges. Phys. Chem.*, 73, 575, 1969.
517. **Horvath, C. and Lin, H.-J.,** Band spreading in liquid chromatography. General plate height equation and a method for the evaluation of the individual plate height contribution, *J. Chromatogr.*, 149, 43, 1978.
518. **Horvath, C., Nahum, A., and Frenz, J. H.,** High performance displacement chromatography, *J. Chromatogr.*, 218, 365, 1981.
519. Howe-Baker Engineers, Inc., Custom Designed Hydrogen Plants SAS III, Box 956, Tyler, Tex, 75710, Dec. 1980.
520. **Hubbard, R.,** Monoclonal antibodies: production, properties and applications, in *Topics in Enzyme and Fermentation Biotechnology,* Vol. 7, Wiseman, A., Ed., Ellis Horwood, Chichester, 1983. chap. 7.
521. **Hughes, J. J. and Charm, S. E.,** Method for continuous purification of biological material using immunosorbent, *Biotechnol. Bioeng.*, 21, 1431, 1979.
522. **Humenick, M. J., Jr.,** *Water and Wastewater Treatment. Calculations for Chemical and Physical Processes,* Marcel Dekker, New York, 1977, chap. 6.
523. **Humphreys. R. L.,** Cyclic adsorption refining, *Petrol. Refiner.*, 32(9), 99, 1953.
524. **Hupe, K. P.,** Design procedures for preparative and production gas chromatography, *J. Chromatogr.*, 203, 41, 1981.
525. **Husband, W. H., Barker, P. E., and Kimi, K. D.,** The separation of liquid mixtures by vapour-phase adsorption in a moving-bed column, *Trans. Inst. Chem. Eng.*, 42, T387, 1964.

526. **Hutchins, R. A.,** Activated-carbon systems for separation of liquids, in *Handbook of Separation Techniques for Chemical Engineers,* Schweitzer, P. A., Ed., McGraw-Hill, New York, 1979, Section 1.13.
527. **Hutchins, R. A.,** Development of design parameters, in *Activated Carbon Adsorption for Wastewater Treatment,* Perrich, J. R., Ed., CRC Press, Boca Raton, Fla., 1981, chap. 4A.
528. HP Chemicals, Inc., Modular preparative HPLC columns, April 1984; Liquid chromatography products, March 1985; A unique approach to liquid chromatography, March, 1985, 4221 Forest Park Blvd., St. Louis, Mo. 63108.
529. **Ikoku, C. U.,** *Natural Gas Engineering. A Systems Approach,* Pennwell, Tulsa, Okla, 1980, chap. 4.
530. Illinois Water Treatment Co., IWT DESEP process, *Making Waves in Liquid Processing,* Vol. 2, No. 1, 4669 Shepherd Trail, Rockford, Ill., 61105, 1984, 1.
531. Illinois Water Treatment Co., IWT Adsep System, *Making Waves in Liquid Processing,* Vol. 1, No. 1, Rockford, Ill., 1984, 1.
532. Institut Francais du Petrole, *N-ISELF,* Rueil-Malmaison, France, 1981.
533. **Ishikawa, H., Tanabe, H., and Usui, K.,** Process of the operation of a simulated moving bed, U.S. Patent 4,182,633, Jan. 8, 1980.
534. **Iwasyk, J. M. and Thodos, G.,** Continuous solute removal from aqueous solutions, *Chem. Eng. Prog.,* 54(4), 69, 1958.
535. ICI Australia Petrochemicals Ltd., Alkali and Chemical Gp., Sirotherm ion exchange resins for desalination. Process Information., Melbourne, Australia, no date.
536. **Jacob, L. and Guiochon, G.,** Theory of Chromatography at finite concentrations, *Chromatogr. Rev.,* 14, 77, 1971.
537. **Jacob, P. and Tondeur, D.,** Nonisothermal adsorption: separation of gas mixtures by modulation of feed temperature, *Sep. Sci. Technol.,* 15, 1563, 1980.
538. **Jacob, P. and Tondeur D.,** Non-isothermal gas adsorption in fixed beds. I. A simplistic linearized equilibrium model, *Chem. Eng. J.,* 22, 187, 1981.
539. **Jacob, P. and Tondeur, D.,** Non-isothermal gas adsorption in fixed beds. II. Nonlinear equilibrium theory and "Guillotine" effect, *Chem. Eng. J.,* 26, 41, 1983.
540. **Jacob, P. and Tondeur, D.,** Adsorption Non-Isotherme de gaz en lit fixe. III. Etude experimental des effets de Guillotine et de focalisation. Separation n-pentane/isopentane sur tamis 5A, *Chem. Eng. J.,* 26, 143, 1983.
541. **Jacob, P. and Tondeur, D.,** Non-isothermal gas adsorption in fixed beds. IV. Feasibility, design, performance and simulation of "Guillotine" process, *Chem. Eng. J.,* 31, 23, 1985.
542. **James, A. T. and Martin, A. J. P.,** Gas liquid partition chromatography: the separation and microestimation of volatile fatty acids. Formic acid to dodecanoic acid, *Biochem. J.,* 50, 679, 1952.
543. **Jamrack, W. D.,** *Rare Metal Extraction by Chemical Engineering Methods,* Pergamon Press, New York, 1963, chap. 3.
544. **Janson, J.-C.,** Columns for large-scale gel filtration on porous gels, *J. Agric. Food Chem.,* 19, 581, 1971.
545. **Janson, J.-C.,** Large scale chromatography of proteins, International Workshop on Technology for Protein Separation and Improvement of Blood Plasma Fractionation, Reston, Va., Sept. 7—9, 1977.
546. **Janson, J.-C.,** Scaling-up of affinity chromatography. Technological and economical aspects, in *Affinity Chromatography and Related Techniques,* Gribnau, T. C. J., Visser, J., and Nivard, R. J. F., Eds., Elsevier, Amsterdam, 1982, 503.
547. **Janson, J.-C.,** Large-scale affinity purification — state of the art and future prospects, *Trends Biotechnol.,* 2 (2), 31, 1984.
548. **Janson, J.-C. and Dunnill, P.,** Factors affecting scale-up of chromatography, in *Industrial Aspects of Biochemistry,* Vol. 30, Part I, Spencer, B., Ed., North-Holland/American Elsevier, Amsterdam, and New York, 1974, 81.
549. **Janson, J.-C. and Hedman, P.,** Large-scale chromatography of proteins, in *Advances in Biochemical Engineering,* Vol. 25, *Chromatography,* Fiechter, A., Ed., Springer-Verlag, Berlin, 1982, 43.
550. **Jaraiz-M., E., Zhang, G.-T., Wang, Y., and Levenspiel, O.,** The magnetic distributor — downcomer (MDD) for fluidized beds, *Powder Technol.,* 38, 53, 1984.
551. **Jenczewski, T. J. and Myers, A. L.,** Parametric pumping separates gas phase mixtures, *AIChE J.,* 14, 509, 1968.
552. **Jenczewski, T. J. and Myers, A. L.,** Separation of gas mixtures by pulsed adsorption, *Ind. Eng. Chem. Fundam.,* 9, 216, 1970.
553. **Jensen, R. A.,** Semi-continuous activated carbon system for wastewater treatment, *AIChE Symp. Ser.,* 76(197), 72, 1970.
554. **Johansson, R. and Neretnieks. I.,** Adsorption on activated carbon in countercurrent flow. An experimental study, *Chem. Eng. Sci.,* 35, 979, 1980.
555. **Johnson, A. S. W. and Turner, J. C. R.,** Distribution of residence-times of the resin in a counter-current continuous ion-exchange column, in *Ion Exchange in the Process Industries,* Society of Chemical Industry, London, 1970, 140.

556. **Joithe, W., Bell, A. T., and Lynn, S.,** Removal and recovery of NO_2 from nitric acid plant tail gas by adsorption on molecular sieves, *Ind. Eng. Chem. Proc. Des. Devel.*, 11, 434, 1972.
557. **Jones, J. L. and Gwinn, J. E.,** Activated carbon regeneration apparatus, *Environ. Prog.*, 2, 226, 1983.
558. **Jones, R. L. and Keller, G. E.,** Pressure-swing parametric pumping — a new adsorption process, *J. Sep. Proc. Technol.*, 2(3), 17, 1981.
559. **Jones, R. L., Keller, G. E., and Wells, R. C.,** Rapid pressure swing adsorption process with high enrichment factor, U.S. Patent 4,194,892, March 15, 1980.
560. **Juhola, A. J.,** Package sorption device system study, EPA Publ. PB-221138, 1973.
561. **Juntgen, H.,** New applications for carbonaceous adsorbents, *Carbon*, 15, 273, 1977.
562. **Kahle, H.,** *Chem. Eng. Technol.*, p.75, 1954.
563. Kahn, Co., Compressed air or gas heaterless dryers, Product Data B-30B, 885 Wells Rd., Wethersfield, Conn. 16109, 1974.
564. **Kaiser, R.,** Temperature and flow programming, in *Preparative Gas Chromatography*, Zlatkis, A. and Pretorius, V., Eds., Wiley-Interscience, New York, 1971, 187.
565. **Kalasz, H. and Horvath, C.,** Preparative-scale separation of polymyxins with an analytical high-performance liquid chromatography system by using displacement chromatography, *J. Chromatogr.*, 215, 295, 1981.
566. **Kapfer, W. H., Malow, M., Happel. J., and Marsel, C. J.,** Fractionation of gas mixtures in a moving-bed adsorber, *AIChE J.*, 2, 456, 1956.
567. **Karasek, F. W.,** Laboratory prep-scale liquid chromatography, *Research/Development*, 28 (7), 32, 1977.
568. **Karlson, E. L. and Edkins, S. P.,** A new continuous ion exchange system, *AIChE Symp. Ser.*, 71 (151), 286, 1975.
569. **Karnofsky, G.,** Let's look at selective adsorption, *Chem. Eng.*, 67(9), 189, 1954.
570. **Kars, R. L., Best, R. J., and Drinkenburg, A. A. H.,** The sorption of propane in slurries of active carbon in water, *Chem. Eng. J.*, 17, 201, 1979.
571. **Kast, W.,** Adsorption from the gas phase — fundamentals and processes, *Ger. Chem. Eng.*, 4, 265, 1981.
572. **Kasten, P. R. and Amundson, N. R.,** Analytical solutions for simple systems in moving bed adsorbers, *Ind. Eng. Chem.*, 44, 1704, 1952.
573. **Katoh, S., Kambayashi, T., Deguchi, R., and Yoshida, F.,** Performance of affinity chromatography columns, *Biotechnol. Bioeng.*, 20, 267, 1978.
574. **Katooka, T. and Yoshida, H.,** Intraparticle mass transfer in thermally regenerable ion exchanger, *Chem. Eng. Commun.*, 34, 187, 1985.
575. **Katz, D. L., Cornell, O., Kobayashi, R., Poettmann, F. H., Vary, J. A., Elenbaas, J. R., and Weinaug, C. F.,** *Handbook of Natural Gas Engineering*, McGraw-Hill, New York, 1959, chap. 16.
576. **Kehde, H., Fairfield, R. G., Frank, J. C., and Zahnstecher, L. W.,** Ethylene recovery. Commercial hypersorption operation, *Chem. Eng. Prog.*, 44, 575, 1948.
577. **Keinath, T. M. and Weber, W. J., Jr.,** A predictive model for the design of fluid-bed adsorbers, *J. Water Polut. Control Fed.*, 40, 741, 1968.
578. **Keller, G. E., II,** Gas-adsorption processes: state of the art, in *Industrial Gas Separations*, Whyte, T. E., Jr., Yon, C. M., and Wagener, E. H., Eds., ACS Symp. Ser., #223, American Chemical Society, Washington, D.C., 1983, 145.
579. **Keller, G. E., II** A new approach to adsorptive separation of gases, Seminar Purdue University, West Lafayette, Ind., Feb. 28, 1985.
580. **Keller, G. E., II and Jones, R. L.,** A new process for adsorption separation of gas streams, in *Adsorption and Ion Exchange with Synthetic Zeolites*, Flank, W. H., Ed., ACS Symp. Ser., #135, American Chemical Society, Washington, D.C., 1980, 275.
581. **Keller, G. E. and Kuo, C.-H.A.,** Enhanced gas separation by selective adsorption, U.S. Patent 4,354,859, Oct. 19, 1982.
582. **Kemmer, F. N., Ed.,** *The NALCO Water Handbook*, McGraw-Hill, New York, 1979, chap. 17.
583. **Kenney, C. N. and Kirkby, N. F.,** Pressure swing adsorption, in *Zeolites: Science and Technology*, Ribeiro, F., Rodriques, A. E., Rollman, L. D. and Naccache, C., Eds., Martinas Nijhoff, The Hague, 1984.
584. **Kersten, R. C. and Asselin, G. F.,** Impact of recently developed processes on refining technology, *Proc. Am. Petrol. Inst.*, 41, (111), 339, 1960.
585. **Kim, D. K.,** Parametric Pumping in a Packed Catalytic Column, Ph.D. thesis, Northwestern University, Evanston, Ill., 1971.
586. **King, C. J.,** *Separation Processes*, 2nd ed., McGraw-Hill New York, 1980.
587. **King, W. H., Jr.,** Generation of clean air to permit gas chromatography without gas cylinders, *Anal. Chem.*, 43, 984, 1971.
588. **Kiovsky, J. R., Koradia, P. B., and Hook, D. S.,** Molecular sieve for SO_2 removal, *Chem. Eng. Prog.*, 72 (8), 98, 1976.

589. **Kipling, J. J.,** *Adsorption from Solutions of Non-Electrolytes,* Academic Press, London, 1965.
590. **Kirkegaard, L. H.,** Gradient sievorptive chromatography. A focusing system for the separation of cellular components, *Biochemistry,* 12, 3627, 1973.
591. **Klaus, R., Aiken, R. C., and Rippen, D. W . T.,** Simulated binary isothermal adsorption on activated carbon in periodic countercurrent column operation, *AIChE J.,* 23, 579, 1977.
592. **Klawviter, J., Kaminski, M., and Kowalczyak, J. S.,** Investigation of the relationship between packing methods and efficiency of preparative columns. I. Characteristics of the tamping method for packing preparative columns, *J. Chromatogr.,* 243, 207, 1982.
593. **Kalwviter, J., Kaminski, M., and Kowalczyak, J. S.,** Investigation of the relationship between packing methods and efficiency of preparative columns. II. Characteristics of the slurry method of packing chromatographic columns, *J. Chromatogr.,* 243, 225, 1982.
594. **Klei, H. E., Sahagian, J., and Sundstrom, D. W.,** The effect of thermal regeneration conditions on the loss of activated carbon, *AIChE Symp. Ser.,* 71(151), 387, 1975.
595. **Klein, G.,** Design and development of cyclic operations, in *Percolation Processes, Theory and Applications,* Rodrigues, A. E. and Tondeur, D. Eds., Sijthoff & Noordhoff, Alpheen aan den Rijn, The Netherlands, 1981, 427.
596. **Klein, G., Nassiri, M., and Vislocky, J. M.,** Multicomponent fixed-bed sorption with variable inlet and feed compositions. Computer prediction of local equilibrium behavior, *AIChE Symp. Ser.,* 80(233), 14, 1984.
597. **Knaebel, K. S.,** Multistage cycling zone adsorption for purification of binary mixtures, *AIChE Symp. Ser.,* 78(219), 128, 1982.
598. **Knaebel, K. S.,** Analysis of complimentary pressure swing adsorption, in *Fundamentals of Adsorption,* Myers, A. L. and Belfort, G., Eds., Engineering Foundation, New York, 1984, 273.
599. **Knabel, K. S. and Hill, F. B.,** Analysis of gas purification by heatless adsorption, AIChE Meeting, Los Angeles, Nov. 1982.
600. **Knaebel, K. S. and Hill, F. B.,** Analysis of gas purification by pressure-swing adsorption: priming the parametric pump, *Separ. Sci. Technol.,* 18, 1193, 1983.
601. **Knaebel, K. S. and Pigford, R. L.,** Equilibrium and dissipative effects in cycling zone adsorption, *Ind. Eng. Chem. Fundam.,* 22, 336, 1983.
602. **Knoblauch, K.,** Pressure-swing adsorption: geared for small-volume users, *Chem. Eng.,* 85, (25), 87, 1978.
603. **Knoblauch, K., Reichenberger, J., and Juntgen, H.,** Molekularsiebkokse zur Gewinung von sauerstoff und stickstoff, *gwf/Gas/Erdgas,* 116, 382, 1975.
604. **Knopf, F. C. and Rice, R. G.,** Adsorptive distillation: optimum solids profiles, *Chem. Eng. Commun.,* 15, 109, 1982.
605. **Knox, J. H.,** Practical aspects of LC theory, *J. Chromatogr., Sci.,* 15, 352, 1977.
606. **Knox, J. H. and Scott, H. P.,** B and C terms in the Van Deemter equation for liquid chromatography, *J. Chromatogr.,* 282, 297, 1983.
607. **Koches, C. F. and Smith, S. B.,** Reactive powdered carbon, *Chem. Eng.,* 79(9), 46, 1972.
608. **Kohl, A. L. and Riesenfeld, F. C.,** *Gas Purification,* 3rd ed., Gulf Publishing, Houston Tex., 1979, chap. 12.
609. **Koloini, T., Rodman, P., and Zumer, M.,** Reversible ion exchange in fluidized beds from dilute ternary solutions, *Chem. Eng. Commun.,* 26, 297, 1984.
610. **Koloini, T. and Zumer, M.,** Ion exchange with irreversible reaction in deep fluidized beds, *Can. J. Chem. Eng.,* 57, 183, 1979.
611. **Koloini, T. and Zumer, M.,** Irreversible ion exchange in deep fluidized beds and in series of shallow liquid fluidized beds, *Can. J. Chem. Eng.,* 57, 770, 1979.
612. **Koo, Y.-M. and Wankat, P. C.,** Size-exclusion parametric pumping, *Ind. Eng. Chem. Fundam.,* 24, 108, 1985.
613. **Kopaciewicz, W. and Regnier, F. E.,** A system for coupled multiple-column separation of proteins, *Anal. Biochem.,* 129, 472, 1983.
614. **Kopperschlager, G., Bohme, H.-H., and Hofmann, E.,** Cibacron Blue F3G-A and related dyes as ligands in affinity chromatography, in *Advances in Biochemical Engineering,* Vol. 25, Fiechter, A., Ed., Springer Verlag, New York, 1982, 102.
615. **Koradia, P. B. and Kiovsky, J. R.,** Drying chlorinated hydrocarbons, *Chem. Eng. Prog.,* 73(4), 105, 1977.
616. **Kovach, J. L.,** Gas phase adsorption, in *Handbook for Separation Techniques for Chemical Engineers,* Schweitzer, P. A., Ed., McGraw-Hill, New York, 1979, Sect. 3.1.
617. **Kowler, D. E. and Kadlec, R. H.,** The optimal control of a periodic adsorber. I. Experiment. II. Theory, *AIChE J.,* 18, 1207 and 1212, 1972.

618. **Krill, H.,** Industrial off-gas purification with activated carbon and inorganic solvents, in *Sorption and Filtration Methods for Gas and Water Purification,* Bonnievie-Svendsen, M., Ed., NATO ASI, Noordhoff, Leyden, The Netherlands, 1977, 485.
619. **Kumar, R., Sircar, S., White, T. R., and Greskovich, E. J.,** Argon Purification, U.S. Patent 4,477,265, Oct. 16, 1984.
620. **Kunii, D. and Levenspiel, O.,** *Fluidization Engineering,* Wiley, New York, 1969.
621. **Kunin, R.,** *Elements of Ion Exchange,* Reinhold, New York, 1960.
622. **Kunin, R.,** *Ion Exchange Resins,* 2nd ed., Robert E. Krieger, Huntington, Ky., 1972.
623. **Kunin, R., Tavares, A., Forman, R., and Wilker, G.,** New developments in the use of ion-exchangers and adsorbents as precoat filters, in *Ion Exchange Technology,* Naden D. and Streat M., Eds., Ellis Horwood, Chichester, 1984, 563.
624. **Kuo, Y.,** Exsorption: a new separation process, *Chem. Eng. Prog.,* 80(12), 37, 1984.
625. **Ladisch, M. R., Voloch, M., Hong, P., Bienkowski, P., and Tsao, G. T.,** Cornmeal adsorber for dehydrating ethanol vapors, *Ind. Eng. Chem. Process Des. Devel.,* 23, 437, 1984.
626. **Ladisch, M. R., Voloch, M., and Jacobson, B.,** Bioseparations: column design factors in liquid chromatography, *Biotechnol. Bioeng. Symp.,* No. 14, 1984.
627. **Landolt, G. R. and Kerr, G. T.,** Methods of separation and purification using the molecular sieving properties of zeolites, *Sep. Purific. Methods,* 2, 283, 1973.
628. **Langer, S. H. and Patton, J. E.,** Chemical reactor application of the gas chromatographic column, in *New Developments in Gas Chromatography,* Purnell, J. H., Ed., Wiley, New York, 1973, 293.
629. **Langmuir, I.,** The adsorption of gases on plane surfaces of glass, mica and platinum, *J. Am. Chem. Soc.,* 40, 1361, 1918.
630. **Lapidus, L. and Amundson, N. R.,** Mathematics of adsorption in beds. VI. The effect of longitudinal diffusion in ion exchange and chromatographic columns, *J. Phys. Chem.,* 56, 984, 1952.
631. **Larsson, P. O.,** High performance liquid affinity chromatography, in *Methods in Enzymology,* Vol. 104, *Enzyme Purification and Related Techniques,* Jakoby, W. B., Ed., Academic Press, Orlando, Fla., 1984, 212.
632. **Lasch, J. and Koelsch, R.,** Enzyme leakage and multipoint attachment of agarose-bound enzyme preparations, *Eur. J. Biochem.,* 82, 181, 1978.
633. **Latty, J. A.,** The Use of Thermally Sensitive Ion Exchange Resins or Electrically Sensitive Liquid Crystals as Adsorbents, Ph.D. thesis, University of California-Berkeley, California, 1974.
634. **Laub, R. J. and Zink, D. L.,** Rotating-disk thin layer chromatography, *Am. Lab.,* 13(1), 55, 1981.
635. **Lavie, R. and Rielly, M. J.,** Limit cycles in fixed beds operated in alternating modes, *Chem. Eng. Sci.,* 27, 1835, 1972.
636. **Lawson, C. T. and Fisher, J. A.,** Limitations of activated carbon adsorption for upgrading petrochemical effluents, *AIChE Symp. Ser.,* 70 (#136), 577, 1974.
637. **Laxen, P. A., Becker, G. S. M., and Rubin, T.,** The carbon-in-pulp gold recovery process — a major breakthrough, *S. Af. Min. Eng. J.,* 90, 4152, 43, 1979.
638. **Leatherdale, J. W.,** Air pollution control by adsorption, in *Carbon Adsorption Handbook,* Cheremisinoff, P. N. and Ellerbusch, F., Eds., Ann Arbor Science, Ann Arbor, Mich., 1978, chap. 11.
639. **Leavitt, F. W.,** Non-isothermal adsorption in large fixed beds, *Chem. Eng. Prog.,* 58(8), 54, 1962.
640. **Ledoux, E.,** Avoiding destructive velocity through adsorbent beds, *Chem. Eng.,* 55(3), 118, 1948.
641. **Lee, H.,** Applied aspects of zeolite adsorbents, in *Molecular Sieves,* Meier, W. M. and Uytterhoeven, J. B., Eds., Advances in Chemistry Series #121, American Chemical Society, Washington, D.C., 1973, 311.
642. **Lee, H. and Cummings, W. P.,** A new design method for silica gel air driers under nonisothermal conditions, *Chem. Eng. Prog. Symp. Ser.,* 63(#74), 42, 1967.
643. **Lee, H. and Stahl, D. E.,** Oxygen rich gas from air by pressure swing adsorption process, *AIChE Symp. Ser.,* 69(134), 1, 1973.
644. **Lee, K. C. and Kirwan, D. J.,** Cycling electrosorption, *Ind. Eng. Chem. Fundam.,* 14, 279, 1975.
645. **Lee, M. N. Y.,** Novel separation with molecular sieves adsorption, in *Recent Developments in Separation Science,* Vol. I, Li, N. N., Ed., CRC Press, Boca Raton, Fla., 1972, 75.
646. **Lerch, R. G. and Ratowsky, D. A.,** Optimum allocation of adsorbent in stagewise adsorption allocations, *Ind. Eng. Chem. Fundam.,* 6, 308, 480 (erratum), 1967.
647. **Lergmigeaux, G. and Roques, H.,** L'echange d'ions continu etat actual de developpement des procedes, *Chim. Ind. Gen. Chim.,* 105 (12), 725, 1972.
648. **Letan, R.,** A continuous ion-exchanger, *Chem. Eng. Sci.,* 28, 982, 1973.
649. **Levendusky, J. A.,** Progress report on the continuous ion exchange process, *CEP Symp. Ser.,* 65 (97), 113, 1969.
650. **Levinson, S. Z. and Orochko, D. I.,** Staged countercurrent multisectional contactors for continuous adsorptive treatment of petroleum products, *Int. Chem. Eng.,* 7, 649, 1967.

651. **Lewis, W. K., Gilliland, E. R., Chertow, B., and Cadogan, W. P.,** Adsorption equilibria: hydrocarbon gas mixtures, *Ind. Eng. Chem.*, 42, 1319, 1950.
652. **LeVan, M. D. and Vermeulen, T.,** Binary Langmuir and Freundlich isotherms for ideal adsorbed solutions, *J. Phys. Chem.*, 85, 3247, 1981.
653. **LeVan, M. D. and Vermeulen, T.,** Channeling and bed-diameter effects in fixed-bed adsorber performance, *AIChE Symp. Ser.*, 80 (233), 34, 1984.
654. **Liapis, A. I. and Rippin, D. W. T.,** The simulation of binary adsorption in activated carbon columns using estimates of diffusional resistance within the carbon particles derived from batch experiments, *Chem. Eng. Sci.*, 33, 593, 1978.
655. **Liapis, A. T. and Rippen, D. W. T.,** The simulation of binary adsorption in continuous countercurrent operation and a comparison with other operating modes, *AIChE J.*, 25, 455, 1979.
656. **Liberti, L. and Helfferich, F. G., Eds.,** *Mass Transfer and Kinetics of Ion Exchange*, Martinus Nijhoff, The Hague, 1983.
657. **Lightfoot, E. N., Sanchez-Palma, R. J., and Edwards, D. O.,** Chromatography and allied fixed-bed separations processes, in *New Chemical Engineering Separation Techniques*, Schoen, H. M., Ed., Interscience, New York, 1962, 99.
658. **Lindblom, H. and Fagerstam, L. G.,** An automatic multidimensional chromatography system for the separation of proteins, *Liquid Chromatogr.*, 3, 360, 1985.
659. **Lindquest, L. O. and Williams, K. W.,** Aspects of whey processing by gel filtration, *Dairy Ind.*, 38, 459, 1973.
660. **Liteanu, C. and Gocan, S.,** *Gradient Liquid Chromatography*, Ellis Horwood, Chichester, 1974.
661. **Little, C. J. and Stahel, O.,** Role of column switching in semi-preparative liquid chromatography. Isolation of the sweetener stevioside, *J. Chromatogr.*, 316, 105, 1984.
662. **Little, C. J. an Stahel, O.,** Increased sample throughput in HPLC using sample switching, *Chromatographic*, 19, 322, 1984.
663. **Little, J. N., Cotter, R. L., Prendergast, J. A., and McDonald, P. D.,** Preparative liquid chromatography using radially compressed columns, *J. Chromatogr.*, 126, 439, 1976.

663a. LKB Produkter AB, Ultrogel and Trisacryl, A Practical Guide for Use in Gel Filtration, Box 305, S-16126, Bromma, Sweden, 1984.

664. **Loven, A. W.,** Perspectives on carbon regeneration, *Chem. Eng. Prog.*, 69 (11), 56, 1973.
665. **Loven, A. W.,** Activated carbon regeneration perspectives, *AIChE Symp. Ser.*, 70(144), 285, 1974.
666. **Lovering, P. E.,** Gas adsorbents for breathing apparatus and respirators, *Processes for Air Pollution Control*, Nonhebel, G., Ed., CRC Press, Boca Raton, Fla., 1972, chap. 17B.
667. **Lovett, W. D. and Cunniff, F. T.,** Air pollution control by activated carbon, *Chem. Eng. Prog.*, 70 (5), 43, 1974.
668. **Lowe, C. R.,** Applications of reactive dyes in biotechnology, in *Topics in Enzyme and Fermentation Biotechnology*, Vol. 9, Wiseman, A., Ed., Ellis Horwood, Chichester, 1984, chap. 3.
669. **Lowe, C. R. and Dean, P. D. G.,** *Affinity Chromatography*, Wiley, New York, 1974.
670. **Lowe, C. R. and Pearson, J. C.,** Affinity chromatography on immobilized dyes, in *Methods in Enzymology*, Vol. 104, *Enzyme Purification and Related Techniques*, Jakoby, W. B., Ed., Academic Press, Orlando, Fla., 1984, 97.
671. **Lowell, S. and Shields, J. E.,** *Powder Surface Area and Porosity*, 2nd ed., Chapman & Hall, New York, 1984.
672. **Luaces, E. L.,** Vapor adsorbent carbons, in *Colloid Chemistry. Theoretical and Applied*, Alexander, J., Ed., Vol. 6, Reinhold, New York, 1946, 840.
673. **Lucas, J. P. and Ratkowsky, D. A.,** Optimization in various multistage adsorption processes, *Ind. Eng. Chem. Fundam.*, 8, 576, 1969.
674. **Lucchesi, P. J., Hatch, W. H., Mayer, F. X., and Rosensweig, R. E.,** Magnetically stabilized beds. New gas solids contracting technology, *Proc. 10th World Petroleum Congress*, Vol. 4, SP-4, Heyden & Sons, Philadelphia, 1979, 49.
675. **Luft, L.,** U.S. Patent 3,016,107, 1962.
676. **Lukchis, G. M.,** Adsorption systems. I. Design by mass transfer-zone concept, *Chem. Eng.*, 80(13), 111, 1973.
677. **Lukchis, G. M.,** Adsorption systems. II. Equipment design, *Chem. Eng.*, 80(16), 83, 1973.
678. **Lukchis, G. M.,** Adsorption systems. III. Adsorbent regeneration, *Chem. Eng.*, 80(18), 83, 1973.
679. **Lyman, W. J.,** Carbon adsorption, in *Unit Operations for Treatment of Hazardous Industrial Wastes*, DeRenzo, D. J., Ed., Noyes Data, Park Ridge, N.J., 1978, 97.
680. **Lyman, W. J.,** Resin adsorption, in *Unit Operations for Treatment of Hazardous Industrial Wastes*, DeRenzo, D. J., Ed., Noyes Data, Park Ridge, N.J., 1978, 138.
681. LKB-Produkter AB, LKB Preparative Scale Columns, #2237, 1984, and LKB Chromatography Column, #2137, 1984, Box 305, S-16126 Bromma, Sweden.

682. **Macnair, R. N. and Arons, G. N.,** Sorptive textile systems containing activated carbon fibers, in *Carbon Adsorption Handbook,* Cheremisinoff, P. N. and Ellerbusch, F., Eds., Ann Arbor Science, Ann Arbor, Mich., 1978. chap. 22.
683. **Madic, C., Kertesz, C., Sontag, R., and Koehly, G.,** Application of extraction chromatography to the recovery of neptunium, plutonium and americum from an industrial waste, *Sep. Sci. Technol.,* 15, 745, 1980.
684. **Majors, R. E.,** Column switching in HPLC, *Liquid Chromatography,* 2, 358, 1984.
685. **Majors, R. E., Barth, H. G., and Lochmuller, C. H.,** Column liquid chromatography, *Anal. Chem.,* 56, 300, 1984.
686. **Mansour, A., Von Rosenberg, D. U., and Sylvester, N. D.,** Numerical solution of liquid-phase multicomponent adsorption in fixed beds, *AIChE J.,* 28, 765, 1982.
687. **Mantell, C. L.,** *Adsorption,* 2nd ed., McGraw-Hill, New York, 1951.
688. **Mantell, C. L.,** Adsorption, in *Chemical Engineer's Handbook,* 3rd ed., Perry, J. H., Ed., McGraw-Hill, New York, 1950, chap. 14.
689. **Manzone, R. R., and Oakes, D. W.,** Profitably recycling solvents from process systems, *Pollut. Eng.,* 5(10), 23, 1973.
690. **Marchello, J. M.,** *Control of Air Pollution Sources,* Marcel Dekker, New York, 1976, 232.
691. **Marchello, J. M. and Davis, M. W.,** Theoretical investigation of agitated ion exchange beds, *Ind. Eng. Chem. Fundam.,* 2, 27, 1963.
692. **Markham, E. D. and Benton, A. F.,** The adsorption of gas mixtures by silica, *J. Am. Chem. Soc.,* 53, 497, 1931.
693. **Marra, R. A. and Cooney, D. O.,** An equilibrium theory of sorption accompanied by sorbent bed shrinking or swelling, *AIChE J.,* 19, 181, 1973.
694. **Martin, A. J. P.,** Summarizing paper, *Discuss. Faraday, Soc.,* 7, 332, 1949.
695. **Martin, A. J. P. and Synge, R. L. M.,** A new form of chromatogram employing two liquid phases, *Biochem. J.,* 35, 1358, 1941.
696. **Martin, D.,** Solvent recovery saves profits from evaporating, *Process Eng.,* p.66, 1976.
697. **Martin, M., Verillon, F., Eon, C., and Guiochon, G.,** Theoretical and experimental study of recycling in high-performance liuqid chromatography, *J. Chromatogr.,* 125, 17, 1976.
698. **Maslan, F.,** Acetylene Separation, U.S. Patent 2,823,763, Feb. 18, 1958.
699. **Maslan, F.,** Adsorption of Gases with a Liquid-Adsorbent Slurry, U.S. Patent 2.823,765, Feb. 18, 1958.
700. **Maslan, F.,** Gas Removal with a Carbon-Water Slurry, U.S. Patent 2,823,766, Feb. 18, 1958.
701. **Mastroianni, M. L. and Rochelle, S. G.,** Improvement in acetone adsorption efficiency, *Environ. Prog.,* 4(1), 7, 1985.
702. **Mathews, A. P. and Fan, L. T.,** Comparison of performance of packed and semifluidized beds for adsorption of trace organics, *AIChE Symp. Ser.,* 79(230), 79, 1983.
703. **Matsuda, H., Yamamoto, T., Goto, S., and Teshima, H.,** Periodic operation for desalination with thermally regenerable ion-exchange resin. Dynamic studies, *Sep. Sci. Technol.,* 16, 31, 1981.
704. **Matsumoto, Z. and Numasaki, K.,** Regenerate activated carbon, *Hydrocarb. Proc.,* 55(5), 157, 1976.
705. **Maugh, T. H.,** A survey of separation techniques, *Science,* 22, 259, 1983.
706. **McAllister, D.,** How to select a system for drying instrument air, *Chem. Eng.,* 80(5), 39, 1973.
707. **McCabe, W. L., Smith, J. C., and Harriott, P.,** *Unit Operations of Chemical Engineering,* 4th ed., McGraw-Hill, New York, 1984.
708. **McCormack, R. H., and Howard, J. F.,** A continuous countercurrent ion-exchange unit, *Chem. Eng. Prog.,* 49, 404, 1953.
709. **McGary, R. S. and Wankat, P. C.,** Improved preparative liquid chromatography: the moving feed point method, *Ind. Eng. Chem. Fundam.,* 22, 10, 1983.
710. **McGill, Inc.,** Processing Systems, P.O. Box 9667, Tulsa, Okla. 74107, 1981.
711. **McGill, Inc.,** The McGill adsorption/adsorption gasoline vapor recovery system, P.O. Box 9667, Tulsa, Okal., 74107, 1980.
712. **McGill, Inc.,** Fuel conditioning units for natural gas fueled engines, 5800 West 68th St., Tulsa, Okla., 74157, 1982.
713. **McGinnis, F. K.,** Infrared furnaces for reactivation, in *Activated Carbon Adsorption for Wastewater Treatment,* Perrich, J. E., Ed., CRC Press, Boca Raton, Fla., 1981, chap. 6B.
714. **McGuire, M. J. and Suffet, I. H., Eds.,** *Treatment of Water by Granular Activated Carbon,* ACS Advances in Chemistry Series #202, American Chemical Society, Washington, D.C., 1983.
715. **McGuire, M. J. and Suffet, I. H., Ed.,** Discussion. III. Biological/adsorptive interactions, in *Treatment of Water by Granular Activated Carbon,* American Chemical Society, Washington, D.C., 1983, 373.
716. **McKay, G.,** Adsorption of acidic and basic dyes onto activated carbon in fluidized beds, *Chem. Eng. Resh. Des. Trans. Inst. Chem. Eng.,* 61, 29, 1983.
717. **McNair, H. M.** Equipment for HPLC. VI, *J. Chromatogr. Sci.,* 22, 521, 1984.

718. **Megy, J. and Hawkes, S. J.,** Zone spreading and optimization of parallel plate continuous chromatographs, *Chem. Eng. Sci.,* 29, 1513, 1974.
719. **Mehrotra, A. K. and Tien, C.,** Further work in species grouping in multicomponent adsorption calculation, *Can. J. Chem. Eng.,* 62, 632, 1984.
720. **Mehta, D. S. and Calvert, S.,** Gas sorption by suspensions of activated carbon in water, *Environ. Sci. Technol.,* 1, 325, 1967.
721. **Meir, D. and Lavie, R.,** Continuous cyclic zone adsorption, *Chem. Eng. Sci.,* 29, 1133, 1974.
722. **Merk, W., Fritz, W., and Schlunder, E. U.,** Competitive adsorption of two dissolved organics onto activated carbon. III. Adsorption kinetics in fixed beds, *Chem. Eng. Sci.,* 36, 743, 1981.
723. **Mersmann, A., Münstermann, U., and Schadl, J.,** Separation of gas mixtures by adsorption, *German Chem. Eng.,* 7, 137, 1984.
724. Met-Pro Corp., Solvent Concentrating System, Bulletin 11B3, Harleysville, Pa., 1980.
725. Met-Pro Corp., Systems Division, Carbon Adsorption Systems, KF Series for Gaseous Applications, 1980; a Guide to the Control of Volatile Organic Emissions, 1981; Series 1000 Air Pollution Control Systems, 1985, 160 Cassell Rd., Box 144, Harleysville, Pa., 19438.
726. **Michaels, A.,** Simplified method of interpreting kinetic data in fixed-bed ion exchange, *Ind. Eng. Chem.,,* 44, 1922, 1952.
727. **Miller, G. H. and Wankat, P. C.,** Moving port chromatography: a method of improving preparative chromatography, *Chem. Eng. Commun.,* 31, 21, 1984.
728. **Miller, R. L., Ogan, K., and Poile, A. F.,** Automated column switching in liquid chromatography, *Int. Lab.,* 11(6), 60, 1981.
729. **Miller, S. A.,** Leaching, in *Chemical Engineer's Handbook,* 5th ed., Perry, R. H. and Chilton, C. H., Eds., McGraw-Hill, New York, 1973, 19-41.
730. **Miller, W. C. and Philcox, J. E.,** *The Purasiv Hg Process for Hg Removal and Recovery,* Union Carbide Corp., New York, no date.
730a. **Millipore Corp.,** The advantages of product purification with process HPLC, 1984; shorten the race to market with pilot-scale HPLC, 1985, Bedford, Mass., 01730.
731. **Mills, B. and Rothery, E.,** Gas drying, *Chem. Eng.,* 402, 19, 1984.
732. **Minty, D. W., McNeil, R., Ross, M., Swinton, E. A., and Weiss, D. E.,** Countercurrent adsorption separation processes. IV. Batch method of fractionation, *Aust. J. Appl. Sci.,* 4, 530, 1953.
733. **Misic, D. M. and Smith, J. M.,** Adsorption of benzene in carbon slurries, *Ind. Eng. Chem. Fundam.,* 10, 380, 1971.
734. **Misra, K. S.,** Design and performance of fixed bed adsorption driers, *Chem. Age India,* 20, 808, 1969.
735. **Mitchell, H. L., Schrenk, W. G., and Silker, R. E.,** Preparation of carotene concentrates from dehydrated alfalfa meal, *Ind. Eng. Chem.,* 45, 415, 1953.
736. **Mitchell, J. E. and Shendalman, L. H.,** A study of heatless adsorption in the model system CO_2 in He. II, *AIChE Symp. Ser.,* 69(134), 25, 1973.
737. **Miwa, K. and Inoue, T.,** Production of nitrogen adsorption separation, *Chem. Econ. Eng. Rev. (Tokyo),* 12(11), 40, 1982.
738. **Miyauchi, T., Kaji, H., and Sato, K.,** Fluid and particle dispersion in fluid-bed reactors, *J. Chem. Eng. Jpn.,* 1, 72, 1968.
739. **Modell, M., Robey, R. J., Krukonis, V., deFilippi, R. P., and Oestreich, D.,** Supercritical fluid regeneration of activated carbon, AIChE 87th National Meeting, Boston, Mass., Aug, 21, 1979.
740. **Moir, G., Ross, M., and Weiss, D. E.,** Countercurrent adsorption separation processes. V. Froth flotation in columns, *Aust. J. Appl. Sci.,* 4, 543, 1953.
741. **Moison, R. L. and O'Hern, H. A., Jr.,** Ion exchange kinetics, *Chem. Eng. Prog. Symp. Ser.,* 55(24), 71, 1959.
742. **Morbidelli, M., Storti, G., Carra, S., Niederjaufner, G. and Pontoglio, A.,** Study of a separation process through adsorption of molecular sieves. Application to a chlorotoluene isomers mixture, *Chem. Eng. Sci.,* 39, 383, 1984.
742a. **Morbidelli, M., Storti, G., Carra, S., Niederjaufner, G., and Pontoglio, A.,** Role of the desorbent in bulk adsorption separations. Applications to a chlorotoluene isomer mixture, *Chem. Eng. Sci.,* 40, 1155, 1985.
743. **Morr, C. V., Coulter, S. T., and Jenness, R.,** Comparison of column and centrifugal Sephadex methods for fractionating whey and skim milk systems, *J. Dairy Sci.,* 51, 1155, 1968.
744. **Morr, C. V., Nielsen, M., and Coulter, T. S.,** Centrifugal Sephadex procedure for fractionation of concentrated skim milk, whey and similar biological systems, *Dairy Sci.,* 50, 305, 1967.
745. **Moseman, M. H. and Bird, G.,** Desiccant dehydration of natural gasoline, *Chem. Eng. Prog.,* 78(2), 78, 1982.
746. **Moshenskaya, M. B. and Vigdergauz, M. S.,** The evolution of the construction and manufacturing of gas chromatographs in the Soviety Union, *J. Chromatogr., Sci.,* 16, 351, 1978.

747. **Mosier, L. C.,** U.S. Patent, 3,078,647, 1963.
748. **Muendel, C. H. and Selke, W. A.,** Continuous ion exchange with an endless belt of phosphorylated cotton, *Ind. Eng. Chem.*, 47, 374, 1955.
749. **Myers, A. L.,** Activity coefficents of mixtures adsorbed on heterogeneous surfaces, *AIChE J.*, 29, 691, 1983.
750. **Myers, A. L. and Belfort, G., Eds.,** *Fundamentals of Adsorption,* Engineering Foundation, New York, 1984.
751. **Myers, A. L. and Moser, F.,** Slurry sorption separations. I. Equilibrium adsorption of gases by suspensions of solid adsorbents in liquids, *Chem. Eng. Sci.*, 32, 529, 1977.
752. **Myers, A. L. and Prausnitz, J. M.,** Thermodynamics of mixed gas adsorption, *AIChE J.*, 11, 121, 1965.
753. **Naden, D., and Streat, M., Eds.,** *Ion Exchange Technology,* Ellis Horwood, Chichester, 1984.
754. **Naden, D. and Willey, G.,** Reduction in copper recovery costs using solid ion exchange, in *The Theory and Practice of Ion Exchange,* Slater, M. J., Ed., Society of Chemical Industry, London, 1979, 44.1.
755. **Naden, D., Willey, G., Bicker, E., and Lunt, D.,** Development of an in-pulp contactor for metals recovery, in *Ion Exchange Technology,* Naden, D. and Streat, M., Eds., Ellis Horwood, Chichester, 1984, 690.
756. **Narraway, R.,** Market switches on to PSA gas generators, *Processing,* 24(1), 29, 1978.
757. **Nasseri, S., Dussap, G., and Bugarel, R.,** Modelisation d'un systeme de pompage parametrique en mode direct fonctionnant en discontinu. II. Etude experimentale: separation d'un melange binarie toluen, n-heptane, *Chem. Eng. J.*, 20, 121, 1980.
758. **Nataraj, S. and Wankat, P. C.,** Multicomponent pressure swing adsorption (MCPSA), *AIChE Symp. Ser.*, 78(219), 29, 1982.
759. **Nelson, W. C., Silarski, D. F., and Wankat, P. C.,** Continuous flow equilibrium staged model for cycling zone adsorption, *Ind. Eng. Chem. Fundam.*, 17, 32, 1978.
760. **Nelson, W. C. and Wankat, P. C.,** Application of cycling zone separation to preparative high-pressure liquid chromatography, *J. Chromatogr.*, 121, 205, 1976.
761. **Nettleton, D. E., Jr.,** Preparative liquid chromatography. I. Approaches utilizing highly compressed beds, *J. Liq. Chromatogr.*, Suppl. 4, 141, 1981.
762. **Nettleton, D. E., Jr.,** Preparative liquid chromatography. II. Approaches on noncompressed beds, *J. Liq. Chromatogr.*, Suppl. 4, 359, 1981.
763. **Neuzil, R. W. and Jensen, R. H.,** Development of the Sarex process for the separation of saccharides, Paper 22d presented at AIChE meeting, Philadelphia, Pa., June 6, 1978.
764. **Neuzil, R. W., and Priegnitz, J. W.,** Process for separating a ketose from a aldose by selective adsorption, U.S. Patent, 4,024,331, May 17, 1977.
765. **Neuzil, R. W., Rosback, D. H., Jensen, R. H., Teague, J. R., and de Rosset, A. J.,** An energy-saving separation scheme, *Chemtech,* 10, 498, 1980.
766. **Newman, J.,** Fluidized moving bed ion exchange for treatment of cooling tower blowdown, Cooling Tower Institution Annual Meeting, Houston, Tex., *Jan. 29—31, 1973.*
767. **Nguyen, H. V. and Potter, O. E.,** Adsorption effects in fluidized beds, in *Fluidization Technology,* Vol. II, Keairns, D. L., Ed., Hemisphere, Washington, D.C., 1976, 193.
768. **Nicholas, R. A. and Fox, J. B.,** Continuous chromatography apparatus. III. Application, *J. Chromatogr.*, 43, 61, 1969.
769. **Nishikawa, A. H.,** Chromatography, affinity, in *Kirk-Othmer Encyclopedia of Chemical Technology,* Vol. 6, 3rd ed., Wiley-Interscience, New York, 1979, 35.
770. **Nord, M.,** Chromatography comes of age, *Chem. Eng.*, 61 (9), 256, 1954. Report on patent of Beard, L. C. Jr., U.S. Patent 2,678,132.
771. **Novgorodov, A. F., Adilbish, M., Zaitseva, N. G., Kovalev, A. S., and Kovach, Z.,** Behavior of nuclear reaction products upon sublimation from irradiated Ag and Au under a dynamic vacuum of 1 to 10^{-1} Pa for O_2 for H_2O (trans.), *Soviet Radiochem.*, 22 (5), 590, 1981.
772. **O'Brien, R. P., Clemens, M. M., and Schuliger, W. G.,** Treatment of contaminated groundwater with granular activated carbon, *AIChE Symp. Ser.*, 79 (230), 44, 1983.
773. **Ohline, R. W. and DeFord, D. D.,** Chromathermography, the application of moving thermal gradients to gas-liquid partition chromatography, *Anal. Chem.*, 35, 227, 1963.
774. **Oren, Y. and Soffer, A.,** Electrochemical parametric pumping, *J. Electrochem. Soc.*, 125, 869, 1978.
775. **Oren, Y. and Soffer, A.,** Water desalting by means of electrochemical parametric pumping. I. The equilibrium properties of a batch unit cell, *J. Appl. Electrochem.*, 13, 473, 1983.
776. **Oren, Y. and Soffer, A.,** Water desalting by means of electrochemical parametric pumping. II. Separation properties of a multistage column, *J. Appl. Electrochem.*, 13, 489, 1983.
777. **Oren, Y. and Soffer, A.,** Theory of isotope separation by electrochemical parametric pumping, *Sep. Sci. Technol.*, 19, 575, 1984.
778. **Ortlieb, H.-J., Bunke, G., and Gelbin, D.,** Separation efficiency in the cyclic steady state for periodic countercurrent adsorption, *Chem. Eng. Sci.*, 36, 1009, 1981.

779. **Otani, S.,** Adsorption separates xylenes, *Chem. Eng.*, 80(21), 106, 1973.
780. **Otozai, K., Tsuneyoshi, T., and Okada, H.,** Continuous preparation by return gas chromatography. I. Construction of a model apparatus, *J. Chromatogr. Sci.*, 23, 151, 1985.
781. **Ouano, A. C.,** Kinematics of gel permeation chromatography, *Adv. Chromatogr.*, 15, 233, 1977.
782. Pall Trinity Micro Corp., Heat-Les Dryers, Bulletin HA 308f, 1970 and Type HV Vacuum Regenerating Heat-Les Desiccant Dryers for Reduced Purge, Bulletin HV 333a, no date, 30 Sea Cliff Ave., Glen Cove, N.Y., 11542.
783. **Palmer, G. H.,** Maloney gas dehydration, *Hydrocarb. Process.*, 56 (4), 103, 1977.
784. **Pan, C. Y. and Basmadjian, D.,** An analysis of adiabatic sorption of single solutes in fixed beds: equilibrium theory, *Chem. Eng. Sci.*, 26, 45, 1971.
785. **Parkhurst, J. D., Dryden F. D., McDermott, G. N., and English, J.,** Ponoma activated carbon pilot plant, *J. Water Pollut. Control Fed.*, 39 (10), R70, 1967.
786. **Parmele, C. S., O'Connell, W. L., and Basdekis, H. S.,** Vapor-phase adsorption cuts pollution-recovers solvent, *Chem. Eng.*, 86 (28), 58, 1979.
787, **Parris, N. A.,** *Instrumental Liquid Chromatography — A Practical Manual on High Performance Liquid Chromatographic Methods,* 2nd ed., Elsevier, Amsterdam, 1984.
788. **Patrick, R. R., Schrodt, J.T., and Kermode, R. I.,** Thermal parametric pumping of air-SO_2, *Sep. Sci.*, 7, 331, 1972.
789. **Pazar, C., Ed.,** *Air and Gas Cleanup Equipment 1970,* Noyes Data Corp., Park Ridge, N.Y., 1970, 628.
790. **Peaden, P. A. and Lee, M. L.,** Supercritical fluid chromatography: methods and principles, *J. Liquid Chromatogr.*, 5 (Suppl. 2), 179, 1982.
791. **Peel, R. G. and Benedek, A.,** Attainment of equilibrium in activated carbon isotherm studies, *Environ. Sci. Technol.*, 14(1), 66, 1980.
792. **Peel, R. G., Benedek, A., and Crowe, C. M.,** A branched pore kinetic model for activated carbon adsorption, *AIChE J.*, 27, 26, 1981.
793. **Perrich, J. R., Ed.,** *Activated Carbon Adsorption for Wastewater Treatment,* CRC Press, Boca Raton, Fla., 1981.
794. **Perrut, M.,** La chromatographic d'elution a L'echelle industrielle, XIIIe Conference internationale des arts chimiques, Paris, Dec. 6—9, 1983, Colloque 2, Genie Chimique, p. 3—1.
795. **Peterson, E. A. and Torres, A. R.,** Displacement chromatography of proteins, in *Methods in Enzymology,* Vol. 104, *Enzyme Purification and Related Techniques,* Jackoby, W. B., Ed., Academic Press, Orlando, Fla., 1984, 113.
796. **Peyrouset, A., Orozco, J., Godbille, E., and Devaux, P.,** Industrial scale preparative high performance liquid chromatography using a 30 cm (I.D.) column, Pittsburgh Conference and Exposition on Analytical Chemistry and Applied Spectroscopy, *Atlantic City, N.J., 1984, Abstr. 517.*
797. **Pfefferle, K.-H. and Piesche, M.,** Purification of exhaust air and recovery of toluene with integrated waste water treatment and heat reclaim system for rotogravure printing processes, International Rotogravure Conference, Tratranska, Lomnice, Czechoslovakia, May 1978.
798. **Pfefferle, K.-H.,** Waste gas purification with simultaneous solvent recovery in the surface coating industry, (trans.), *Kunststoffe German Plastics,* 3, 136, 1979.
799. Pharmacia Fine Chemicals, Pharmacia FPLC: Data File, HR prepacked columns. Mono Q: Mono S: Mono P, Piscataway, N.J., 08854, 1983.
800. Pharmacia Fine Chemicals AB, *Pharmacia Sectional Comumn KS 370, the Stack, Instructional Manual,* Uppsala, Sweden, 1982.
801. Pharmacia Fine Chemicals, *Sophomatic Gel Filters, For Large Scale Gel Filtration Chromatography,* Uppsala, Sweden, 1979.
802. Pharmacia Fine Chemicals, *The Large Scale Purification of Insulin by Gel Filtration Chromatography,* Uppsala, Sweden, 1983.
803. Pharmacia Fine Chemicals AB, *Scale Up to Process Chromatography,* Uppsala, Sweden, 1983.
804. **Picht, R. D., Dillman, T. R., Burke, D. J., and deFilippi, R. P.,** Regeneration of adsorbents by a supercritical fluid, *AIChE Symp. Ser.*, 78 (219), 136, 1982.
805. **Pigford, R. L., Baker, B., and Blum, D. E.,** An equilibrium theory of the parametric pump, *Ind. Eng. Chem. Fundam.*, 8, 144, 1969. *Note:* this solution is for a special case. See Reference 58.
806. **Pigford, R. L., Baker, B., and Blum, D. E.,** Cycling zone adsorption, a new separation process, *Ind. Eng. Chem. Fundam.*, 8, 848, 1969.
807. **Polezzo, S. and Taramasso, M.,** Rotating unit for preparative-scale gas chromatography. II. Theoretical aspects, *J. Chromatogr.*, 11, 19, 1963.
808. **Pollock, A. W., Brown, M. F., and Dempsey, C. W.,** Machine solution of a boundary value problem for a continuous Arosorb process, *Ind. Eng. Chem.*, 50, 725, 1958.
809. **Ponec, V., Knor, Z., and Cerny, S.,** *Adsorption on Solids,* Butterworths, London, 1974, Chap. 8 to 10.

810. **Prabhudesai, R. K.,** Leaching, in *Chemical Engineer's Handbook,* 5th ed., Perry, R. H. and Chilton, C. H., Eds., McGraw-Hill, New York, 1973, 17—3.
811. **Prahacs, S. and Barclay, H. G.,** Session II discussion and some studies of the regeneration of powdered activated carbon, *Chem. Eng. Prog. Symp. Ser.,* 70(144), 378, 1974.
812. **Priestly, A. J.,** Continuous ion exchange with magnetic shell resins. Dealkalisation. V. Process development, *J. Chem. Technol. Biotechnol.,* 29, 273, 1979.
813. **Pritchard, D. W., Probert, M. E., and Tiley, P. F.,** Experimental and theoretical studies of countercurrent solvent-gas extraction. I. Continuous gas-chromatography in a laboratory moving-bed column, *Chem. Eng. Sci.,* 26, 2063, 1971.
814. **Pritchard, D. W. and Tiley, P. F.,** Experiments and theoretical studies of countercurrent solvent-gas extraction. II. The construction and operation of a laboratory differential vapoun-absorption column, *Chem. Eng. Sci.,* 28, 1839, 1973.
815. **Prober, R., Pyeha, J. J., Helfgott, T., and Rader, B. B.,** Chemical implications in activated carbon regeneration, *Chem. Eng. Prog. Symp. Ser.,* 70(144), 339, 1974.
816. **Prout, W. E. and Fernandez, L. P.,** Performance of anion resins in agitated beds, *Ind. Eng. Chem.,* 53, 449, 1961.
817. **Rachez, D., Delaveau, G., Grevillot, G., and Tondeur, D.,** Stagewise liquid-liquid extraction parametric pumping. Equilibrium analysis and experiments, *Sep. Sci. Technol.,* 17, 589, 1982.
818. **Taghavan, N. S. and Ruthven, D. M.,** Dynamic behavior of an adiabatic adsorption column. II. Numerical solution and analysis of experimental data, *Chem. Eng. Sci.,* 39, 1201, 1984.
819. **Raghavan, N. S., Hassan, M. M., and Ruthven, D. M.,** Numerical simulation of a PSA system. I. Isothermal trace component system with linear equilibrium and finite mass transfer resistance, *AIChE J.,* 31, 385, 1985.
820. **Raghurman, J. and Varma, Y. B. G.,** Aspects of multistage fluidization, *Chem Process. Eng.,* 53(7), 48, 1972.
821. **Rahman, K. and Streat, M.,** Mass transfer in lqiuid fluidized beds of ion exchange particles, *Chem. Eng. Sci.,* 36, 293, 1981.
822. **Ramaswamy, S. R. and Gerhard, E. R.,** A gas-granular solids contacting apparatus, *Br. Chem. Eng.,* 11, 1210, 1966.
823. **Ramshaw, C.,** Higee distillation — an example of process intensification, *Chem. Eng.,* 389, 13, 1983.
824. Ray Solv. Inc., The experiences and complete recovery systems supplier, 225 Old New Brunswick Rd., Piscataway, N.J., 08854, no date.
825. **Ray, M. S.,** The separation and purification of gases using solid adsorbents, *Sep. Sci. Technol.,* 18, 95, 1983.
826. **Reed, A. K., Tewksbury, T. L., and Smithson, G. R., Jr.,** Development of a fluidized-bed technique for the regeneration of powdered activated carbon, *Environ. Sci. Technol.,* 4, 432, 1970.
827. **Regnier, F. E.,** High-performance liquid chromatography of biopolymers, *Science,* 222, 245, 1983.
828. **Regnier, F. E.,** High-performance ion exchange chromatography, in *Methods in Enzymology,* Vol. 104, *Enzyme Purification and Related Techniques,* Jakoby, W. B., Ed., Academic Press, Orlando, Fla., 1984, 170.
829. **Rendell, M.,** The real future for large-scale chromatography, *Process Eng.,* April, 66, 1975.
830. Research-Cotttrell, New Oxycat CA-66 Solvent Recovery System, Oxy-Catalyst, West Chester, Pa., no date. *Note:* this company no longer manufactures carbon adsorption systems.
831. **Reynolds, T. D.,** *Unit Operations and Processes in Environmental Engineering,* Brooks/Cole, Monterey, Calif., 1982, chap. 6.
832. **Rhee, H.-K.,** Equilibrium theory of multicomponent chromatography, in *Percolation Processes, Theory and Applications,* Rodrigues, A. E., and Tondeur, D., Eds., Sijthoff and Noordhoff, Alphen aan de Rijn, Netherlands, 1981, 285.
833. **Rhee, H.-K. and Amundson, N. R.,** Correspondence, an equilibrium theory of the parametric pump, *Ind. Eng. Chem. Fundam.,* 9, 303, 1970.
834. **Rhee, H.-K. and Amundson, N. R.,** Asymptotic solution to moving-bed exchange operations, *Chem. Eng. Sci.,* 28, 55, 1973.
835. **Rhee, H.-K. and Amundson, N. R.,** Analysis of multicomponent separation by displacement development, *AIChE J.,* 28, 423, 1982.
836. **Rhee, H.-K., Aris, R., and Amundson, N. R.,** On the theory of chromatography, *Phil. Trans. R. Soc. London,* A267, 419, 1970.
837. **Rice, R. G.,** Disperson and ultimate separation in the parametric pump, *Ind. Eng. Chem. Fundam.,* 12, 406, 1973.
838. **Rice, R. G.,** Transport resistances influencing the estimation of optimum frequencies in parametric pumps, *Ind. Eng. Chem. Fundam.,* 14, 202, 1975.

839. **Rice, R. G.,** The effect of purely sinusoidal potentials on the performance of equilibrium parapumps, *Ind. Eng. Chem. Fundam.,* 14, 362, 1975.
840. **Rice, R. G.,** Progress in parametric pumping, *Sep. Purif. Methods,* 5, 139, 1976.
841. **Rice, R. G.,** Adsorptive distillation, *Chem. Eng. Commun.,*, 10, 111, 1981.
842. **Rice, R. G.,** Adsorptive fractionation: cyclic recycle cascades, in *Fundamentals of Adsorption,* Myers, A. L. and Belfort, G., Eds., Engineering Foundation, New York, 1984, 481.
843. **Rice, R. G. and Foo, S. C.,** Thermal diffusion effects and optimal frequencies in parametric pumps, *Ind. Eng. Chem. Fundam.,* 13, 396, 1974.
844. **Rice, R. G. and Foo, S. C.,** Continuous desalination using cyclic mass transfer on bifunctional resins, *Ind. Eng. Chem. Fundam.,* 20, 150, 1981.
845. **Rice, R. G., Foo, S. C., and Gough, G. G.,** Limiting separations in parametric pumps, *Ind. Eng. Chem. Fundam.,* 18, 117, 1979.
846. **Rice, R. G. and Mackenzie, M.,** A curious anomaly in parametric pumping, *Ind. Eng. Chem. Fundam.,* 12, 486, 1973.
847. **Richardson, J. F. and Szekely, J.,** Mass transfer in a fluidized bed, *Trans. Inst. Chem. Eng.,* 39, 212, 1961.
848. **Richey, J. and Beadling, L.,** Chromatofocusing: a method for high resolution protein purification, *Int. Lab.,* 12(1), 56, 1982.
849. **Richter, E., Strunk, J., Knoblauch, K., and Jüntgen, H.,** Modelling of desorption by depressurization as partial step in gas separation by pressure swing adsorption, *German Chem. Eng.,* 5, 147, 1982.
850. **Rieke, R. D.,** Cycling zone adsorption: variable-feed mode of operation, *Sep. Sci. Technol.,* 19, 261, 1984.
851. **Ripperger, S. and Germerdonk, R.,** Binary adsorption equilibria of organic compounds and water on active carbon, *German Chem. Eng.,* 6, 249, 1983.
852. **Robinson, K. S. and Thomas, W. J.,** The adsorption of methane/ethane mixtures on a molecular sieve, *Trans. Inst. Chem. Eng.,* 58(4), 219, 1980.
853. **Robinson, P. J., Wheatly, M. A., Janson, J.-C., Dunnill, P., and Lilly, M. D.,** Pilot scale affinity purification: purification of β-galactosidase, *Biotech. Bioeng.,* 16, 1103, 1974.
854. Rohm & Haas, Helpful hints in ion exchange technology, Philadelphia, Pa., 1972.
855. **Roland, L. D.,** Ion exchange-operational advantages of continuous plants, *Processing,* 22(1), 11, 1976.
856. **Rolke, R. W. and Wilhelm, R. H.,** Recuperative parametric pumping model development and experimental evaluation, *Ind. Eng. Chem. Fundam.,* 8, 235, 1969.
857. **Rosensweig, R. E.,** Magnetic stabilization of the state of uniform fluidization, *Ind. Eng. Chem. Fundam.,* 18, 260, 1979.
858. **Rosensweig, R. E.,** Fluidization: hydrodynamic stabilization with a magnetic field, *Science,* 204, 57, 1979.
859. **Rosensweig, R. E., Jerauld, G. R., and Zahn, M.,** Structure of magnetically stabilized fluidized solids, in *Continuous Models of Discrete Systems-4,* Brulin, O. and Hsieh, R. K. T., Eds., North-Holland, Amsterdam, 1981, 137.
860. **Rosensweig, R. E., Siegell, J. H., Lee, W. K., and Mikus, T.,** Magnetically stabilized fluidized solids, *AIChE Symp. Ser,* 77(205), 8, 1981.
861. **Rosensweig, R. E., Zahn, M., Lee, W. K., and Hagan, P. S.,** Theory and experiments in the mechanics of magnetically stabilized fluidized solids, in *Theory of Dispersed Multiphase Flow,* Academic Press, New York, 1983, 359.
862. **Ross, J. R. and George, D. R.,** Recovery of Uranium from Natural Mine Waters by Countercurrent Ion Exchange, U.S. Bureau of Mines, 1971, RI7471.
863. **Rowson, H. M.,** Fluid bed adsorption of carbon disulphide, *Br. Chem. Eng.,* 8(3), 180, 1963.
864. **Roz, B., Bonmati, R., Hagenbach, G., Valentin, P., and Guiochon, G.,** Practical operation of prep-scale gas chromatographic units, *J. Chromatogr. Sci.,* 14, 367, 1976.
865. **Ruthven, D. M.,** *Principles of Adsorption and Adsorption Processes,* John Wiley & Sons, New York, 1984.
866. **Ruthven, D. M.,** The axial dispersed plug flow model for continuous counter-current adsorbers, *Can. J. Chem. Eng.,* 61, 881, 1983.
867. **Ryan, J. M. and Dienes, G. L.,** *Drug Cosmet. Ind.,* 99 (4), 60, 1966.
868. **Ryan, J. M., Timmins, R. S., and O'Donnell, J. F.,** Production scale chromatography, *Chem. Eng. Prog.,* 64 (8), 53, 1968.
869. **Sabadell, J. E. and Sweed, N. H.,** Parametric pumping with pH, *Sep. Sci.,* 5, 171, 1970.
870. **Said, A. S.,** *Theory and Mathematics of Chromatography,* Huthig, Heidelberg, 1981.
871. **Said, A. S.,** Calculations of a continuous and preparative gas chromatograph. I, *Int. Lab.,* 13(6), 10, 1983.
872. **Said, A. S.,** Calculations of a continuous and preparative gas chromatograph. II, *Int. Lab.,* 13(8), 28, 1983.

873. **Sakodynskii, K. I., Volkov, S. A., Kovan'ko, Yu. A., Zel'venskii, V. Yu, Reznikov, V. I., and Averin, V. A.,** Design of and experience in operating technological preparative installations, *J. Chromatogr.*, 204, 167, 1981.
874. **Sawyer, D. T. and Hargrove, G. L.,** Preparative gas chromatography, in *Progress in Gas Chromatography*, Purnell, J. H., Ed., Interscience, New York, 1968, 325.
875. **Scamehorn, J. F.,** Removal of vinyl chloride from gaseous streams by adsorption on activated carbon, *Ind. Eng. Chem. Process Des. Devel.*, 18, 210, 1979.
876. **Scawen, M. D., Atkinson, A., and Darbyshire, J.,** Large scale enzyme purification, in *Applied protein Chemistry*, Grant R. A., Ed., Applied Science, London, 1980, Chap. 11.
877. **Scheibel, E. G.,** Liquid-liquid extraction in *Kirk-Othmer Encyclopedia of Chemical Technology*, Vol. 8, 2nd ed., Wiley-Interscience, New York, 1965, 719.
878. **Schoenmaker, P. J.,** Thermodynamic model for supercritical fluid chromatography, *J. Chromatogr.*, 315, 1, 1984.
879. **Schroeder, H. G. and Hamrin, C. E.,** Separation of boron isotopes by direct mode thermal parametric pumping, *AIChE J.*, 21, 807, 1975.
880. **Schuliger, W. G. and MacCrum, J. M.,** Granular activated carbon reactivation system design and operating conditions, *AIChE Symp. Ser.*, 70(144), 352, 1974.
881. **Schulz, H.,** Continuous counter-current separation under conditions of elution gas chromatography, in *Gas Chromatography 1962*, van Swaay, M., Ed., Butterworths, London, 1962, 225.
882. **Schultz, W. W., Wheelwright, E. J., Godbee, H., Mallory, C. W., Burney, G. A., and Wallace, R. M.,** Ion exchange and adsorption in nuclear chemical engineering, *AIChE Symp. Ser.*, 80(233), 96, 1984.
883. **Schweich, D. and Villermaux, J.,** The preparative chromatographic reactor revisited, *Chem. Eng. J.*, 24, 99, 1982.
884. **Schweich, D., Villermaux, J., and Sardin, M.,** An introduction to the nonlinear theory of adsorptive reactors, *AIChE J.*, 26, 477, 1980.
885. **Sciance, C. T. and Crosser, O. K.,** Column height required for continuous chromatographic separation: a probalistic model, *AIChE J.*, 12, 100, 1966.
886. **Scott, C. D., Spence, R. D., and Sisson, W. G.,** Pressurized, annular chromatograph for continuous separations, *J. Chromatogr.*, 126, 381, 1976.
887. **Scott, F.,** Stepping up to kilogram/hr production rates with HPLC, *Process Eng.*, 65 (2), 26, 1984.
888. **Scouten, W.,** *Affinity Chromatography: Bioselective Adsorption on Inert Matrices*, John Wiley & Sons, New York, 1981.
889. **Seko, M., Miyake, T., and Inada, K.,** Economical *p*-xylene and ethylbenze separated from mixed xylene, *Ind. Eng. Chem. Product Res. Devel.*, 18, 263, 1979.
890. **Seko, M., Miyake, T., and Inada, K.,** Sieves for mixed xylenes separation, *Hydrocarb. Process.*, 59(1), 133, 1980.
891. **Seko, M., Miyake, T., and Takeda, H.,** The chromatographic uranium enrichment process by Asahi-Chemical, *AIChE Symp. Ser.*, 78(221), 41, 1982.
892. **Seko, M., Takeachi, H., and Inada, T.,** Scale-up for chromatographic separation of *p*-xylene and ethylbenzene, *Ind. Eng. Chem. Product Res. Devel.*, 21, 656, 1982.
893. **Selke, W. A. and Bliss, H.,** Continuous countercurrent ion exchange, *Chem. Eng. Prog.*, 47, 529, 1951.
894. Separations Technology, Pilot/Production Preparative HPLC, ST/Lab 800A Preparative HPLC, and HPLC Prep Column, Wakefield, R. I., 02879, no date.
895. **Shaffer, A. G. and Hamrin, C. E.,** Enzyme separation by parametric pumping, *AIChE J.*, 21, 782, 1975.
896. **Shaltiel, S.,** Hydrophobic chromatography, in *Methods in Enzymology*, Vol. 104, Enzyme Purification and Related Techniques, Jakoby, W. B., Ed., Academic Press, Orlando, Fla., 1984, 69.
897. **Sharples, P. M. and Bolto, B. A.,** Desalting in Australia — the development of a new process for brackish water, *Desalination*, 20, 391, 1977.
898. **Shearon, W. H. and Gee, O. F.,** Carotene and chlorophyll commercial chromatographic production, *Ind. Eng. Chem.*, 42, 218, 1949.
899. **Shell, G. L.,** Powdered carbon regeneration in a fluid-bed furnace, *Chem. Eng. Prog. Symp. Ser.*, 70(144), 371, 1974.
900. **Shendalman, L. H. and Mitchell, J. E.,** A study of heatless adsorption on the model system CO_2 in He, *Chem. Eng. Sci.*, 27, 1449, 1972.
901. **Sherwood, T. K., Pigford, R. L., and Wilke, C. R.,** *Mass Transfer*, McGraw-Hill, New York, 1975, chap. 10.
902. **Shih, C. K., Snavely, C. M., Molnar, T. E., Meyer, J. L., Caldwell, W. B., and Paul, E. L.,** Large-scale liquid chromatography system, *Chem. Eng. Prog.*, 79(10), 53, 1983.
903. **Shih, T. T. and Pigford, R. L.,** Removal of salt from water by thermal cycling of ion exchange resins, in *Recent Developments in Separation Science*, Vol. III, Part A, Li, N. N., Ed., CRC Press, Boca Raton, Fla., 1977, 129.

904. **Shingari, M. K., Conder, J. R., and Fruitwala, N. A.,** Construction and operation of a pilot scale production gas chromatograph for separating heat-sensitive materials, *J. Chromatogr.*, 285, 409, 1984.
905. **Short, C. S.,** Removal of organic compounds, in *Developments in Water Treatment-2,* Lewis, W. M., Ed., Applied, Science, London, 1980, chap. 2.
906. **Shulman, H. L., Youngquist, G. R., Allen, J. R., Ruths, D. W., and Press, S.,** Development of a continuous countercurrent fluid-solid contactor. Adsorption, *Ind. Eng. Chem. Process Des. Devel.*, 5, 359, 1966.
907. **Shulman, H. L., Youngquist, G. R., and Covert, J. R.,** Development of a continuous countercurrent fluid-solids contactor. Ion exchange, *Ind. Eng. Chem. Process Des. Devel.*, 5, 257, 1966.
908. **Shuman, F. R. and Brace, D. G.,** *Petrol. Eng.*, 25(4), C-9, 1953.
908a. **Siegell, J. H. and Coulaloglou, C. A.,** Crossflow magnetically stabilized fluidized beds, *AIChE Symp. Ser.*, 80(241), 129, 1984.
908b. **Siegell, J. H., Pirkle, J. C., Jr., and Dupre, G. D.,** Crossflow magnetically stabilized bed chromatography, *Sep. Sci. Technol.*, 19, 977, 1984.
909. **Siegmund, C. W., Munro, W. D., and Amundson, N. R.,** Two problems on moving beds, *Ind. Eng. Chem.*, 48, 43, 1956.
910. **Sircar, S.,** Separation of multicomponent gas mixtures, U.S. Patent 4, 171,206, Oct. 16, 1979.
911. **Sircar, S.,** Separation of multicomponent gas mixtures by pressure swing adsorption, U.S. Patent 4,171,207, Oct. 16, 1979.
912. **Sircar, S. and Kumar, R.,** Adiabatic adsorption of bulk binary gas mixtures: analysis by constant pattern model, *Ind. Eng. Chem. Process Des. Devel.*, 22, 271, 1983.
913. **Sircar, S. and Kumar, R.,** Equilibrium theory for adiabatic desorption of bulk binary gas mixtures by purge, *Ind. Eng. Chem. Process Des. Devel.*, 24, 358, 1985.
914. **Sircar, S. and Myers, A. L.,** Surface potential theory of multilayer adsorption from gas mixtures, *Chem. Eng. Sci.*, 28, 489, 1973.
915. **Sircar, S. and Zondlo, J. W.,** Fractionation of air by adsorption, U.S. Patent 4,013,429, March 22, 1977.
916. **Sitrin, R. D., Chan, G., De Phillips, P., Dingerdissen, J., Valenta, J., and Snader, K.,** Preparative reversed phase high performance liquid chromatography. A recovery and purification process for nonextractable polar antibiotics, in *Purification of Fermentation Products. Applications to Large-Scale Processes,* Le Roth, D., Shiloach, J., and Leahy, T. J., Eds., ACS Symp. Ser., #271, American Chemical Society, Washington, D.C., 1985, 71.
917. **Skarstrom, C. W.,** Use of adsorption phenomena in automatic plant-type gas analyzers, *Ann. N.Y. Acad. Sci.*, 72(13), 751, 1959.
918. **Skarstrom, C. W.,** U.S. Patent 2,944,627, July 12, 1960.
919. **Skarstrom, C. W.,** Heatless fractionation of gases over solid adsorbents, in *Recent Developments in Separation Science,* Vol. II, Li, N. N., Ed., CRC Press, Boca Raton, Fla., 1972, 95.
920. **Slater, M. J.,** Comparison of the hydrodynamics of two continous countercurrent ion-exchange contactors, in *Ion Exchange in the Process Industries,* Society of Chemical Industry, London, 1970, 127.
921. **Slater, M. J.,** Continuous ion exchange in fluidized beds, *Can. J. Chem. Eng.*, 52, 43, 1974.
922. **Slater, M. J.,** Recent industrial-scale applications of continuous resin ion exchange systems, *J. Sep. Proc. Technol.*, 2(3), 2, 1981.
923. **Slater, M. J.,** The relative sizes of fixed bed and continuous countercurrent flow ion exchange equipment, *Trans. Inst. Chem. Eng.*, 60, 54, 1982.
924. **Slater, M. J. and Lucas, B. H.,** Flow patterns and mass transfer rates in fluidized bed ion exchange equipment, *Can. J. Chem. Eng.*, 54, 264, 1976.
925. **Small, D. A. P., Atkinson, T., and Lowe, C. R.,** Preparative high performance liquid affinity chromatography, *J. Chromatogr.*, 266, 151, 1983.
926. **Smisek, M. and Cerny, S.,** *Active Carbon, Manufacture, Properties and Applications,* Elsevier, Amsterdam, 1970.
927. **Smith, D. R.,** Solven laden air is cash in the wind, *Processing,* 23(4), 59, 1977.
928. **Smith, J. C., Michalson, A. W., and Roberts, J. T.,** Ion exchange equipment, in *Chemical Engineer's Handbook,* 4th ed., Perry, R. H., Chilton, C. H., and Kirkpatrick, S. D., Eds., McGraw-Hill, New York, 1963, 19—18.
929. **Snowdon, C. B. and Turner, J. C. R.,** Mass transfer in liquid-fluidized beds of ion exchange resin beads, in *Proc. Int. Symp. Fluidization,* Drinkenburg, A. A. H., Ed., Netherlands, University Press, Amsterdam, 1967, 599.
930. **Snyder, L. R. and Kirkland, J. J.,** *Introduction to Modern Liquid Chromatography,* 2nd ed., Wiley, New York, 1979.
931. **Snyder, L. R.,** Gradient elution, in *High Performance Liquid Chromatography. Advances and Perspectives,* Vol. 1, Horvath, C., Ed., Academic Press, New York, 1980, 208.

932. **Snyder, L. R., Dolan, J. W., and Van der Waal, Sj.,** Boxcar chromatography. A new approach to increased analysis rate and very large column plate numbers, *J. Chromatogr.*, 203, 3, 1981.
933. Society of Chemical Industry, *Ion Exchange in the Process Industries,* London, 1970.
934. Society of Chemical Industry, *Ion Exchange and Its Applications,* London, 1955.
935. **Solms, J.,** Kontinuierliche Papierchromatographic, *Helv. Chim. Acta,* 38, 1127, 1955.
936. **Solt, G. S.,** Continuous countercurrent ion exchange: the C.I. process, *Br. Chem. Eng.*, 12, 1582, 1967.
937. **Spedding, F. H., Fulmer, E. I., Butler, T. A., Gladrow, E. M., Gobush, M., Porter, P. E., Powell, J. E., and Wright, J. M.,** The separation of rare earths by ion exchange. III. Pilot plant scale separations, *J. Am. Chem. Soc.*, 69, 2812, 1947.
938. **Sterba, M. J.,** Recovery of normal paraffins by the Molex process, *Proc. Am. Petrol. Inst.*, 45, 111, 209, 1965.
939. **Stevenson, D. G.,** Development of a continuous ion exchange process, in *Ion Exchange in the Process Industries,* Society of Chemical Industry, London, 1970, 114.
940. **Stokes, J. D. and Chen, H. T.,** Design and scale-up of a continuous thermal parametric pumping system, *Ind. Eng. Chem. Process Des. Devel.*, 18, 147, 1979.
941. **Storm, D. W.,** Contacting systems, in *Activated Carbon Adsorption for Wastewater Treatment,* Perrich, J. R., Ed., CRC Press, Boca Raton, Fla., 1981, chap. 5.

941a. **Storti, G., Santacasaria, E., Morbidelli, M., and Carra, S.,** Separation of xylenes on Y zeolite in the vapor phase. 3. Choice of the suitable desorbent, *Ind. Eng. Chem. Process Des. Develop.*, 24, 89, 1985.

942. **Streat, M., Ed.,** *The Theory and Practice of Ion Exchange,* Society of Chemical Industry, London, 1976.
943. **Streat, M.,** Recent developments in continuous ion exchange, *J. Sep. Process. Technol.*, 1(3), 10, 1980.
944. **Suffet, I. H. and McGuire, M. J., Eds.,** *Activated Carbon Adsorption of Organics from the Aqueous Phase,* Vols. 1 and 2, Ann Arbor Science, Ann Arbor, Mich., 1980.
945. **Summers, R. S. and Roberts, P. V.,** Dynamic behavior of organics in full-scale granular activated-carbon columns, in *Treatment of Water by Granular Activated Carbon,* McGuire, M. J. and Suffet, I. H., Eds., American Chemical Society, Washington, D.C., 1983, chap. 22.
946. **Sung, E., Han, C. D., and Rhee, H.-K.,** Optimal design of multistage adsorption-bed systems, *AIChE J.*, 25, 86, 1979.
947. **Sussman, M. V.,** Status report: continuous chromatography, *Chem. Tech.*, 6, 260, 1976.
948. **Sussman, M. V., Astill, K. N., and Rathore, R. N. S.,** Continuous gas chromatography, *J. Chromatogr. Sci.*, 12, 91, 1974.
949. **Sussman, M. V., Astill, K. N., Rombach, R., Cerullo, A., and Chen, S. S.,** Continuous surface chromatography, *Ind. Eng. Chem. Fundam.*, 11, 181, 1972.
950. **Sussman, M. V. and Huang, C. C.,** Continuous gas chromatography, *Science,* 156, 974, 1967.
951. **Sussman, M. V. and Rathore, R. N. S.,** Continuous modes of chromatography, *Chromatographia,* 8, 55, 1975.
952. **Sutikno, T. and Himmelstein, K. J.,** Desorption of phenol from activated carbon by solvent regeneration, *Ind. Eng. Chem. Fundam.*, 22, 420, 1983.

952a. **Suzuki, M.,** Continuous counter-current flow approximation for steady-state profile of pressure swing adsorption, *AIChE Symp. Ser.*, 81(242), 67, 1985.

953. **Suzuki, M. and Doi, H.,** Effect of adsorbed water on adsorption of oxygen and nitrogen on molecular sieving carbon, *Carbon,* 20, 441, 1982.
954. **Svedberg, U. G.,** Numerical solution of multicolumn adsorption processes under periodic countercurrent operation, *Chem. Eng. Sci.*, 31, 345, 1976.
955. **Svensson, H., Agrell, C.-E., Dehlen, S.-O., and Hagadahl, L.,** An apparatus for continuous chromatographic separation, *Sci. Tools,* 2(2), 17, 1955.
956. **Sweed, N. H.,** Parametric pumping, in *Progress in Separation and Purification,* Vol. 4, Perry, E. S. and van Oss, C. J., Eds., Wiley-Interscience, New York, 1971, 171-2
957. **Sweed, N. H.,** Parametric pumping, in *Recent Developments in Separation Science,* Vol. 1, Li, N. N., Ed., CRC Press, Boca Raton, Fla., 1972, 59.
958. **Sweed, N. H.,** Nonisothermal and nonequilibrium fixed bed sorption, in *Percolation Processes, Theory and Applications,* Rodrigues, A. E. and Tondeur, D., Eds., Sijthoff and Noordhoff, Alphen aan den Rijn, Netherlands, 1981, 329.
959. **Sweed, N. H.,** Parametric pumping and cycling zone adsorption. A critical analysis, *AIChE Symp. Ser.*, 80(233), 44, 1984.
960. **Sweed, N. H. and Gregory, R. A.,** Parametric pumping: modeling direct thermal separations of sodium chloride-water in open and closed systems, *AIChE J.*, 17, 171, 1971.
961. **Sweed, N. H. and Rigaudeau, J. M.,** Equilibrium theory and scale-up of parametric pump, *AIChE Symp. Ser.*, 71(152), 1, 1975.
962. **Sweed, N. H. and Wilhelm, R. H.,** Parametric pumping, separations via direct thermal mode, *Ind. Eng. Chem. Fundam.*, 8, 221, 1969.

963. **Swinton, E. A., Bolto, B. A., Eldridge, R. J., Nadebaum, P. R., and Coldrey, P. C.,** The present status of continuous ion exchange using magnetic micro-resins, in *Ion Exchange Technology,* Naden, D. and Streat, M., Eds., Ellis Horwood, Chichester, 1984, 542.
964. **Swinton, E. A., Nadebaum, P. R., Monkhouse, P., and Poulos, A.,** Continuous ion exchange using magnetic microbeads — field trials of a transportable pilot plant, Australian Water and Wastewater Association 10th Federal Convention, Sydney, Australia, 1983.
965. **Swinton, E. A., Nadebaum, P. R., Murtagh, R. W., and O'Beirne, R. J.,** Continuous ion exchange using magnetic microresins, I, *Water (Australia),* 8(2), 15, 1981.
966. **Swinton, E. A., Nadebaum, P. R., Murtagh, R. W., and O'Beirne, R. J.,** Continuous ion exchange using magnetic microresins. II, *Water (Australia),* 8(3), 16, 1981.
967. **Swinton, E. A. and Weiss, D. E.,** Counter-current adsorption separation processes, *Aust. J. Appl. Sci.,* 4, 316, 1953.
968. **Szepesy, L., Sebestyen, Zs., Feher, I., and Nagy, Z.,** Continuous liquid chromatography, *J. Chromatogr.,* 108, 285, 1975.
969. **Szolcsanyi, P., Horvath, G., Kotsis, L., and Szanya, T.,** Mathematical modeling of the pressure swing adsorption process, 5th Congr. CHISA, Prague, Czechoslovakia, 1975, paper 7.2.6.
970. **Takeuchi, K. and Uraguchi, Y.,** Separation conditions of the reactant and the product with a chromatographic moving bed reactor, *J. Chem. Eng. Jpn.,* 9, 164, 1976.
971. **Takeuchi, K. and Uraguchi, Y.,** Basic design of chromatographic moving-bed reactors for product refining, *J. Chem. Eng. Jpn.,* 9, 246, 1976.
972. **Tan, V. A., Astakhor, V. A., Romankor, P. G., Sukin, V. D., and Mokhatkin, R. A.,** A continuous fluid bed adsorber with centrifugal separation of the solid phase, *Br. Chem. Eng.,* 15, 1295, 1970.
973. **Taramasso, M. and Dinelli, D.,** Preparative scale gas chromatography by a laboratory rotating unit, *J. Gas Chromatogr.,* 2, 150, 1964.
974. **Thomas, H. C.,** Heterogeneous ion exchange in a flowing system, *J. Am. Chem. Soc.,* 66, 1664, 1944.
975. **Thomas, H. C.,** Chromatography: a problem in kinetics, *Ann. N.Y. Acad. Sci.,* 49, 161, 1948.
976. **Thompson, D. W. and Abu Goukh, M. E.,** Operation of cyclic electrodialysis in a continuous (open) system, 24th Canadian Chemical Eng. Conf., Ottawa, Canada, Oct. 20—23, 1974.
977. **Thompson, D. W. and Bass, D.,** Cyclic electrodialysis: experimental results in a closed system. *Can. J. Chem. Eng.,* 52, 345, 1974.
978. **Thompson, D. W., Bass D., and Abu-Goukh, M. E.,** Rate models for cyclic separation processes — comparison with experimental results in cyclic electrodialysis, *Can. J. Chem. Eng.,* 52, 479, 1974.
979. **Thompson, D. W. and Bowen, B. D.,** Equilibrium theory of the parametric pump. Effect of boundary conditions, *Ind. Eng. Chem. Fundam.,* 11, 415, 1972.
980. **Thornton, D. P.,** Extract *p*-Xylene with Parex, *Hydrocarb. Process.,* 49(11), 151, 1970.
981. **Tien, C.,** Recent advances in the calculation of multicomponent adsorption in fixed beds, in *Treatment of Water by Granular Activated Carbon,* McGuire, M. J. and Suffet, I. H., Eds., American Chemical Society, Washington, D.C., 1983, chap. 8.
982. **Tien, C., Hsieh, J. S. C., and Turian, R. M.,** Application of h-transformation for the solution of multicomponent adsorption in fixed bed, *AIChE J.,* 22, 498, 1976.
983. **Tiley, P. F.,** Multi-stage computations for continuous gas chromatography, *J. Appl. Chem.,* 17, 131, 1967.
984. **Timmins, R. S., Mir, L., and Ryan, J. M.,** Large-scale chromatography: new separation tool, *Chem. Eng.,* 76(11), 170, 1969.
985. **Toei, R. and Akao, T.,** Multi-stage fluidized bed apparatus with perforated plates, *Inst. Chem. Eng. Symp. Ser.,* 30, 34, 1968.
986. **Tondeur, D.,** Theory of ion-exchange columns, *Chem. Eng. J.,* 4, 337, 1970.
987. **Tondeur, D., Jacob, P., Schweich, D., and Wankat, P. C.,** The Guillotine effect: a new concept in chromatographic separations, *Proc. 2nd World Congr. Chem. Eng.,* Vol. 4, Canadian Society for Chemical Engineering, Ottawa, 1981, 226.
988. **Tondeur, D. and Wankat, P. C.,** Gas purification by pressure-swing adsorption, *Sep. Purif. Methods,* 14, 157, 1985.
989. **Traut, D. E., Nichols, I. L., and Seidel, D. C.,** Design Requirements for Uranium Ion Exchange From Acidic Solutions in a Fluidized System, U.S. Bureau of Mines, RI8282, 1978.
990. **Treybal, R. E.,** *Mass Transfer Operations,* 3rd ed., McGraw-Hill, New York, 1980.
991. **Tsai, M. C., Wang, S.-S., and Yang, R. T.,** Pore-diffusion model for cyclic separation: temperature swing separation of hydrogen and methane at elevated pressures, *AIChE J.,* 29, 966, 1983.
992. **Tsai, M. C., Wang, S. S., Yang, R. T., and Desai, N. J.,** Temperature -swing separation of hydrogen-methane mixtures, *Ind. Eng. Chem. Process Des. Devel.,* 24, 57, 1985.
993. **Tudge, A. P.,** Studies in chromatographic transport. III. Chromatothermography, *Can. J. Phys.,* 40, 557, 1962.

994. **Turina, S., Krajovan, V., and Kostomaj, T.,** Eine einfache apparatur für die kontinuierliche stofftrennug aus der gasphase, *Z. Anal. Chem.,* 189, 100, 1962.
995. **Turina, S. and Marjanovic-Krajovan, V.,** Continuous separation by the method of thin layer chromatography, *Anal. Chem.,* 36, 1905, 1964.
996. **Turk, A.,** Source control by gas-solid adsorption and related processes, in *Air Pollution,* Vol. 3, 2nd ed., Stern, A. C., Ed., Academic Press, New York, 1968, 497.
997. **Turkova, J.,** *Affinity Chromatography,* Elsevier, Amsterdam, 1978.
998. **Turner, J. C. R. and Church, M. R.,** A continuous ion exchange column, *Trans. Inst. Chem. Eng.,* 41, 283, 1963.
999. **Turnock, P. H. and Kadlec, R. H.,** Separation of nitrogen and methane via periodic adsorption, *AIChE J.,* 17, 335, 1971.
1000. **Tuthill, E. J.,** A new concept for the continuous chromatographic separation of chemical species, *J. Chromatogr. Sci.,* 8, 285, 1970.
1001. **Umehara, T., Harriott, P., and Smith, J. M.,** Regeneration of activated carbon. I. Thermal decomposition of adsorbed sodium dodecylbenzene sulfonate. II. Gasification kinetics with steam, *AIChE J.,* 29, 732, 737, 1983.
1002. **Umehara, T. and Smith, J. M.,** Regeneration of carbon containing sodium dodecylbenzene sulfonate — cyclic regeneration with steam in fluidized beds, *AIChE J.,* 30, 177, 1984.
1003. **Unger, K.,** High-performance size-exclusion chromatography, in *Methods in Enzymology,* Vol. 104, *Enzyme Purification and Related Techniques,* Jakoby, W. B., Ed., Academic Press, Orlando, Fla., 1984, 154.
1004. Union Carbide Corp., Purasiv HR, for hydrocarbon recovery, Bulletin F-48668 15M, no date, and Purasir HR. A solvent recovery and income producing system from the solvent management company, Bulletin F-48668A 5M, 1983, Union Carbide Corp., Danbury, Conn.,
1005. **Urano, K., Yamamoto, E., and Takeda, H.,** Regeneration rates of granular activated carbons containing adsorbed organic matter, *Ind. Eng. Chem. Process Des. Devel.,* 21, 180, 1982.
1006. **Valentin, P.,** Design and optimization of preparative chromatographic separations, in *Percolation Processes, Theory and Applications,* Rodrigues, A. E. and Tondeur, D., Eds., Sijthoff and Noordhoff, Alphen aan den Rijn, Netherlands, 1981, 141.
1007. **Valentin, P., Hagenbach, G., Roz, B., and Guiochon, G.,** New advances in the operation of large-scale gas chromatographic units, in *Gas Chromatography—1972,* Perry, S. G. and Adlord, E. R., Eds., Applied Science, Barking, Essex, England, 1973, 157.
1008. **Van Deemter, J. J., Zuiderweg, F. J., and Klinkenberg, A.,** Longitudinal diffusion and resistance to mass transfer as causes of nonideality in chromatography, *Chem. Eng. Sci.,* 5, 271, 1956.
1009. **van der Meer, A. P., Woerde, H. M., and Wesselingh, J. A.,** Mass transfer in a countercurrent ion-exchange plate column, *Ind. Eng. Chem. Process. Des. Devel.,* 23, 660, 1984.
1010. **Van der Vlist, E.,** Oxygen and nitrogen enrichment in air by cycling zone adsorption, *Sep. Sci.,* 6, 727, 1971.
1011. **Vanderschuren, J.,** The plate efficiency of multistage fluidized bed adsorbers, *Chem. Eng. J.,* 21, 7, 1981.
1012. Varex, Corp., Varex PSLC 100, Rockville, Md., no date.
1013. **Vatavuk, W. M. and Neveril, R. B.,** Costs of carbon adsorbers., XIV, Chem. Eng., *90(2), 131, 1983.*
1014. **Vaughan, M. F. and Dietz, R.,** Preparative gel permeation chromatography (GPC), in *Chromatography of Synthetic and Biological Polymers, Vol. 1,* Epton R., Ed., Ellis Horwood, Chichester, 1978, 199.
1015. **Vermeulen, T.,** Separation by adsorption methods, in *Advances in Chemical Engineering,* Vol. II, Drew, T. B. and Hoopes, J. W., Jr., Eds., Academic Press, New York, 1958, 14.
1016. **Vermeulen, T., Klein, G., and Hiester, N. K.,** Adsorption and ion exchange, in *Chemical Engineers' Handbook,* Perry, R. H., and Chilton, C. H., Eds., 5th ed., Section 16, McGraw-Hill, New York, 1973.
1017. **Vermeulen, T., LeVan, M. D., Hiester, N. K., and Klein, G.,** Adsorption and ion exchange in *Perry's Chemical Engineers' Handbook,* 6th ed., Perry, R. H. and Green, D., Eds., McGraw-Hill, New York, 1984, Section 16.
1018. **Verzele, M.,** Preparative gas chromatography, in *Progress in Separation and Purification,* Vol. 1, Perry, E. S., Ed., Interscience, New York, 1968, 83.
1019. **Verzele, M. and Dewaele, C.,** Optimization of column packing and column packing materials in HPLC, 23rd Eastern Analytical Symposium, New York, Nov. 16, 1984.
1020. **Verzele, M. and Dewaele, C.,** Preparative liquid chromatography. A critical review: 1980—1984, *Liquid Chromatogr. HPLC,* 3(1), 22, 1985.
1021. **Verzele, M., Dewaele, C., Van Dijck, J., and Van Haver, D.,** Preparative-scale high-performance liquid chromatography on analytical columns, *J. Chromatogr.,* 249, 231, 1982.
1022. **Verzele, M. and Geeraert, E.,** Preparative liquid chromatography, *J. Chromatogr. Sci.,* 18, 559, 1980.
1023. **Viswanathan, S. and Aris, R.,** Countercurrent moving bed chromatographic reactor, *Adv. Chem. Ser.,* 134, 191, 1974.

1024. **von Dreusche, C.,** Regeneration systems, in *Activated Carbon Adsorption for Wastewater Treatment,* Perrich, J. R., Ed., CRC Press, Boca Raton, Fla., 1981, chap. 6A.
1025. **von Szirmay, L.,** Process for gas separation on a moving adsorbent bed, *AIChE Symp. Ser.,* 71 (152), 104, 1975.
1026. **Vulliez-Sermet, P. R. E. and Fiorentino, E.,** Adsorption Process, U.S. Patent 3,979,287, Sept. 7, 1976.
1027. VIC Manufacturing Co., VIC Hydrocarbon vapor adsorption systems, (no date); Carbon adsorption/emission control. Benefits and limitations, (no date); VIC Air Pollution Control systems for coating processes, (1979), 1620 Central Ave. N.E., Minneapolis, Minn., *55413.*
1028. VIC Manufacturing Co., Energy conservation through combination adsorption-incineration technologies, 1620 Central Ave. NE, Minneapolis, Minn., 55413, no date.
1029. **Wagner, J. L.,** U.S. Patent 3,430,418, March, 4, 1969.
1030. **Wakao, N., Matsumoto, H., Suzuki, K., and Kawahara, A.,** Adsorption separation of liquids by means of parametric pumping, *Kogaku kogaku,* 32, 169, 1968 (in Japanese).
1031. **Walter, J. E.,** Multiple adsorption from solutions, *J. Chem. Phys.,* 13, 229, 1945.
1032. **Wang, S.-C. and Tien, C.,** Further work on multicomponent liquid phase adsortion in fixed beds, *AIChE J.,* 28, 565, 1982.
1033. **Wang, S.-C. and Tien, C.,** Fluidized bed expansion as a result of biomass growth, *Can. J. Chem. Eng.,* 61, 64, 1983.
1034. **Wang, S.-C. and Tien, C.,** Bilayer film model for the interaction between adsorption and bacterial activity in granular activated carbon columns. I. Formulation of equations and their numerical solutions. II. Experiment, *AIChE J.,* 30, 786, 794, 1984.
1035. **Wang, S. I., Nicholas, D.M., and DiMartino, S. P.,** Selection and optimization of hydrogen purification processes, Paper 66d presented at AIChE Denver meeting, Aug, 29, 1983.
1036. **Wang, S.-S. and Yang, R. T.,** Multicomponent separation by cyclic processes — a process for combined hydrogen/methane separation and acid gas removal in coal conversion, *Chem. Eng. Commun.,* 20, 183, 1983.
1037. **Wankat, P. C.,** Two-dimensional development in staged systems, *Sep. Sci.,* 7, 345, 1972.
1038. **Wankat, P. C.,** Liquid-liquid extraction parametric pumping, *Ind. Eng. Chem. Fundam.,* 12, 372, 1973.
1039. **Wankat, P. C.,** Cycling zone extraction, *Sep. Sci.,* 8, 473, 1973.
1040. **Wankat, P. C.,** Review: cyclic separation processes, *Sep. Sci.,* 9, 85, 1974.
1041. **Wankat, P. C.,** Thermal wave cycling zone separation: a preparative separation technique for countercurrent distribution and chromatography, *J. Chromatogr.,* 88, 211, 1974.
1042. **Wankat, P. C.,** Theory of affinity chromatography, *Anal. Chem.,* 46, 1400, 1974.
1043. **Wankat, P. C.,** Multicomponent cycling zone separations, *Ind. Eng. Chem. Fundam.,* 14, 96, 1975.
1044. **Wankat, P. C.,** Fractionation by cycling zone adsorption, *Chem. Eng. Sci.,* 32, 1283, 1977.
1045. **Wankat, P. C.,** Improved efficiency in preparative chromatographic columns using a moving feed, *Ind. Eng. Chem. Fundam.,* 16, 408, 1977.
1046. **Wankat, P. C.,** The relationship between one-dimensional and two-dimensional separatio· ›rocesses, *AIChE J.,* 23, 859, 1977.
1047. **Wankat, P. C.,** Increasing feed throughput in preparative two-dimensional separations, *Sep. Sci.,* 12, 553, 1977.
1048. **Wankat, P. C.,** Continuous recuperative mode parametric pumping, *Chem. Eng. Sci.,* 33, 723, 1978.
1049. **Wankat, P. C.,** Calculations for separations with three phases. I. Staged systems, *Ind. Eng. Chem. Fundam.,* 19, 358, 1980.
1050. **Wankat, P. C.,** An analogy between countercurrent and two-dimensional separation cascades, *Sep. Sci. Technol.,* 15, 1599, 1980.
1051. **Wankat, P. C.,** Cyclic separation techniques, in *Percolation Processes, Theory and Applications,* Rodrigues, A. E. and Tondeur, D., Eds., Sijthoff and Noordhoff, Alphen aan den Rijn, Netherlands, 1981, 443.
1052. **Wankat, P. C.,** Operational techniques for adsorption and ion exchange, *Proceedings, Corn Refiner's Association 1982 Scientific Conference,* June 16—18, 1982, p.119.
1053. **Wankat, P. C.,** New adsorption methods, *Chem. Eng. Educ.,* 18, 20, 1984.
1054. **Wankat, P. C.,** Improved preparative chromatography: moving port chromatography, *Ind. Eng. Chem. Fundam.,* 23, 256, 1984.
1055. **Wankat, P. C.,** Two-dimensional separation processes, *Sep. Sci. Technol.,* 19, 801, 1984.
1056. **Wankat, P. C.,** Large scale chromatography, in *Handbook of Separation Process Technology,* Rousseau, R., Ed., Wiley, New York, chap. 14, in press.
1057. **Wankat, P. C.,** An engineer's perspective — increasing the efficiency of packed bed chromatographic separations, in *Solving Problems with Preparative Liquid Chromatography,* Bidlingmeyer, B. H., Ed., Elsevier, Amsterdam, in press.
1058. **Wankat, P. C., Dore, J. C., and Nelson, W. C.,** Cycling zone separations, *Sep. Purif. Methods,* 4, 215, 1975.

1059. **Wankat, P. C., Middleton, A. R., and Hudson, B. L.,** Steady-state continuous, multicomponent separations in regenerated two-dimensional cascades, *Ind. Eng. Chem. Fundam.*, 15, 309, 1976.
1060. **Wankat, P. C. and Noble, R. D.,** Calculations for separations with three phases. II. Continuous contact systems, *Ind. Eng. Chem. Fundam.*, 23, 137, 1984.
1061. **Wankat, P. C. and Ortiz, P. M.,** Moving feed point gel permeation chromatography. An improved preparative technique, *Ind. Eng. Chem. Process Design Devel.*, 21, 416, 1982.
1062. **Wankat, P. C. and Partin, L. R. ,** Process for recovery of solvent vapors with activated carbon, *Ind. Eng. Chem. Process Des. Devel.*, 19, 446, 1980.
1063. **Wankat, P. C. and Ross, J. W.,** Communication, partial fractionation of dyes by cycling zone separation, *Sep. Sci.*, 11, 207, 1976.
1064. **Wankat, P. C. and Tondeur, D.,** Use of multiple sorbents in pressure swing adsorption, parametric pumping and cycling zone adsorption, *AIChE Symp. Ser.*, 81(242), 74, 1985.
1065. Waters Div., Millipore, New Waters Kiloprep Process Scale Separation Systems, Millford, Mass., 1983.
1066. **Watson, A. M.,** Use pressure swing adsorption for lowest cost hydrogen, *Hydrocarb. Process.*, 62(3), 91, 1983.
1067. **Weaver, K. and Hamrin, C. E.,** Separation of hydrogen isotopes by heatless adsorption, *Chem. Eng. Sci.*, 29, 1873, 1974.
1068. **Weber, T. W. and Chakravorti, R. K.,** Pore and solid diffusion models for fixed-bed adsorbers, *AIChE J.*, 20, 228, 1974.
1069. **Weber, W. J., Jr., Ed.,** *Physicochemical Processes for Water Quality Control*, Wiley-Interscience, New York, 1972.
1070. **Weber, W. J., Jr., Freeman, L. D., and Bloom, R.,** Biologically extended physiochemical treatment, in *Advances in Water Pollution Research — Proceedings 6th International Conference*, Jenkins, S. H., Ed., Pergamon Press, London, 1972.
1071. **Weber, W. J., Jr., Hopkins, C. B., and Bloom, R.,** Physico-chemical treatment of wastewater, *J. Water Pollut. Control Fed.*, 42, 83, 1970.
1072. **Weber, W. J., Jr. and Keinath, T. M.,** Mass transfer of perdurable pollutants from dilute aqueous solutions in fluidized adsorbers, *Chem. Eng. Prog. Symp. Ser.*, 63(74), 79, 1967.
1073. **Weber, W. J., Jr. and Liang, S.** A dual particle-diffusion model for porous adsorbents in fixed beds, *Environ. Prog.*, 2(3), 167, 1983.
1074. **Weber, W. J., Jr. and Morris, J. C.,** Kinetics of adsorption in columns of fluidized media, *J. Water Pollut. Control Fed.*, 37, 425, 1965.
1075. **Weber, W. J., Jr. and Thaler, J. O.,** Modeling for process scale-up of adsorption and ion exchange systems, in *Scale-up of Water and Wastewater Treatment Processes*, Schmidtke, N. W. and Smith, D. W., Eds., Butterworths, Boston, 1983, 233.
1076. **Wehr, C. T.,** Commercially available columns, in *CRC Handbook of HPLC for the Separation of Amino Acids, Peptides, and Proteins*, Vol. 1, Hancock, W. S., Ed., CRC Press, Boca Raton, Fla., 1984, 31.
1077. **Wehr, C. T.,** High-performance liquid chromatography: care of columns, in *Methods in Enzymology*, Vol. 104, *Enzyme Purification and Related Techniques*, Jakoby, W. B., Ed., Academic Press, Orlando, Fla., 1984, 133.
1078. **Wehrli, A., Hermann, U., and Huber, J. F. K.,** Effect of phase system selectivity in preparative column liquid chromatography, *J. Chromatogr.*, 125, 59, 1976.
1079. **Weiner, A. L.,** Drying gases and liquids. Dynamic fluid drying, *Chem. Eng.*, 81 (19)92, 1974.
1080. **Weiss, D. E.,** Industrial fractional adsorption techniques, *Aust. Chem. Inst. J. Proc.*, p. 141, April 1950.
1081. **Weiss, J.,** On the theory of chromatography, *J. Chem. Soc.*, p. 297, 1943.
1082. Westvaco, Corp., Westvaco pulsed bed carbon adsorption system, Covington, Va., 24426, 1983.
1083. Whatman Magnum Production LC, Whatman Chemical Separation, Inc., Clifton, N.J., 1983.
1084. **Wheelwright, E. J.,** Recovery and purification of promethium, in *Promethium Technology*, Wheelwright, E. J., Ed., American Nuclear Society, Hinsdale, Ill., 1973, chap. 2.
1085. **Wheelwright, E. J.,** Kilogram-scale purification of americium by ion exchange, *Sep. Sci. Technol.*, 15, 783, 1980.
1086. **White, R. J., Klein, F., Chan, J. A., and Stroshane, R. M.,** Large-scale production of human interferons, in *Annual Reports on Fermentation Processes*, Vol. 4, Tsao, G. T., Ed., Academic Press, New York, 1980, chap. 8.
1087. **Whitley, M. D. and Hamrin, C. E.,** Separation of hydrogen sulfide-hydrogen mixtures by heatless adsorption, in *Adsorption and Ion Exchange with Synthetic Zeolites*, Flank, W. H., Eds., ACS Symp. Ser., #134, American Chemical Society, Washington, D.C., 1980, 261.
1088. **Wilchek, M., Miron, T., and Kohn, J.,** Affinity chromatography, in *Methods in Enzymology*, Vol. 104, *Enzyme Purification and Related Techniques*, Jakoby, W. B., Ed., Academic Press, Orlando, Fla., 1984.

1089. **Wilhelm, R. H.,** Parametric pumping, a model for active transport, in *Intracellular Transport,* Warren, K. B., Ed., Symposia of the International Society for Cell Biology, Vol. 5, Academic Press, New York, 1966, 199.
1090. **Wilhelm, R. H., Rice, A. W., and Bendelius, A. R.,** Parametric pumping: a dynamic principle for separating fluid mixtures, *Ind. Eng. Chem. Fundam.,* 5, 141, 1966.
1091. **Wilhelm, R. H., Rice, A. W., Rolke, R. W., and Sweed, N. H.,** Parametric pumping, *Ind. Eng. Chem. Fundam.,* 7, 337, 1968.
1092. **Wilhelm, R. H. and Sweed, N. H.,** Parametric pumping: separation of mixtures of toluene and *n*-heptane, *Science,* 159, 522, 1968.
1093. **Willmott, F. W., Mackenzie, I., and Dolphin, R. J.,** Microcomputer controlled column switching system for high performance liquid chromatography, *J. Chromatogr.,* 167, 31, 1978.
1094. **Wilson, J. N.,** A theory of chromatography, *J. Am. Chem. Soc.,* 62, 1583, 1940.
1095. Witco Chemical Corp., Inorganic Specialties Div., Witco activated carbon: granules and pellets for superior adsorption, 520 Madison Ave., New York, 10022.
1096. **Wolf, E. and Vermeulen, T.,** A multiple-layer cross flow configuration for preparative chromatography of multicomponent mixtures, *Ind. Eng. Chem. Process Des. Devel.,* 15, 485, 1976.
1097. **Wolf, W.,** PSA system can reduce hydrogen costs, *Oil Gas J.,* 88, Feb. 23, 1976.
1098. **Wong, Y. W. and Hill, F. B.,** Separation of hydrogen isotopes via single column pressure swing adsorption, *Chem. Eng. Commun.,* 15, 343, 1982; *2nd World Congress Chemical Engineering,* Vol. IV, 230, Montreal Canada, Oct. 1981.
1099. **Wong, Y. W., Hill, F. B., and Chan, Y. N. I.,** Studies of the separation of hydrogen isotopes by a pressure swing adsorption process, *Sep. Sci. Technol.,* 15, 423, 1980.
1100. **Wu, X.-Z. and Wankat, P. C.,** Continuous multicomponent parametric pumping, *Ind. Eng. Chem. Fundam.,* 22, 172, 1983.
1101. **Yang, C.-M. and Tsao, G. T.,** Packed bed adsorption theories and their applications to affinity chromatography, in *Advances in Biochemical Engineering,* Vol. 25, Fiechter, A., Ed., Springer-Verlag, New York, 1982, 1.
1102. **Yang, C.-M. and Tsao, G. T.,** Affinity chromatography, in *Advances in Biochemical Engineering,* Vol. 25, Fiechter, A., Ed., Springer-Verlag, New York, 1982, 19.
1103. **Yang, R. T. and Cen, P.-L.,** Pressure swing adsorption processes for gas separation: by heat exchange between adsorbers and by high-heat-capacity inert additives, *Ind. Eng. Chem. Process Des. Devel.,* 25, 54, 1986.
1105. **Yang, R. T., Doong, S. J., and Cen, P. L.,** Bulk gas separation of binary and ternary mixtures by pressure swing adsorption, *AIChE Symp. Ser.,* 81(242), 84, 1985.
1106. **Yang, R. T. and S.-J. Doong,** Gas separation by pressure swing adsorption: a pore diffusion model for bulk separation, *AIChE J.,* 31, 1829, 1985.
1107. **Yanovskii, S. M., Silaeva, I. E., and Alksnis, O. N.,** A chromathermographic method for the determination of adsorption isotherms at high adsorbate concentrations, *J. Chromatogr.,* 93, 464, 1974.
1108. **Yates, J. G.,** *Fundamentals of Fluidized-Bed Chemical Processes,* Butterworths, London, 1983, chap. 1.
1109. **Yau, W. W., Kirkland, J. J., and Bly, D. D.,** *Modern Size-Exclusion Chromatography,* Wiley, New York, 1979.
1110. **Yehaskel, A.,** *Activated Carbon — Manufactured and Regenerations,* Noyes Data Corp., Park Ridge, N.J., 1979.
1111. **Yomiyama, A., Nanke, Y., and Mizuma, N.,** Method and Apparatus for Continuously Transferring Ion Exchange Resins, U.S. Patent 3,152,072, 1964.
1112. **Yon, C. M. and Turnock, P. H.,** Multicomponent adsorption equilibrium on molecular sieves, *Chem. Eng. Prog. Symp. Ser.,* 67(117), 67, 1971.
1113. **Yoon, S. M. and Kunii, D.,** Physical absorption in a moving bed of fine adsorbents, *Ind. Eng. Chem. Process. Des. Devel.,* 10, 64, 1971.
1114. **Young, D. M. and Crowell, A. D.,** *Physical Adsorption of Gases,* Butterworths, London, 1962.
1115. **Young G. W., Korvsky, J. R., and Koradia, P. B.,** Liquid phase drying application of zeolites, in *Adsorption and Ion Exchange with Synthetic Zeolites,* Flank, W. H., Ed., ACS Symp. Ser., #135, American Chemical Society, Washington, D. C., 1980, 201.
1116. YMC, Inc. Process Liquid Chromatography, no date; Liquid Chromatography Product Guide, no date; Liquid Chromatography Price List (Feb. 1985), P.O. Box 492, Mt. Freedom, N.J., 07970.
1117. **Zalewski, W. C. and Hanesian, D.,** A study of the dynamics of gas adsorption in fixed and fluidized beds, *AIChE Symp. Ser.,* 69(128), 58, 1973.
1118. **Zemskov, I. F.,** Mass Transfer in a fluidized bed of activated charcoal, *Int. Chem. Eng.,* 5, 596, 1965.
1119. **Zhukhovitsky, A. A.,** Some developments in gas chromatography in the USSR, in *Gas Chromatography, 1960,* Scott, R. P. W., Ed., Butterworths, London, 1960, 293.

1120. **Zhukhovitskii, A. A. and Turkel'taub, N. M.,** The chromathermographic method of separation and analyzing gases, *Usp. Khim.*, 25, 859, 1956.
1121. **Zhukhovitskii, A. A. and Yanovskii, S. M.,** Applications of chromatographic distillation (review), *Ind. Lab.*, 47(2), 114, 1981 (trans.).
1122. **Zhukkovitskii, A. A., Yanovskii, S. M., and Shvarzman, V. P.,** Chromadistillation, *J. Chromatogr.*, 119, 591, 1976.
1123. **Zhukhovitskii, A. A., Zolotareva, O. V., Sokolov, V. A., and Turkel'taub, N. M.,** A new gas-chromatographic method, *Dokl. Akad. Nauk SSSR*, 77, 435, 1951.
1124. **Zlatkis, A. and Pretorius, V., Eds.,** *Preparative Gas Chromatography*, Wiley-Interscience, New York, 1971.
1125. **Zolandz, R. R. and Myers, A. L.,** Adsorption on heterogeneous surfaces, in *Progress in Filtration and Separation*, Vol. 1, Wakeman, R. J., Ed., Elsevier, Amsterdam, 1979, chap. 1.
1126. **Zweig, G. and Sherma, J., Eds.,** *CRC Handbook of Chromatography*, Vols. I and II, CRC Press, Boca Raton, Fla., 1972.

INDEX

A

B

C

D

E

F

G

H

I

K

L

M

N

O

P

R

S

T

U

V

W

Z